# A Gentle Introduction to S

## 4th Edition

# A Gentle Introduction to Stata

## 4th Edition

ALAN C. ACOCK
*Oregon State University*

A Stata Press Publication
StataCorp LP
College Station, Texas

Published by Stata Press, 4905 Lakeway Drive, College Station, Texas 77845
Typeset in LaTeX $2_\varepsilon$
Printed in the United States of America

10 9 8 7 6 5 4 3 2 1

ISBN-10: 1-59718-142-0
ISBN-13: 978-1-59718-142-6

Library of Congress Control Number: 2014935652

# Contents

# Figures

# Tables

# Boxed tips

# Preface

This book was written with a particular reader in mind. This reader is learning social statistics and needs to learn Stata but has no prior experience with other statistical software packages. When I learned Stata, I found there were no books written explicitly for this type of reader. There are certainly excellent books on Stata, but they assume extensive prior experience with other packages, such as SAS or IBM SPSS Statistics; they also assume a fairly advanced working knowledge of statistics. These books moved quickly to advanced topics and left my intended reader in the dust. Readers who have more background in statistical software and statistics will be able to read chapters quickly and even skip sections. The goal is to move the true beginner to a level of competence using Stata.

With this target reader in mind, I make far more use of the menus and dialog boxes in Stata's interface than do any other books about Stata. Advanced users may not see the value in using the interface, and the more people learn about Stata, the less they will rely on the interface. Also, even when you are using the interface, it is still important to save a record of the sequence of commands you run. Although I rely on the commands much more than the dialog boxes in the interface in my own work, I still find value in the interface. The dialog boxes in the interface include many options that I might not have known or might have forgotten.

To illustrate the interface as well as graphics, I have included more than 100 figures, many of which show dialog boxes. I present many tables and extensive Stata "results" as they appear on the screen. I interpret these results substantively in the belief that beginning Stata users need to learn more than just how to produce the results—users also need to be able to interpret them.

I have tried to use real data. There are a few examples where it is much easier to illustrate a point with hypothetical data, but for the most part, I use data that are in the public domain. For example, I use the General Social Surveys for 2002 and 2006 in many chapters, as well as the National Survey of Youth, 1997. I have simplified the files by dropping many of the variables in the original datasets, but I have kept all the observations. I have tried to use examples from several social-science fields, and I have included a few extra variables in several datasets so that instructors, as well as readers, can make additional examples and exercises that are tailored to their disciplines. People who are used to working with statistics books that have contrived data with just a few observations, presumably so work can be done by hand, may be surprised to see more than 1,000 observations in this book's datasets. Working with these files provides better experience for other real-world data analysis. If you have your own data and the dataset has a variety of variables, you may want to use your data instead of the data provided with this book.

The exercises use the same datasets as the rest of the book. Several of the exercises require some data management prior to fitting a model because I believe that learning data management requires practice and cannot be isolated in a single chapter or single set of exercises.

This book takes the student through much of what is done in introductory and intermediate statistics courses. It covers descriptive statistics, charts, graphs, tests of significance for simple tables, tests for one and two variables, correlation and regression, analysis of variance, multiple regression, logistic regression, reliability, factor analysis, and path analysis. There are chapters on constructing scales to measure variables and on using multiple imputation for working with missing values.

By combining this coverage with an introduction to creating and managing a dataset, the book will prepare students to go even further on their own or with additional resources. More advanced statistical analysis using Stata is often even simpler from a programming point of view than what we will cover here. If an intermediate course goes beyond what we do with logistic regression to multinomial logistic regression, for example, the programming is simple enough. The `logit` command can simply be replaced with the `mlogit` command. The added complexity of these advanced statistics is the statistics themselves and not the Stata commands that implement them. Therefore, although more advanced statistics are not included in this book, the reader who learns these statistics will be more than able to learn the corresponding Stata commands from the Stata documentation and help system.

I would like to point out the use of punctuation after quotes in this book. While the standard U.S. style of punctuation calls for periods and commas at the end of a quote to always be enclosed within the quotation marks, Stata Press follows a style typically used in mathematics books and British literature. In this style, any punctuation mark at the end of a quote is included within the quotation marks only if it is part of the quote. For instance, the pleased Stata user said she thought that Stata was a "very powerful program". Another user simply said, "I love Stata."

I assume that the reader is running Stata 13, or a later version, on a Windows-based PC. Stata works equally as well on Mac and on Unix systems. Readers who are running Stata on one of those systems will have to make a few minor adjustments to some of the examples in this book. I will note some Mac-specific differences when they are important. In preparing this book, I have used both a Windows-based PC and a Mac.

Corvallis, OR                                                    Alan C. Acock
March 2014

# Acknowledgments

I acknowledge the support of the Stata staff who have worked with me on this project. Special thanks goes to Lisa Gilmore, the Stata Press production manager, and Deirdre Skaggs, the Stata Press technical editor. I also thank my students who have tested my ideas for the book. They are too numerous to mention, but Shauna Tominey deserves special recognition for going through the entire draft of the second edition to find errors.

Stata has many outstanding technical support people. I was lucky have Kristin MacDonald be assigned the task as a technical support for the fourth edition. After I made an initial draft of changes and additions to this edition of the book, Kristin found several errors. She helped make sure these were fixed. The remaining errors are my responsibility. This edition is a vastly better book in terms of the statistical analysis, efficient use of Stata coding, and ease of reading because of Kristin's work. You, the reader, will benefit from her amazingly helpful technical support.

My education benefited from the knowledge of many, but I would like to acknowledge two of my former professors who were especially important. Henry (Bud) Kass, one of my undergraduate professors, was a role model for how to work with students; he encouraged me to pursue my graduate education. Louis Gray, one of my graduate school professors, taught me the power of quantitative analysis and shared his enthusiasm for research.

Finally, I thank my wife, Toni Acock, for her support and for her tolerance of my endless excuses for why I could not do things. She had to pick up many tasks I should have done, and she usually smiled when told it was because I had to finish this book.

# Support materials for the book

All the datasets and do-files for this book are freely available for you to download. In the Command window, type

```
. net from http://www.stata-press.com/data/agis4/
. net describe agis4
. net get agis4
```

Notice that each of these commands is preceded by a period (.) and a space. This is a convention used by Stata. When you enter the command, you just type the command without the . and space that precede it in the instructions.

Stata comes in several varieties. Small Stata is limited to analyzing datasets with a maximum of 99 variables and 1,200 observations. If you are using Small Stata, you will be able to do everything in this book, but you will need to download a different set of datasets that meet these restrictions. In the Command window, type

```
. net from http://www.stata-press.com/data/agis4/
. net describe agis4_small
. net get agis4_small
```

Stata will place the datasets in a directory where you can access them. On a Windows machine, this is probably `C:\Users\`*userid*`\Documents`. You may want to create a new directory and copy the materials there. If you have several projects, it may be useful to have a separate folder for each project. For simplicity, throughout this book, we will use `C:\data` as the data directory.

To open one of the datasets that you downloaded from the commands above, for example, `relate.dta`, type `use relate` in the Command window. If you are using Small Stata, `_small` is appended to the dataset name (`relate_small.dta`), so you would type `use relate_small`. Those readers using Small Stata will need to append `_small` whenever I mention a dataset in this book.

If your computer is connected to the Internet, you can also load the dataset by specifying the complete URL of the dataset. For example,

```
. use http://www.stata-press.com/data/agis4/firstsurvey
```

This text complements the material in the Stata manuals but does not replace it. For example, chapters 5 and 6, respectively, show how to generate graphs and tables, but these are only a few of the possibilities described in the Stata *Reference* manuals. All reference material is available in PDF format. In the Stata menu, click on Help ▷ PDF

*HELP*

Documentation. One of the best aspects of the Stata documentation is that it provides several real-data examples for most commands. An entry will start with a fairly simple example and then give examples that are more complex. Looking at the examples is how I have learned much of what I know about Stata. You will find that the capabilities for many of the commands I discuss far exceed what I was able to cover here.

If you remember the name of a command, you can type `help` *command_name* in the Command window. For example, typing `help summarize` would display a Viewer window with brief information and examples of how to run the command. If you do not know the exact name of the command, you could just enter the first part. For example, typing `help sum` opens a window with two options, one of which is `summarize`. If you enter the wrong name for a command, say, you type `help summary`, Stata opens a Viewer window with a list of files where the word "summary" was listed as a keyword. You scroll through the list and find the `summarize` command. If you click on `summarize`, the help file for the `summarize` command opens in the Viewer window.

The help file does not give you all the detailed explanation and examples that you get from the PDF documentation, but it is often all you need. You can open the PDF document for a specific command by clicking on the command name in the *Title* section or in the **Also See** menu of the help file.

My hope in writing this book is to give you sufficient background so that you can use the manuals effectively.

# 1 Getting started

## 1.1 Conventions

Listed below are the conventions that are used throughout the book. I thought it might be convenient to list them all in one place should you want to refer to them quickly.

**Typewriter font.** I use this font when something would be meaningful to Stata as input. I also use it to indicate Stata output.

I use a typewriter font to indicate the text to type in the Command window. Because Stata commands do not have any special characters at the end, any punctuation mark at the end of a command in this book is not part of the command. Sometimes, to be consistent with Stata manuals, I will put a command on a line by itself with the dot preceding it, as in

```
. sysuse cancer, clear
```

All of Stata's dialog boxes generate commands, which will be displayed in the Review window and in the Results window. In the Results window, each command will be preceded by the dot prompt. If you make a point of looking at the command Stata prints each time you use the dialog boxes, you will quickly learn the commands. I may include the equivalent command in the text after explaining how to navigate to it through the dialog boxes.

**Why do we show the dot prompt with these commands?**

When we show a listing of Stata commands, we place a dot and a space in front of each command. When you enter these commands in the Command window, you enter the command itself and not the dot prompt or space. We include these because Stata always shows commands this way in the Results window. Stata manuals and many other books about Stata follow this convention.

When you type a Stata command in the Command window, you execute the command when you press the Enter key. The command may wrap onto more than one line, but if you press the Enter key in the middle of entering a command, Stata will interpret that as the end of the command and will probably generate an error. The rule is that you should just keep typing when entering a command in the Command window, no matter how long the command is. Press Enter only when you want to execute the command.

I also use the typewriter font for variable names, for names of datasets, and to show Stata's output. In general, I use the typewriter font whenever the text is something that can be typed into Stata or when the text is something that Stata might print as output. This approach may seem cumbersome now, but you will catch on quickly.

Folder names, filenames, and filename extensions, as in "The `survey.dta` file is in the `C:\data` directory (or folder)", are also denoted in the typewriter font. Stata assumes that `.dta` will be the extension, so you can use just the filename without an extension, if you prefer.

Sans serif font. I use this font to indicate menu items (in conjunction with the ▷ symbol), button names, dialog-box tab names, and particular keys:

- Menu items, such as "Select Data ▷ Data utilities ▷ Rename groups of variables from the Stata menu" (see figure 1.1).

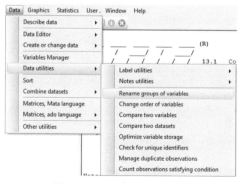

Figure 1.1. Stata menu

- Buttons that can be clicked on, as in "Remember, if you are working on a dialog box, it will now be up to you to click on OK or Submit, whichever you prefer."
- Keys on your keyboard, as in "The Page Up and Page Down keys will move you backward and forward through the commands in the Review window." Some functions require the use of the Shift, Ctrl, or Alt key, which will be held down while the second key is pressed. For example, Alt+f will open the File menu.

*Slant font.* I use this font for dialog-box titles and when I talk about labeled elements of a dialog box, with both items capitalized as they are on the dialog box.

*Italics font.* I use this font when I refer to a word that is to be replaced.

**Quotes**. I use double quotes when I am talking about labels in a general way, but I will use the typewriter font to indicate a specific label in a dataset. For example, if we decided to label the variable `age` "Age at first birth", we would enter `Age at first birth` in the textbox.

**Capitalization**. Stata is case sensitive, so `summarize` is a Stata command, whereas `Summarize` is not and will generate an error if you use it. Stata also recognizes capitalization in variable names, so `agegroup`, `Agegroup`, and `AgeGroup` will be three different variables. Although you can certainly use capital letters in variable names, you will probably find yourself making more typographical errors if you do. I have found that using all lowercase letters when creating variable names is usually the best practice.

I will capitalize the names of the various Stata windows, but I do not set them off by using a different font. For example, we will type commands in the Command window and look at the output in the Results window.

**Setting how much output is in the Results window**

The default size for the scrollback buffer size for the Results window is 200 kilobytes, approximately 200,000 characters. If you have many results being displayed in the Results window, the default is to drop the oldest lines once you use up the 200 kilobyte buffer. If you want to be able to scroll back further, you can make the buffer size larger, up to 2,000 kilobytes. Select Edit ▷ Preferences ▷ General Preferences... and click on the Windowing tab. Stata for Mac users can make this change by selecting Stata ▷ Preferences ▷ General Preferences... and clicking on the Windows tab. You might change the scrollback buffer size from the default 200 kilobytes to 500 kilobytes. This change will not take effect until you restart Stata.

Stata for Unix users cannot make this change from the *Preferences* dialog box; they must type the command set scrollbufsize 500000 directly in the Command window.

Typing the command sets the scrollback buffer size in bytes by default, whereas using the menu method sets the size in kilobytes.

Many Stata users find having to click on the —more— message when it appears in the Results window irritating. It is designed to make it easier to read the results of a single command, but if you do not like this feature, you can type the command set more off or set more off, permanently. The permanently option specifies that the setting be remembered for each future Stata session until you reverse the action by typing set more on or set more on, permanently.

## 1.2   Introduction

The best way to learn data analysis is to actually do it with real data. These days, doing statistics means doing statistics with a computer and a software package. There is no other software package that can match the internal consistency of Stata, which makes it easy to learn and a joy to use. Stata empowers users more effectively than any other statistical package.

**Work along with the book**

Although it is not necessary, you will probably find it helpful to have Stata running while you read this book so that you can follow along and experiment for yourself. Having your hands on a keyboard and replicating the instructions in this book will make the lessons that much more effective, but more importantly, you will get in the habit of just trying something new when you think of it and seeing what happens. In the end, experimentation is how you will really learn how Stata works. The other great advantage to following along is that you can save the examples we do for future use.

Stata is a powerful tool for analyzing data. Stata makes statistics and data analysis fun because it does so much of the tedious work for you. A new Stata user should start by using the dialog boxes. As you learn more about Stata, you will be able to do more sophisticated analyses with Stata commands. Learning Stata well now is an investment that will pay off in saved time later. Stata is constantly being extended with new capabilities, which you can install using the Internet from within Stata. Stata is a program that grows with you.

Stata is a command-driven program. It has a remarkably simple command structure that you use to tell it what you want it to do. You can use a dialog box to generate the commands (this is a great way to learn the commands or prompt yourself if you do not remember one exactly), or you can enter commands directly. If you enter the `summarize` command, you will get a summary of all the variables in your dataset (mean, standard deviation, number of observations, minimum value, and maximum value). Enter the command `tabulate gender`, and Stata will make a frequency distribution of the variable called `gender`, showing you the number and percentage of men and women in your dataset.

After you have used Stata for a while, you may want to skip the dialog box and enter these commands directly. When you are just beginning, however, it is easy to be overwhelmed by all the commands available in Stata. If you were learning a foreign language, you would have no choice but to memorize hundreds of common words right away. This is not necessary when you are learning Stata because the dialog boxes are so easy to use.

**Searching for help**

Stata can help when you want to find out how to do something. You can use the `search` command along with a keyword. For example, you believe that a *t* test is what you want to use to compare two means. Enter `search t test`; Stata searches its own resources and others that it finds on the Internet. The first entry of the results is

```
[R]     ttest . . . . . . . . . . . . . . . t tests (mean-comparison tests)
        (help ttest)
```

The `[R]` at the beginning of the line means that details and examples can be found in the *Stata Base Reference Manual*. Click on the blue `ttest` to go to the help file for the `ttest` command. If you think this help is too cryptic, repeat the `search t test` command and look farther down the list. Scroll past the lines starting with `Video`, and look for the lines starting with `FAQ` (frequently asked questions). One of these is "What statistical analysis should I use?" Click on the blue URL to go to a UCLA webpage that will help you decide whether the *t* test is the best choice for what you are doing. You might click on some of the other resources to see how much support you get from a wide variety of resources.

When using the `search` command, you need to pick a keyword that Stata knows. You might have to try different keywords before you get one that works. Searching these Internet locations is a remarkable capability of Stata. If you are reading this book and want to know more about a command, the online help is the first place to start. Suppose that we are discussing the `summarize` command and you want to know more options for this command. Type `help summarize` and you will get an informative help screen. To obtain complete information for a command, you should see the PDF documentation. The PDF documentation can be opened from the Stata menu by selecting Help ▷ PDF Documentation. Bookmarks to all the Stata manuals are available; click on the plus sign (+) next to each manual to see bookmarks to sections therein.

Stata has done a lot to make the dialog boxes as friendly as possible so that you feel confident using them. The dialog boxes often show many options, which control the results that are shown and how they are displayed. You will discover that the dialog boxes have default values that are often all you need, so you may be able to do a great deal of work without specifying any options.

As we progress, you will be doing more complex analyses. You can do these using the dialog boxes, but Stata lets you create files that contain a series of commands you can run all at once. These files, called do-files, are essential once you have many commands to run. You can reopen the do-file a week or even several months later and repeat exactly what you did. Keeping a record of what you do is essential; otherwise, you will not be able to replicate results of elaborate analyses. Fortunately, Stata makes this easy.

You will learn more about replicating results in chapter 4. The do-files that reproduce most of the tables, graphs, and statistics for each chapter are available on the webpage for this book (http://www.stata-press.com/data/agis4/).

Because Stata is so powerful and easy to use, I may include some analyses that are not covered in your statistics textbook. If you come to a procedure that you have not already learned in your statistics text, give it a try. If it seems too daunting, you can skip that section and move on. On the other hand, if your statistics textbook covers a procedure that I omit, you might search the dialog boxes yourself. Chances are that you will find it there.

Depending on your needs, you might want to skip around in the book. Most people tend to learn best when they need to know something, so skipping around to the things you do not know may be the best use of the book and your time. Some topics, though, require prior knowledge of other topics, so if you are new to Stata, you may find it best to work through the first four chapters carefully and in order. After that, you will be able to skip around more freely as your needs or interests demand.

## 1.3   The Stata screen

When you open Stata, you will see a screen that looks something like figure 1.2.

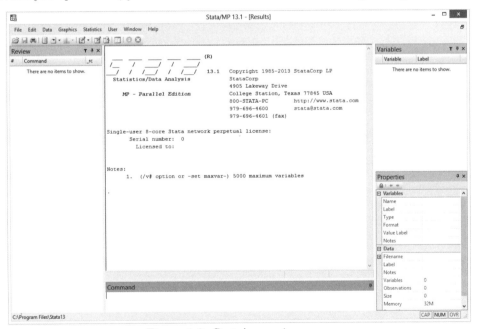

Figure 1.2. Stata's opening screen

You can rearrange the windows to look the way you want them, although many users are happy with the default layout. If you are satisfied with the defaults, you might skip the next couple paragraphs and come back to them if you change your mind later. Many experienced Stata users have particular ways to arrange these screens. Feel free to experiment with the layout.

Selecting Edit ▷ Preferences gives you several options. One thing you might want to do is change the size of the buffer for the Results window. The factory default of 200 kilobytes may be too small to be able to scroll through all your results. To change the size of the buffer, select Edit ▷ Preferences ▷ General Preferences... and then click on the tab labeled Windowing. Depending on how much memory your computer has available, you might want to raise the default value to as much as 500 kilobytes. You can resize the Stata interface as you would any other Windows product. There are other options you can try under Windowing and each of the other tabs. It is nice to personalize your interface in a way that is attractive to you. I will use the generic "factory settings" for this book, however. If you make several changes and want to return to the starting point, select Edit ▷ Preferences ▷ Load Preference Set ▷ Widescreen Layout (default). If you are using Stata for Mac, select Stata ▷ Preferences ▷ Manage Preferences ▷ Factory Settings.

When you open a file that contains Stata data, which we will call a Stata dataset, a list of the variables will appear in the Variables window. The Variables window reports the name of the variable (for example, `abortion`) and a label for the variable (for example, `Attitude toward abortion`). Other information about the variable is shown in the Properties window, such as the type of variable (for example, `float`) and the format of the variable (for example, `%8.0g`). For now, just consider the name and label. You can vary the width of each column in the Variables window by placing your cursor on the vertical line between the name and label, clicking on it, and then dragging your cursor to the right or left.

When Stata executes a command, it prints the results or output in the Results window. First, it prints the command preceded by a . (dot) prompt, and then it prints the output. The commands you run are also listed in the Review window. If you click on one of the commands listed in the Review window, it will appear in the Command window. If you double-click on one of the commands listed in the Review window, it will be executed. You will then see the command and its output, if any, in the Results window.

When you are not using the interface, you enter commands in the Command window. You can use the Page Up and Page Down keys on your keyboard to recall commands from the Review window. On a Mac that does not have the Page Up and Page Down keys, you can use the fn key with the arrow up or arrow down key. You can also edit commands that appear in the Command window. I will illustrate all these methods in the coming chapters.

The gray bar at the bottom of the screen, called the status bar, displays the current working directory (folder). This directory may be different on different computers depending on how Stata was installed. The working directory is where Stata will look for a file or save a file unless you specify the full path to a different directory that contains the file. If you have a project and want to store all files related to that project in a particular directory, say, `C:\data\thesis`, you could enter the command `cd C:\data\thesis`. This command assumes that this directory already exists on your computer.

On a Mac, the gray bar at the bottom looks slightly different. To change the working directory on a Mac or Unix computer from the current working directory to a Documents folder in your home directory, you would type `cd "~/Documents"`. Stata recognizes the tilde to represent your home directory. If you had a folder in your Documents folder called `Learning Stata`, you would type `cd "~/Documents/Learning Stata"`. Also on a Mac, you have help if you cannot remember where you saved a file containing data: You can click on the magnifying glass in the upper right corner of your screen to search the name of the file, and then click on the file to open it. You may want to type the `clear` command first.

Stata has the usual Windows title bar across the top, on the right side of which are the three buttons (in order from left to right) to minimize, to expand to full-screen mode, and to close the program. Immediately below the Stata title bar is the menu bar, where the names of the menus appear. Some of the menu items (File, Edit, and Window) will look familiar because they are used in other programs. The Data, Graphics, and Statistics menus are specific to Stata, but their names provide a good idea of what you will find under them.

Figures 1.3 and 1.4 show the Stata toolbar as it appears in Windows and Mac, respectively. The icons provide alternate ways to perform some of the actions you would normally do with the menus. If you hold the cursor over any of these icons for a couple of seconds, a brief description of the function appears. For a complete list of the toolbar icons and their functions, see the *Getting Started with Stata* manual.

Figure 1.3. The toolbar in Stata for Windows

Figure 1.4. The toolbar in Stata for Mac

## 1.4  Using an existing dataset

Chapter 2 discusses how to create your own dataset, save it, and use it again. You will also learn how to use datasets that are on the Internet. For now, we will use a simple dataset that came with Stata. Although we could use the dialog box to do this, we will enter a simple command. Click once in the Command window to put the cursor there, and then type the command `sysuse cancer, clear`; the Command window should look like the one in figure 1.5.

Figure 1.5. Stata command to open `cancer.dta`

The `sysuse` command we just used will find the sample dataset on your computer by name alone, without the extension; in this case, the dataset name is `cancer`, and the file that is found is actually called `cancer.dta`. The `cancer` dataset was installed with Stata. This particular dataset has 48 observations and 4 variables related to a cancer treatment.

What if you forget the command `sysuse`? You could open a file that comes with Stata by using the menu File ▷ Example Datasets.... A new window opens in which you click on *Example datasets installed with Stata*. The next window then lists all the datasets that come with Stata. You can click on **use** to open the dataset.

Now that we have some data read into Stata, type `describe` in the Command window. That is it: just type `describe` and press the Enter key. `describe` will yield a brief description of the contents of the dataset.

```
. describe
Contains data from C:\Program Files\Stata13\ado\base/c/cancer.dta
  obs:           48                          Patient Survival in Drug Trial
 vars:            8                          3 Mar 2011 16:09
 size:          576

              storage   display    value
variable name   type    format     label      variable label

studytime       int     %8.0g                 Months to death or end of exp.
died            int     %8.0g                 1 if patient died
drug            int     %8.0g                 Drug type (1=placebo)
age             int     %8.0g                 Patient's age at start of exp.
_st             byte    %8.0g
_d              byte    %8.0g
_t              byte    %10.0g
_t0             byte    %10.0g

Sorted by:
```

The description includes a lot of information: the full name of the file, `cancer.dta` (including the path entered to read the file); the number of observations (48); the number of variables (8); the amount of memory the data consume (576 bytes); a brief description of the dataset (*Patient Survival in Drug Trial*); and the date the file was last saved. The body of the table displayed shows the names of the variables on the far left and the labels attached to them on the far right. We will discuss the middle columns later.

Now that you have opened `cancer.dta`, note that the Variables window lists the eight variables `studytime`, `died`, `drug`, `age`, `_st` `_d`, `_t`, and `_t0`.

### Internet access to datasets

Stata can use data stored on the Internet just as easily as data stored on your computer. If you did not have the `cancer.dta` file installed on your computer, you could read it by entering `webuse cancer`. However, you are not limited to data stored at the Stata site. Typing `use http://www.ats.ucla.edu/stat/stata/notes/hsb2` will open a dataset stored at the UCLA website.

Stata does not discard changes to the dataset currently in memory unless you tell it to do so. That is, if you have a dataset in memory and you have modified it, you will receive an error message if you try to load another dataset. You need to save the dataset in memory, type the `clear` command to discard the changes, or type the `clear` option of the `use` command to discard the changes. You can then load the new dataset.

Stata provides all the datasets for every example in its manuals. For example, click on File ▷ Example Datasets.... A new window opens in which you click on Stata 13 manual datasets. There you might click on *Base Reference Manual [R]*; scroll down to `correlate`, and click on `use` to open any of the datasets or `describe` to see what variables are in the dataset.

## 1.5 An example of a short Stata session

If you do not have `cancer.dta` loaded, type the command `sysuse cancer`. We will execute a basic Stata analysis command. Type `summarize` in the Command window and then press Enter.

Rather than typing in the command directly, you could use the dialog box by selecting Data ▷ Describe data ▷ Summary statistics to open the corresponding dialog box. Simply clicking on the OK button located at the bottom of the dialog box will produce the `summarize` command we just entered. Because we did not enter any variables in the dialog box, Stata assumed that we wanted to summarize all the variables in the dataset.

You might want to select specific variables to summarize instead of summarizing them all. Open the dialog box again and click on the pulldown menu within the *Variables* box, located at the top of the dialog box, to display a list of variables. Clicking on a variable name will add it to the list in the box. Dialog boxes allow you to enter a variable more than once, in which case the variable will appear in the output more than once. You can also type variable names in the *Variables* box. A last alternative is to click on the variable name in the Variables window. Figure 1.6 shows the dialog box with the drop-down variable list displaying the variables in your dataset:

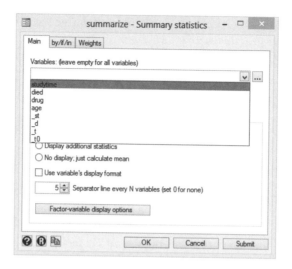

Figure 1.6. The `summarize` dialog box

In the bottom left corner of the dialog box, there are three icons: ❓, ®, and 🖹. The ❓ icon gives us a help screen explaining the various options. The explanations are brief, but there are examples at the bottom of the Viewer window. The ® icon resets the dialog box. Just to the right of the ® icon is an icon that looks like two pages. If you click on this icon, the command is copied to the Clipboard.

If you enter the `summarize` command directly in the Command window, simply follow it with the names of the variables for which you want summary statistics. For example, typing `summarize studytime age` will display only statistics for the two variables named `studytime` and `age`.

In the Results window, the `summarize` command will display the number of observations (also called cases or $N$), the mean, the standard deviation, the minimum value, and the maximum value for each variable.

```
. summarize
```

| Variable | Obs | Mean | Std. Dev. | Min | Max |
|---|---|---|---|---|---|
| studytime | 48 | 15.5 | 10.25629 | 1 | 39 |
| died | 48 | .6458333 | .4833211 | 0 | 1 |
| drug | 48 | 1.875 | .8410986 | 1 | 3 |
| age | 48 | 55.875 | 5.659205 | 47 | 67 |
| _st | 48 | 1 | 0 | 1 | 1 |
| _d | 48 | .6458333 | .4833211 | 0 | 1 |
| _t | 48 | 15.5 | 10.25629 | 1 | 39 |
| _t0 | 48 | 0 | 0 | 0 | 0 |

The first line of output displays the dot prompt followed by the command. After that, the output appears as a table. As you can see, there are 48 observations in this dataset. *Observations* is a generic term. These could be called participants, patients, subjects, organizations, cities, or countries depending on your field of study. In Stata, each row of data in a dataset is called an observation. The average, or mean, age is 55.875 years with a standard deviation of 5.659,[1] and the subjects are all between 47 (the minimum) and 67 (the maximum) years old.

If you have computed means and standard deviations by hand, you know how long this can take. Stata's virtually instant statistical analysis is what makes Stata so valuable. It takes time and skill to set up a dataset so that you can use Stata to analyze it, but once you learn how to set up a dataset (chapter 2), you will be able to compute a wide variety of statistics in little time.

We will do one more thing in this Stata session: we will make the histogram for the `age` variable, shown in figure 1.7.

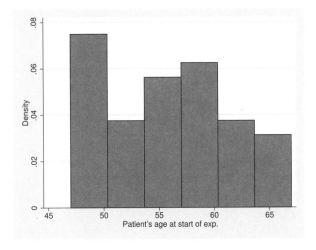

Figure 1.7. Histogram of `age`

A histogram is just a graph that shows the distribution of a variable, such as `age`, that takes on many values.

Simple graphs are simple to create. Just type the command **histogram age** in the Command window, and Stata will produce a histogram using reasonable assumptions. I will show you how to use the dialog boxes for more complicated graphs shortly.

At first glance, you may be happy with this graph. Stata used a formula to determine that six bars should be displayed, and this is reasonable. However, Stata starts the lowest bar (called a bin) at 47 years old, and each bin is 3.33 years wide (this information is displayed in the Results window) even though we are not accustomed to measuring years in thirds of a year. Also notice that the vertical axis measures density, but we

---

1. I may round numbers in the text to fewer digits than shown in the output unless it would make finding the corresponding number in the output difficult.

might prefer that it measure the frequency, that is, the number of people represented by each bar.

Using the dialog box can help us customize our histogram. Let's open the `histogram` dialog box shown in figure 1.8 by selecting Graphics ▷ Histogram from the menu bar.

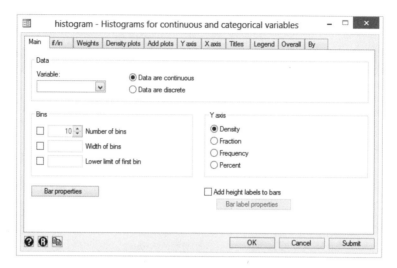

Figure 1.8. The `histogram` dialog box

Let's quickly go over the parts of the dialog box. There is a textbox labeled *Variable* with a pulldown menu. As we saw on the `summarize` dialog, you can pull down the list of variables and click on a variable name to enter it in the box, or you can type the variable's name yourself. Only one variable can be used for a histogram, and here we want to use `age`. If we stop here and click on OK, we will have re-created the histogram shown in figure 1.7.

There are two radio buttons visible to the right of the *Variable* box: one labeled *Data are continuous* (which is shown selected in figure 1.8) and one labeled *Data are discrete*. Radio buttons indicate mutually exclusive items—you can choose only one of them. Here we are treating `age` as if it were continuous, so make sure that the corresponding radio button is selected. On the right side of the Main tab is a section labeled *Y axis*. Click on the radio button for *Frequency* so that the histogram shows the frequency of each interval. In the section labeled *Bins*, check the box labeled *Width of bins* and type `2.5` in the textbox that becomes active (because the variable is `age`, the 2.5 indicates 2.5 years). Also check the box labeled *Lower limit of first bin* and type `45`, which will be the smallest age represented by the bar on the left.

The dialog box shows a sequence of tabs just under its title bar, as shown in figure 1.9. Different categories of options will be grouped together, and you make a different set of options visible by clicking on each tab. The options you have set on the current tab will not be canceled by clicking on another tab.

Figure 1.9. The tabs on the `histogram` dialog box

Graphs are usually clearer when there is a title of some sort, so click on the Titles tab and add a title. Here we type `Age Distribution of Participants in Cancer Study` in the *Title* box. Let's add the text `Data: Sample cancer dataset` to the *Note* box so that we know which dataset we used for this graph. Your dialog box should look like figure 1.10.

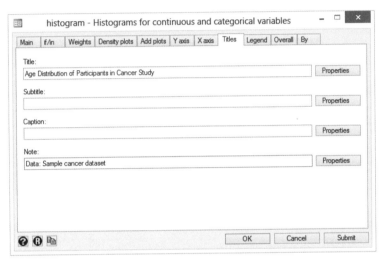

Figure 1.10. The Titles tab of the `histogram` dialog box

Now click on the Overall tab. Let's select *s1 monochrome* from the pulldown menu on the *Scheme* box. Schemes are basically templates that determine the standard attributes of a graph, such as colors, fonts, and size; which elements will be shown; and more.

From the Legend tab, under the *Legend behavior* section, click on the radio button for *Show legend*. Whether a legend will be displayed is determined by the scheme that is being used, and if we were to leave *Default* checked, our histogram might have a legend or it might not, depending on the scheme. Choosing *Show legend* or *Hide legend* overrides the scheme, and our selection will always be honored.

Now that we have made these changes, click on Submit instead of OK to generate the histogram shown in figure 1.11. The dialog box does not close. To close the dialog box, click on the X (close) button in the upper right corner, but we are not ready to do that yet.

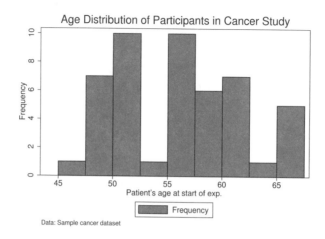

Figure 1.11. First attempt at an improved histogram

If you look at the complex command that the dialog box generated, you will see why even experienced Stata programmers will often rely on the dialog box to create **graph** commands. In reading this command, you will want to ignore the opening dot (Stata prints this in front of commands in the Results window, but the dot is not part of the command and you do not type it). Stata prints the > sign at the start of the second and third line, which might be confusing. Stata uses the Enter key to submit a command. Because of this, Stata sees the entire command as one line. To print the entire line in the confines of the Results window, Stata inserts the > for a line break. If you wanted to enter this command in the Command window, you would simply type the entire thing without the > and let Stata do the wrapping as needed in the Command window. Never press the Enter key until you have entered the entire command.

```
. histogram age, width(2.5) start(45) frequency
> title(Age Distribution of Participants in Cancer Study)
> note(Data: Sample cancer dataset) legend(on) scheme(s1mono)
```

### Clearing the Results window: The cls command

As you run commands, the results are displayed in the Results window. There may be times when you want to clear the Results window, so that, for example, seeing the top of the results of a command is easier, especially if your commands and results are lengthy. Beginning with Stata 13, you can type the **cls** command (with no options) to clear the Results window.

It is much more convenient to use the dialog box to generate that command than to try to remember all its parts and the rules of their use. If you do want to enter a long command in the Command window, remember to type it as one line. Whenever you press Enter, Stata assumes that you have finished the command and are ready to submit it for processing.

**When to use Submit and when to use OK**

Stata's dialogs give you two ways to run a command: by clicking on OK or by clicking on Submit. If you click on OK, Stata creates the command from your selections, runs the command, and closes the dialog box. This is just what you want for most tasks. At times, though, you know you will want to make minor adjustments to get things just right, so Stata provides the Submit button, which still runs the command but leaves the dialog open. This way, you can go back to the dialog box and make changes without having to reopen the dialog box.

The resulting histogram in figure 1.11 is an improvement, but we might want fewer bins. Here we are making small changes to a Stata command, then looking at the results, and then trying again. The Submit button is useful for this kind of interactive, iterative work. If the dialog box is hidden, we can use the Alt+Tab (Windows) or Cmd+Tab (Mac) key combination to move through Stata's windows until the one we want is on top again.

Instead of a width of 2.5 years, let's use 5 years, which is a more common way to group ages. If you clicked on OK instead of on Submit, you need to reopen the `histogram` dialog box as you did before. When you return to a dialog that you have already used in the current Stata session, the dialog box reappears with the last values still there. So all you need to do is change 2.5 to 5 in the *Width of bins* box on the Main tab and click on Submit. The result is shown in figure 1.12.

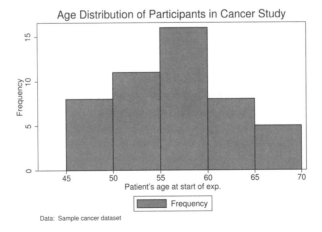

Figure 1.12. Final histogram of age

Notice how different the three graphs appear. You need to use judgment to pick the best combination and avoid using graphs that misrepresent the distribution. A good graph will give the reader a true picture of the distribution, but a poor graph may be quite deceptive. When people say that you can lie with statistics, they are often thinking about graphs that do not provide a fair picture of a distribution or a relationship. Can you think of any more improvements? The legend at the bottom center of the graph is unnecessary. You might want to go back to the dialog box, click on the Legend tab, and click on *Hide legend* to turn off the legend.

To finish our first Stata session, we need to close Stata. Do this with File ▷ Exit. If you are using Stata for Mac, select Stata ▷ Quit Stata.

## 1.6   Summary

We covered the following topics in this chapter:

- The font and punctuation conventions I will use throughout the book

- The Stata interface and how you can customize it

- How to open a sample Stata dataset

- The parts of a dialog box and the use of the OK and Submit buttons

- How to summarize the variables

- How to create and modify a simple histogram

## 1.7   Exercises

Some of these exercises involve little or no writing; they are simply things you can do to make sure you understand the material. Other exercises may require a written answer.

1. You can copy and paste text to and from Stata as you wish. You should try highlighting some text in Stata's Results window, copying it to the Clipboard, and pasting it into another program, such as your word processor. To copy highlighted text, you can use the Edit ▷ Copy menu or, as indicated on the menu, Ctrl+c. You will probably need to change the font to a monospaced font (for example, Courier), and you may need to reduce its font size (for example, to 9 point) after pasting it to prevent the lines from wrapping. You may wish to experiment with copying Stata output into your word processor now so that you know which font size and typeface work best. It may help to use a wider margin, such as 1 inch, on each side.

2. After you highlight material in the Results window, right-click on it. You can copy this output in several formats, including Copy, Copy Table (only works with some

commands), Copy Table as HTML, and Copy as Picture (copies a graphic image
of what you highlighted). The Copy option works nicely, but you will need to
use a monospaced font, such as Courier, and may need to use a smaller font size
when you paste it into your word processing document. The Copy Table option
is limited because it only works with a few commands. The Copy Table as HTML
option will create a table that looks like what you would see on a webpage. Using
Microsoft Word, you can edit the table by making columns wider or narrower and
by aligning the columns so that each number has the same number of decimal
places. Just copy the tabular results and not the command when using the HTML
option. The Copy as Picture option works nicely in Windows, but you cannot edit
it in Word because it is a graphic image. In Word, you can resize the image.

Run the `summarize` command and copy the results to a Word document by using
each of the options. Highlight the table. Right-click on it and then select the
option you want. Switch to your word processor. Press Ctrl+v to paste what you
copied. In your word processor, make the table as nice as you can by adjusting
the font, font size, margins, etc.

3. Stata has posted all the datasets from its manuals that were used to illustrate
how to do procedures. You can access the manual datasets from within Stata by
going to the File ▷ Example Datasets... menu, which will open a Viewer window.
Click on *Stata 13 manual datasets* and then click on *User's Guide [U]*.

The Viewer window works much like a web browser, so you can click on any of
the links in the list of datasets. Scroll down to chapter 25, and select the `use` link
for `censusfv.dta`, which opens a dataset that is used for chapter 25 of the *User's
Guide*. Run two commands, `describe` and `summarize`. What is the variable
`divorcert` and what is the mean (average) divorce rate for the 50 states?

4. Open `cancer.dta`. Create histograms for `age` using bin widths of 1, 3, and 5. Use
the right mouse button to copy each graph to the Clipboard, and then paste it
into your word processor. Does the overall shape of the histogram change as the
bins get wider? How?

5. UCLA has a Stata portal containing a lot of helpful material about Stata. You
might want to browse the collection now just to get an idea of the topics covered
there. The URL for the main UCLA Stata page is

http://www.ats.ucla.edu/stat/stata/

In particular, you might want to look at the links listed under Learning Stata.
On the *Stata Starter Kit* page, you will find a link to *Class notes with movies*.
These movies demonstrate using Stata's commands rather than the dialog box.
The topics we will cover in the first few chapters of this book are also covered on
the UCLA webpage using the commands. Each movie is about 25 minutes long.
Some of these movies are for older versions of Stata, but they are still useful.

# 2 Entering data

## 2.1  Creating a dataset

In this chapter, you will learn how to create a dataset. Data entry and verification can be tedious, but these tasks are essential for you to get accurate data. More than one analysis has turned out wrong because of errors in the data.

In the first chapter, you worked with some sample datasets. Most of the analyses in this book use datasets that have already been created for you and are available on the book's webpage, or they may already be installed on your computer. If you cannot wait to get started analyzing data, you can come back to this chapter when you need to create a dataset. You should look through the sections in this chapter that deal with documenting and labeling your data because not all prepared datasets will be well documented or labeled.

The ability to create and manage a dataset is invaluable. People who know how to create a dataset are valued members of any research team. For small projects, collecting, entering, and managing the data can be straightforward. For large, complex studies that extend over many years, managing the data and the documentation can be as complex as any of the analyses that may be conducted.

Stata, like most other statistical software, almost always requires data that are set in a grid, like a table, with rows and columns. Each row contains data for an observation (which is often a subject or a participant in your study), and each column contains the measurement of some variable of interest about the observation, such as age. This system will be familiar to you if you have used a spreadsheet.

## Variables and items

In working with questionnaire data, an item from the questionnaire almost always corresponds to a variable in the dataset. So if you ask for a respondent's age, then you will have a variable called `age`. Some questionnaire items are designed to be combined to make scales or composite measures of some sort, and new variables will be created to contain those items, but there is no single item corresponding to the scale or construct on the questionnaire. A questionnaire may also have an item where the respondent is asked to "mark all that apply", and commonly there is a variable for each category that could be marked under that item. The terms *item* and *variable* are often used interchangeably, but they are not synonyms. An item always refers to one question. A variable may be the score on an item or a composite score based on several items.

There is more to a dataset than just a table of numbers. Datasets usually contain labels that help the researcher use the data more easily and efficiently. In a dataset, the columns correspond to variables, and variables must be named. We can also attach to each variable a descriptive label, which will often appear in the output of statistical procedures. Because most data will be numbers, we can also attach value labels to the numbers to clarify what the numbers mean.

It is extremely helpful to pick descriptive names for each variable. For example, you might call a variable that contains people's responses to a question about their state's schools q23, but it is often better to use a more descriptive name, such as `schools`. If there were two questions about schools, you might want to name them `schools1` and `schools2` rather than q23a and q23b. If a set of items are to be combined into scales and are not intended to be used alone, you may want to use names that correspond to the questionnaire items for the original variables and reserve the descriptive names for the composite scores. This is useful with complex datasets where several people will be using the same dataset. Each user will know that q23a refers to question 23a, whereas "friendly" names like `schools1` may make sense to one user but not to another user.

There are a few variable names that Stata reserves: `_all`, `_cons`, `_N`, `using`, `with`, etc. For example, Stata uses the variable name `_N` to always be the size of your sample and uses the variable name `_cons` to refer to the constant (intercept). See [U] **11.3 Naming conventions** for the complete list of reserved names.

No matter what the logic of your naming, try to keep names short. You will probably have to type them often, and shorter names offer fewer opportunities for typing errors.

A respondent's age can be named `age` or `age_of_respondent` (note that no blank spaces are allowed in a name).

Even with relatively descriptive variable names, it is usually helpful to attach a longer and, one hopes, more descriptive label to each variable. We call these *variable labels* to distinguish them from *variable names*. For example, you might want to further describe the variable `schools` with the label "Public school rating". The variable label gives us a clearer understanding of the data stored in that variable.

For some variables, the meaning of the values in the data is obvious. If you measure people's height in inches, then when you see the values, it is clear what they mean. For other variables, the meaning of the values needs to be specified with a value label. Most of us have run across questions that ask us if we "Strongly agree", "Agree", "Disagree", or "Strongly disagree", or that ask us to answer some question with "Yes" or "No". This sort of data is usually coded as numbers, such as 1 for "Yes" and 2 for "No", and it can make understanding tables of data much easier if you create *value labels* to be displayed in the output in addition to (or instead of) the numbers.    *How?*

If your project is just a short homework assignment, you could consider skipping the labeling (though your instructor would no doubt appreciate anything that makes your work easier to understand). For any serious work though, clear labeling will make your work much easier in the long run. Remember, we need a variable name, a descriptive label for the variable, and sometimes labels for the values the variable can have.

## 2.2   An example questionnaire

We have discussed datasets in general, so now let's create one. Suppose that we conducted a survey of 20 people and asked each of them six questions, which are shown in the example questionnaire in figure 2.1.

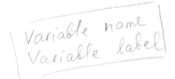
*Variable name*
*Variable label*

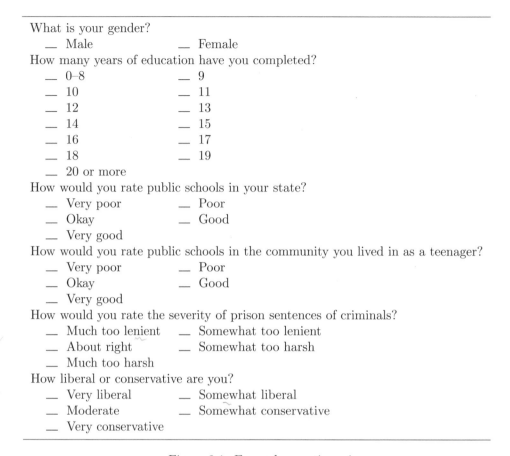

Figure 2.1. Example questionnaire

Our task is to convert the questionnaire's answers into a dataset that we can use with Stata.

## 2.3   Developing a coding system

Statistics is most often done with numbers, so we need to have a numeric coding system for the answers to the questions. Stata can use numbers or words. For example, we could type Female if a respondent checked "Female". However, it is usually better to use some sort of numeric coding, so you might type 1 if the respondent checked "Male" on the questionnaire and 2 if the respondent checked "Female". We will need to assign a number to enter for each possible response for each of the items on the survey.

You will also need a short variable name for the variable that will contain the data for each item. Variable names can contain uppercase and lowercase letters, numerals, and the underscore character, and they can be up to 32 characters long. No blank spaces are allowed in variable names. The variable name `mother age` would be interpreted as two variables, `mother` and `age`. Generally, you should keep your variable names to 10 characters or fewer, but 8 or fewer is best. Variable names should start with a letter.

If appropriate, you should explain the relationship between any numeric codes and the responses as they appeared on the questionnaire. For an example, see the example codebook (not to be confused with the Stata command `codebook`, which we will use later) for our questionnaire in table 2.1.

Table 2.1. Example codebook

| Question | Variable name | Value labels | Code |
|---|---|---|---|
| Identification number | | | |
| | id | Record in order | 1 to 20 |
| What is your gender? | | | |
| | gender | Male | 1 |
| | | Female | 2 |
| | | No answer | −9 |
| How many years of education have you completed? | | | |
| | education | 0–8 | 8 |
| | | 9–20 | 9 to 20 |
| | | No answer | −9 |
| Rate public schools in your state. | | | |
| | sch_st | Very poor | 1 |
| | | Poor | 2 |
| | | Okay | 3 |
| | | Good | 4 |
| | | Very good | 5 |
| | | No answer | −9 |
| Rate public schools in the community you lived in as a teenager. | | | |
| | sch_com | Very poor | 1 |
| | | Poor | 2 |
| | | Okay | 3 |
| | | Good | 4 |
| | | Very good | 5 |
| | | No answer | −9 |
| Rate severity of prison sentences of criminals. They are ... | | | |
| | prison | Much too lenient | 1 |
| | | Too lenient | 2 |
| | | About right | 3 |
| | | Too harsh | 4 |
| | | Much too harsh | 5 |
| | | No answer | −9 |
| How liberal or conservative are you? | | | |
| | conserv | Very liberal | 1 |
| | | Liberal | 2 |
| | | Moderate | 3 |
| | | Conservative | 4 |
| | | Very conservative | 5 |
| | | No answer | −9 |

A codebook translates the numeric codes in your dataset back into the questions you asked your participants and the choices you gave them for answers. Regardless of whether you gather data with a computer-aided interviewing system or with paper questionnaires, the codebook is essential to help make sense of your data and your analyses. If you do not have a codebook, you might not realize that everyone with eight or fewer years of education is coded the same way, with an 8. That may have an impact on how you use that variable in later analyses.

We have added an `id` variable to identify each respondent. In simple cases, we just number the questionnaires sequentially. If we have a sample of 5,000 people, we will number the questionnaires from 1 to 5,000, write the identification number on the original questionnaire, and record it in the dataset. If we discover a problem in our dataset (for example, somebody with a coded value of 3 for `gender`), we can go back to the questionnaire and determine what the correct value should be. Some researchers will eventually destroy the link between the ID and the questionnaire as a way to protect human participants. It is good to keep the original questionnaires in a safe place that is locked and accessible only by members of the research team.

Some data will be missing—people may refuse to answer certain questions, interviewers forget to ask questions, equipment fails; the reasons are many, and we need a code to indicate that data are missing. If we know why the answer is missing, we will record the reason too, so we may want to use different codes that correspond to different reasons the data are missing.

On surveys, respondents may refuse to answer, may not express an opinion, or may not have been asked a particular question because of the answer to an earlier question. Here we might code "invalid skip" (interviewer error) as -5, "valid skip" (not applicable to this respondent) as -4, "refused to answer" as -3, "don't know" as -2, and "missing for any other reason" as -1. For example, adolescent boys should not be asked when they had their first menstrual period, so we would type -4 for that question if the respondent is male. We should pick values that can never be a valid response. In chapter 3, we will redefine these values to Stata's missing-value codes. In this chapter, we will use only one missing-value code, -9.

We will be entering the data manually, so after we administered the questionnaire to our sample of 20 people, we prepared a coding sheet that will be used to enter the data. The coding sheet originates from the days when data were entered by professional keypunch operators, but it can still be useful or necessary. When you create a coding sheet, you are converting the data from the format used on the questionnaire to the format that will actually be stored in the computer (a table of numbers). The more the format of the questionnaire differs from a table of numbers, the more likely it is that a coding sheet will help prevent errors.

There are other reasons you may want to use a coding sheet. For example, your questionnaire may include confidential information that should be shown to as few people as possible; also, if there are many open-ended questions for which extended answers were collected but will not be entered, working from the original questionnaire can be unwieldy.

In general, if you transcribe the data from the questionnaire to the coding sheet, you will need to decide which responses go in which columns. Deciding this will reduce errors from those who perform the data entry, who may not have the information needed to make those decisions properly.

Whether you enter the data directly from the questionnaire or create a coding sheet will depend largely on the study and on the resources that are available to you. Some experience with data entry is valuable because it will give you a better sense of the problems you may encounter, whether you or someone else enters the data. For our example questionnaire, we have created a coding sheet, shown in table 2.2.

Table 2.2. Example coding sheet

| id | gender | education | sch_st | sch_com | prison | conserv |
|----|--------|-----------|--------|---------|--------|---------|
| 1  | 2      | 15        | 4      | 5       | 4      | 2       |
| 2  | 1      | 12        | 2      | 3       | 1      | 5       |
| 3  | 1      | 16        | 3      | 4       | 3      | 2       |
| 4  | 1      | 8         | -9     | 1       | -9     | 5       |
| 5  | 2      | 12        | 3      | 3       | 3      | 3       |
| 6  | 2      | 18        | 4      | 5       | 5      | 1       |
| 7  | 1      | 17        | 3      | 4       | 2      | 4       |
| 8  | 2      | 14        | 2      | 3       | 1      | 5       |
| 9  | 2      | 16        | 5      | 5       | 4      | 1       |
| 10 | 1      | 20        | 4      | 4       | 5      | 2       |
| 11 | 1      | 12        | 2      | 2       | 1      | 5       |
| 12 | 2      | 11        | -9     | 1       | 1      | 3       |
| 13 | 2      | 18        | 5      | 5       | -9     | -9      |
| 14 | 1      | 16        | 5      | 5       | 5      | 1       |
| 15 | 2      | 16        | 5      | 5       | 4      | 2       |
| 16 | 1      | 17        | 4      | 3       | 4      | 3       |
| 17 | 1      | 12        | 2      | 2       | -9     | 1       |
| 18 | 2      | 12        | 2      | 2       | 2      | 2       |
| 19 | 2      | 14        | 4      | 5       | 4      | 4       |
| 20 | 1      | 13        | 3      | 3       | 5      | 5       |

Because we are not reproducing the 20 questionnaires in this book, it may be helpful to examine how we entered the data from one of them. We will use the ninth questionnaire. We have assigned an id of 9, as shown in the first column. Reading from left to right, for gender, we have recorded a 2 to indicate a woman, and for education, we have recorded a response of 16 years. This woman rates schools in her state and in the community in which she lived as a teenager as very good, which we can see from the 5s in the fourth and fifth columns. She thinks prison sentences are too harsh and she considers herself to be very liberal (4 and 1 in the sixth and seventh columns, respectively).

## 2.4   Entering data using the Data Editor

Stata has a Data Editor, shown in figure 2.2, in which we can enter our data from the coding sheet in table 2.2. The Data Editor provides an interface that is similar to that of a spreadsheet, but it has some features that are particularly suited to creating a Stata dataset.

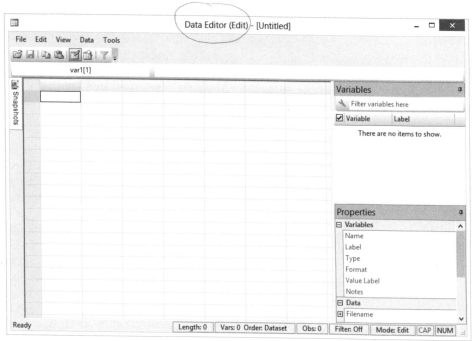

Figure 2.2. The Data Editor

Before we begin, let's type the command `clear` in the Command window. This step will give us a fresh Data Editor in which to enter data. To open the Data Editor in edit mode, type `edit` in the Command window. Alternatively, click on the icon on the toolbar that looks like a spreadsheet with a pencil pointing to it (see figure 2.3) or click on Data ▷ Data Editor ▷ Data Editor (Edit). The Data Editor in edit mode is a convenient way to enter or change data, but if you just want to see the data, you should use the Data Editor in browse mode. The icon for the Data Editor (Browse) is just to the right of the icon for the Data Editor (Edit); it is a spreadsheet with a magnifying glass on top (see figure 2.3). You can also type the command `browse` in the Command window.

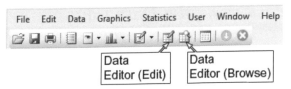

Figure 2.3. Data Editor (Edit) and Data Editor (Browse) icons on the toolbar

Data are entered in the white columns, just as in other spreadsheets. Let's enter the data for the first respondent from the coding sheet in table 2.2. In the first cell under the first column, enter the identification number, which is a 1 for the first observation. Press the Tab key to move one cell to the right, and enter the value for the person's gender, which is a 2 for the first observation. Keep entering the values and pressing Tab until you have entered all the values for the first participant. Now you can press the Enter key, which will move the cursor to the second row of the Data Editor (Edit), and then press the Home key, which will move the cursor to the first column of this row. You can proceed to enter the rest of the data. The Tab key is smart, so after you enter the last value for the second row of data, pressing Tab will move the cursor to the first column of the third row. However, before doing this it would be good to enter more informative variable names to replace the default names of var1, var2, etc. To change a variable's name, click anywhere in the column for that variable or click on its name in the Variables pane at the top right of the Data Editor, for example, var1. Then in the Properties pane at the lower right of the Data Editor, change *Name* from var1 to id. The name can be up to 32 characters, but it is usually best to keep the name short. Stata is case sensitive, so it is best to use lowercase names. By using lowercase names, you will not have to remember what case to use because the names will all be lowercase. From the Properties pane, we can also change the variable's label to be "Respondent's identification" by typing this text in the *Label* box. Our Data Editor (Edit) will look like figure 2.4.

Figure 2.4. Variable name and variable label

Once we click outside the changed field in the Properties pane or press **Enter**, we will see the new variable name displayed. Change each of the remaining generic variable names to rename and label them as listed in table 2.3.

Table 2.3. New variable names and labels

| Generic variable name | New variable name | Variable label |
| --- | --- | --- |
| var2 | gender | Participant's gender |
| var3 | education | Years of education |
| var4 | sch_st | Ratings of schools in your state |
| var5 | sch_com | Ratings of schools in your community of origin |
| var6 | prison | Ratings of prison sentences |
| var7 | conserv | Conservatism/liberalism |

As well as allowing us to specify the variable name and variable label, the Properties pane includes a place to specify the format of the variable. The current format is %9.0g, which is the default format for numeric data in the Data Editor. This format instructs Stata to try to print numbers within nine columns. Stata will adjust the format if you enter a number wider than nine columns.

**I typed the letter l for the number 1**

A common mistake is to type a letter instead of a numeric value, such as typing the letter l rather than the number 1 or simply pressing the wrong key. When this happens for the first value you enter for a variable, Stata will make the variable a "string" variable, meaning that it is no longer a numeric variable. When this happens for a subsequent value you enter for a variable, Stata will display an error message. If you made the mistake of inputting a variable as a string variable instead of a numeric variable, you need to correct this. Simply changing the letter to a number will not work because a string variable may include numbers, such as a variable for a street address. Stata will still think it is a string variable. There is a simple command to force Stata to change the variable from a string variable back to a numeric variable: `destring` *varlist*, `replace`. For example, to change the variable `id` from string to numeric, you would type `destring id, replace`.

If you have string data (that is, data that contain letters or symbols), you will see a format that looks like %9s, which indicates a variable that is a string (the s) with nine characters. For more detailed information about formats, see the *Stata Data-Management Reference Manual*, which is available under Help ▷ PDF Documentation. Some people like to enter string data where they enter yes or no, male or female, and so on. It is usually best here to simply enter numbers; I will show how to add value labels for the numbers in the next section. Entering numbers will benefit you later because some Stata commands work fine with string variables, but many of them require numeric values. After entering the rest of your data, the Data Editor (Edit) should now look like figure 2.5.

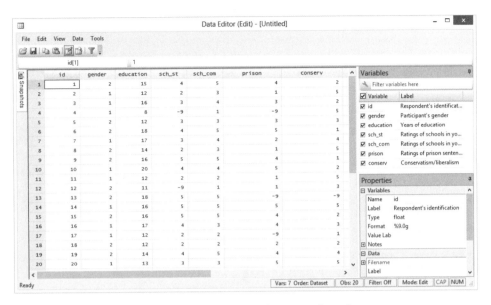

Figure 2.5. Data Editor with a complete dataset

### 2.4.1   Value labels

It would be nice to have labels for the values on most of these variables. For example, we would like **gender** labeled, so a code of 1 was labeled "Male" and a code of 2 was labeled "Female". We would not want to label values for **id** or **education**, but the rest of them should have value labels. In Stata, labeling values is a two-step process.

1. First, we define a label and give it a nickname. Some defined labels may apply to just one variable, such as **gender**, but other defined labels may apply to multiple variables that share identical value labels, such as **sch_st** and **sch_com**.

2. Second, we assign the defined label to each variable. You may not appreciate this being a two-step process with a small dataset, but in a major study, this is a valuable feature of Stata. For example, you might have 30 variables that are all coded as 1 for yes and 2 for no. You could define one value label and give it a nickname of **agree**. Then we could apply this defined label to each of the 30 variables.

## 2.5   The Variables Manager

We could use the Data Editor to define and apply value labels, but we will use another feature of Stata, the Variables Manager. The Stata for Windows and Stata for Unix toolbars have an icon for the Variables Manager, as shown in figure 2.6. By default, this icon is not shown in the Stata for Mac toolbar. You can add the Variables Manager icon by right-clicking on the toolbar and selecting **Customize Toolbar**.... You can then drag the Variables Manager icon to the toolbar.

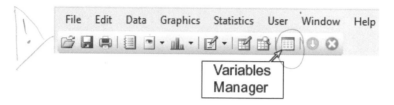

Figure 2.6. The Variables Manager icon on the Stata toolbar

Let's open the Variables Manager and define a value label for `gender`. We could have entered the variable names and variable labels here, but that was simple enough in the Data Editor. The Variables Manager is a convenient place to make any changes we might want. For example, when we click on `gender`, the Variable Properties pane on the right shows whatever information we already have for the variable: its name, label, type, format, value label, and notes.

For the moment, we will use the Variables Manager to define the value labels. Click on the **Manage**... button next to the *Value Label* field, which opens a new dialog box called *Manage Value Labels*; see figure 2.7. Because we have no value labels yet, nothing is listed. Click on **Create Label**, and the *Create Label* dialog box opens as shown in figure 2.7. For *Label name*, type `sex`. Notice that this is not the name of the variable but a nickname for the set of labels we will use for the variable. For *Value*, type 1; tab to *Label* and type `Male`; and then click on **Add**. Next type 2 under *Value*; tab to *Label* and type `Female`; and then click on **Add**. Figure 2.7 shows the Variables Manager and what we have done so far.

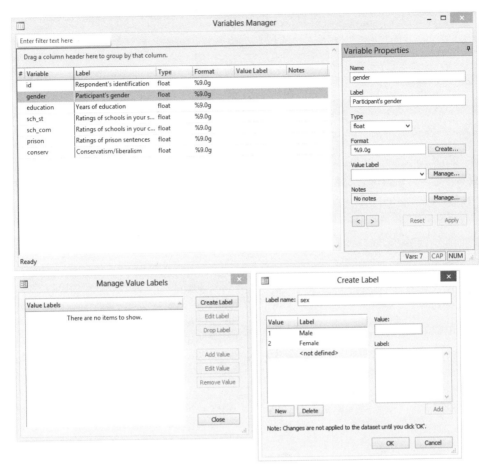

Figure 2.7. Using the Variables Manager to add a label for `gender`

Because we have no more values, click on **OK** to close the *Create Label* dialog box and add the label `sex` to the *Manage Value Labels* dialog box. Now click on the **Close** button of the *Manage Value Labels* dialog box to return to the Variables Manager. In the Variable Properties pane, we can click on the drop-down menu for *Value Label* and select our label of `sex`. Then click on **Apply**. You will notice that the value label for `gender` is now `sex` in the list of variables on the left.

Let's define three more value labels. We do not need to click on a variable. We can simply click on the **Manage...** button next to *Value Label*, and then click on **Create Label**, which opens the *Create Label* dialog box. Let's make a label nickname of `rating` that we will apply to `sch_st` and `sch_com`. For *Value*, type 1; tab to *Label* and type **Very poor**; and then click on **Add**. Repeat this step to add the following values and labels for the `rating` value label:

| Value | Label |
|-------|-------|
| 2 | Poor |
| 3 | Okay |
| 4 | Good |
| 5 | Very good |
| -9 | Missing |

This label needs to have a final value of $-9$ with the label "Missing" because not everybody answered the relevant items. Click on OK in the *Create Label* dialog box.

Repeat these steps, starting by clicking on Create Label, and create value labels named `harsh` and `conserv` with the following values and labels:

| Label name | Value | Label |
|------------|-------|-------|
| harsh | 1 | Much too lenient |
|  | 2 | Too lenient |
|  | 3 | About right |
|  | 4 | Too harsh |
|  | 5 | Much too harsh |
|  | -9 | Missing |
| conserv | 1 | Very liberal |
|  | 2 | Liberal |
|  | 3 | Moderate |
|  | 4 | Conservative |
|  | 5 | Very conservative |
|  | -9 | Missing |

Once you have finished creating the last value label, click on Close in the *Manage Value Labels* dialog box. Now we are back to the Variables Manager. Here we can click on `sch_st`, and then click on the drop-down menu next to *Value Label*. We select our just created nickname, `rating`. Clicking on Apply will assign this set of value labels to the `sch_st` variable. We can repeat this for `sch_com` because that variable uses the same set of value labels, that is, `rating`.

Now click on `prison`, and then click on the drop-down menu next to *Value Label*. Select the nickname `harsh`. Clicking on Apply will assign this set of values labels to the `prison` variable. Repeat these steps to assign the nickname `conserv` to the `conserv` variable.

Going through a series of dialog boxes can be confusing. You should experiment with these dialog boxes until you are confident in using them. Try to change some of the value labels, for example, use the *Edit Label* dialog box that appears as an option when you click on the Manage... button next to *Value Label* in the Variable Properties pane.

Fortunately, working with value labels is the most confusing part of data management that we will cover. If you can get comfortable with this, the rest of the material on data management will be less confusing. The key to remember is that you use the *Value Label* dialog box to define a nickname (sex, harsh, rating, conserv) for sets of value labels. Then you apply these sets to the appropriate variables. As you do this, Stata will issue the commands you created, and they will appear in the Results window. Study these commands to see what is involved, and as you get more experience, you will use the commands directly. Here are the commands we generated using the Variables Manager:

```
. label define sex 1 "Male" 2 "Female"
. label values gender sex
. label define rating 1 "Very poor" 2 "Poor" 3 "Okay" 4 "Good" 5 "Very good"
> -9 "Missing"
. label define harsh 1 "Much too lenient" 2 "Too lenient" 3 "About right"
> 4 "Too harsh" 5 "Much too harsh" -9 "Missing"
. label define conserv 1 "Very liberal" 2 "Liberal" 3 "Moderate"
> 4 "Conservative" 5 "Very conservative"
. label values sch_st rating
. label values sch_com rating
. label values prison harsh
. label values conserv conserv
```

We will see how to enter these commands in a do-file in chapter 4. A brief explanation of commands may help now. The label define command says that we are going to define some value labels. Then we have a nickname (sex, harsh, rating, conserv). After the nickname, we have the values followed by their labels. We put quotes around each of the labels. After we have defined the sets of labels, the label values command applies these sets of labels to the variables. For example, the command label values gender sex means to apply value labels to the gender variable and use the set of value labels that have the nickname of sex.

What happened to the Variables Manager? It now looks like figure 2.8:

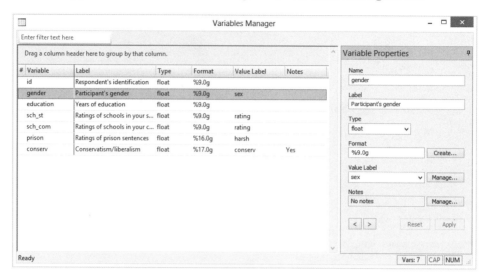

Figure 2.8. Variables Manager with value labels added

We can use the Variables Manager in more ways. Let's add a note to the variable `conserv`. We want to make sure anyone using our dataset knows that this is a single-item measure and not a scale of conservatism or liberalism that is based on a series of items. First, highlight the `conserv` variable by clicking on it. Next click on the **Manage**... button next to *Notes* in the Variable Properties pane. A dialog box opens telling us that there are no items (notes) to show, so we click on **Add**. Enter `This is a single-item measure` and click on **Submit**. Click on **Close** to close this dialog box.

There are several reasons for applying a note to a variable. When several people are working on a large dataset and making changes, it is nice to have a record of those changes. For example, you might find that both observation 222 and observation 318 were coded with an out-of-range value, say, 8 on `sch_com`. You check the original questionnaires they answered and find that observation 222 should have been coded as a 4 and observation 318 should have been coded a missing value of −9. You could attach a note that "Observation 222 changed to 4 and 318 changed to −9 on 6/7/2012." You can have several notes assigned to each variable.

Another thing you might want to do is add a note to the entire dataset, not just to a specific variable. If you right-click on any variable, a drop-down menu appears with several useful options including *Manage Notes for Dataset...*. When you click on this option, you get a dialog box that looks just like the one of *Notes for variable*, but it is for *Notes for Data*. You can click on **Add** and then type `These data were collected in January of 2012`. Now click on **Submit** and then **Close** to close the dialog box. Dataset notes can be very useful for keeping track of general changes in the dataset. For example, if Aidan merged some new variables from a different dataset into this dataset, it would be good to know what dataset was merged and when.

To see how the notes look in output, enter two commands. First, type `describe` in the Command window, and then type `notes`.

```
. describe
Contains data
   obs:             20
   vars:             7
   size:           560                    (_dta has notes)

               storage   display    value
variable name   type     format     label     variable label

id              float    %9.0g                 Respondent's identification
gender          float    %9.0g      sex        Participant's gender
education       float    %9.0g                 Years of education
sch_st          float    %9.0g      rating     Ratings of schools in your state
sch_com         float    %9.0g      rating     Ratings of schools in your
                                                 community of origin
prison          float    %16.0g     harsh      Ratings of prison sentences
conserv         float    %17.0g     conserv  * Conservatism/liberalism
                                             * indicated variables have notes

Sorted by:
     Note:   dataset has changed since last saved

. notes

_dta:
   1.  These data were collected in January of 2012

conserv:
   1.  This is a single-item measure
```

The `describe` command gives us a nice description of the dataset including the number of observations, the number of variables, the size of the dataset, the date the dataset was originally constructed (displayed once the dataset is saved), and a comment that _dta has notes. An asterisk appears by the variable label for `conserv` in the list of variables, and a note at the bottom of the list tells us that an asterisk means there is a note attached to that variable. The `notes` command has one note for the dataset and one note for the `conserv` variable. When a large group of researchers is working on a large and complex research project, the use of notes is extremely important.

There are still other things we can do with the Variables Manager. Suppose that we have a large dataset with 1,500 variables. We are looking for a variable and cannot remember its exact name, but we remember that it starts with the letters so. We can click on *Variable*—the title of the column that includes all the variable names—to sort our variable list alphabetically. (Sorting our variable list in the Variables Manager does not change the order in which the variables are stored.) An alphabetical listing can help us find all the variables beginning with so very quickly. We find the variable; it is soc_sat1. If we want that variable included in a Stata command, such as `summarize`, we could type `summarize` in the Command window, and then right-click in the Variables Manager and choose *Send Varlist to Command Window*.

Still using this large dataset, we remember that a particular set of five variables we want were coded as yes_no variables. We could click on *Value Label*—the title of that column—to sort the variables by value labels. Instead of searching through 1,500 variables in the large dataset, we might have just 20 or 30 variables coded with the yes_no nickname. If we then remember that the variable we want is one of several that start with the variable label of "Peer", we could click on *Label* at the top of its column to sort the variables by variable labels.

At this point, we probably should restore the order of the variables as they are in the dataset. To return to the order of the variables in the dataset, click on the hash mark (#), which is to the left of the *Variable* title. You should experiment with other features in the Variables Manager. It is a valuable tool that was first made available in Stata 11.

## 2.6   The Data Editor (Browse) view

In section 2.4, we entered data using the Data Editor in edit mode. Now that the data have been entered, we want to check our work. We can use the Data Editor in browse mode to do this. The purpose of using the Data Editor in browse mode is to look at data without altering it.

Type browse in the Command window to open the Data Editor (Browse). In the Data Editor window that opens, you see the numeric data where you did not add value labels, but for all the variables to which you did assign value labels, you now see the value label of each person, as shown in figure 2.9.

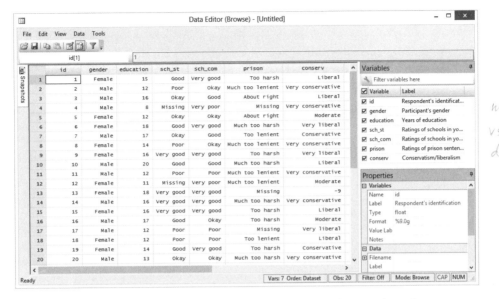

Figure 2.9. Dataset shown in the Data Editor (Browse) mode

Viewing the data in Data Editor (Browse) mode can be quite helpful. For example, person 11 has an ID of 11, is male, has 12 years of education, rates the schools in his state as poor, rates the schools in his community of origin as poor, feels that prison sentences are much too lenient, and considers himself to be very conservative. By seeing it laid out this way, you do not have to remember how each variable was coded. However, if you want to see just numeric values using the Data Editor (Edit), then you need to type the `edit, nolabel` command in the Command window. If you want to see just numeric values using the Data Editor (Browse), then you need to type the `browse, nolabel` command. Some researchers, when working on a small dataset, combine variables or change a variable and instantly see what happens in Browse mode. With a large dataset, this ability is usually not useful.

## 2.7   Saving your dataset

If you look at the Results window in Stata, you can see that the dialog box has done a lot of the work for you. The Results window shows that a lot of commands have been run to label the variables. These commands also appear in the Review window. The Variables window lists all your variables. We have now created our first dataset.

Until now, Stata has been working with our data in memory, and we still need to save the dataset as a file, which we can do from the File ▷ Save As... menu. Doing this will open the familiar dialog box where we can choose where we would like to save our file. Once the data have been saved to a file, we can exit Stata or use another dataset and not lose our changes.

Notice that Stata will give the file the .dta extension if you do not specify one. You must use this extension, because it identifies the file as a Stata dataset. You can give it any first name you want, but do not change the extension. Let's call this file firstsurvey and let Stata add the .dta extension for us. If you look in the Results window, you will see that Stata has printed something like this:

```
. save "C:\data\firstsurvey.dta"
file C:\data\firstsurvey.dta saved
```

It is a good idea to make sure that you spelled the name correctly and that you saved the file into the directory you intended. Note the quotation marks around the filename. If you type a save command in the Command window, you must use the quotes around the filename if the name contains a space; however, if you always use the quotes around the filename, you will never get an error, whether or not the name contains a space.

### Saving data and different versions of Stata

This book is written for users of Stata 13 and later. Stata is careful so that new versions can read data files that were saved using earlier versions of Stata. For example, if you were using an older version of Stata and upgraded, your new version should be able to read datasets you stored when you were using the older version. Stata is backward compatible in this sense.

Stata is not always forward compatible. New versions store data in different formats. Datasets saved in the new data format cannot be opened by older versions of Stata. One solution is to use the saveold command. For instance, if you are using Stata 13 and want to share a dataset with another user who is using Stata 12, you would type the following command:

```
. saveold "C:\data\firstsurvey.dta"
```

This command would store the dataset in the Stata 12 data format.

Stata works with one dataset at a time. When you open a new dataset, Stata will first clear the current one from memory. Stata knows when you have changed your data and will prompt you if you attempt to replace unsaved data with new data. It is, however, a good idea to get in the habit of saving before reading new data because Stata will not prompt you to save data if the only things that have changed are labels or notes.

A Stata dataset contains more than just the numbers, as we have seen in the work we have done so far. A Stata dataset contains the following information:

- Numerical data

- Variable names

- Variable labels

- Missing values

- Formats for printing data

- A dataset label

- Notes

- A sort order and the names of the sort variables

In addition to the labeling we have already seen, you can store notes in the dataset, and you can attach notes to particular variables, which can be a convenient way to document changes, particularly if more than one person is working on the dataset. To see all the things you can insert into a dataset, type the command `help label`. If the dataset has been sorted (observations arranged in ascending order based on the values of a variable), that information is also stored.

We have now created our first dataset and saved it to disk, so let's close Stata by selecting File ▷ Exit.

## 2.8   Checking the data

We have created a dataset, and now we want to check the work we did defining the dataset. Checking the accuracy of our data entry is also our first statistical look at the data. To open the dataset, use the File ▷ Open... menu. Locate `firstsurvey.dta`, which we saved in the preceding section, and click on Open.

Let's run a couple of commands that will characterize the dataset and the data stored in it in slightly different ways. We created a codebook to use when creating our dataset. Stata provides a convenient way to reproduce much of that information, which is useful if you want to check that you entered the information correctly. It is also useful if you share the dataset with someone who does not have access to the original codebook. Use the Data ▷ Describe data ▷ Describe data contents (codebook) menu to open the `codebook` dialog box. You can specify variables for the codebook entry you want, or you can get a complete codebook by not specifying any variables. Try it first without specifying any variables.

**Scrolling the results**

When Stata displays your results, it does so one screen at a time. When there is more output than will fit in the Results window, Stata prints what will fit and displays —more— in the lower left corner of the Results window. Stata calls this a *more condition* in its documentation. To continue, you can click on —more— in the Results window, but most users find it easier to just press the Spacebar. If you do not want to see all the output—that is, you want to cancel displaying the rest of it—you can press the q key. If you do not like having the output pause this way, you can turn this paging off by typing set more off in the Command window. All the output generated by the command will print without pause until you either change this setting back or quit and restart Stata. To restore the paging, use the set more on command.

This setting will only apply to your current session. If you want to make this change permanent, type set more off, permanently.

To see a list of other available settings, type help set.

Let's look at the codebook entry for **gender** as Stata produced it. For this example, you can either scroll back in the Results window or use the Data ▷ Describe data ▷ Describe data contents (codebook) menu and type **gender** in the *Variables* box, as I have done in the examples that follow.

```
. codebook gender
```

| gender | | | | | Participant's gender |
|---|---|---|---|---|---|

```
              type:  numeric (float)
             label:  sex

             range:  [1,2]                        units:  1
     unique values:  2                         missing .:  0/20

       tabulation:  Freq.   Numeric  Label
                      10          1  Male
                      10          2  Female
```

The first line lists the variable name, **gender**, and the variable label, Participant's gender. Next the type of the variable, which is numeric (float), is shown. The value-label mapping associated with this variable is sex. The range of this variable (shown as the lowest value and then the highest value) is from 1 to 2, there are two unique values, and there are 0 missing values in the 20 cases. This information is followed by a table showing the frequencies, values, and labels. We have 10 cases with a value of 1, labeled Male, and 10 cases with a value of 2, labeled Female.

Using codebook is an excellent way to check your categorical variables, such as **gender** and **prison**. If, in the tabulation, you saw three values for **gender** or six values for `prison`, you would know that there are errors in the data. Looking at these kinds of summaries of your variables will often help you detect data-entry errors, which is why it is crucial to review your data after entry.

Let's also look at the display for `education`:

```
. codebook education
```

| education | | | | | Years of education |
|---|---|---|---|---|---|

|  |  |  |  |  |  |
|---|---|---|---|---|---|
| type: | numeric (float) | | | | |
| range: | [8,20] | | | units: | 1 |
| unique values: | 10 | | | missing .: | 0/20 |
| mean: | 14.45 | | | | |
| std. dev: | 2.94645 | | | | |
| percentiles: | 10% | 25% | 50% | 75% | 90% |
|  | 11.5 | 12 | 14.5 | 16.5 | 18 |

Because this variable takes on more values, Stata does not tabulate the individual values. Instead, Stata displays the range (8 to 20) along with the mean, which is 14.45, and the standard deviation, which is 2.95. It also gives the percentiles. The 50th percentile is usually called the median, and here the median is 14.5.

In the `codebook` dialog box, if we leave the box where we enter variables blank, then `codebook` will give us a detailed codebook for every variable in the dataset. This detailed codebook can be combined with other options. By clicking on the Options tab, we can ask to include any notes that are attached to variables. Clicking on *Report potential problems in dataset* can be informative. Using this option, we might be told that `sch_st`, `sch_com`, and `conserv` have incomplete value labels. Why would this happen? If we did not assign a value label to the value of $-9$ for these variables, this option would inform us that there is a value that does not have an assigned value label. We will learn about how to manage missing values in the next chapter.

Sometimes we just want a brief summary of variables like we would get using the `describe` command. Enter `codebook, compact` in the Command window or click on *Display compact report on the variables* on the Options tab of the `codebook` dialog box. This option gives you seven columns of information including the variable name, the number of observations, the number of unique values the variable has in your data, the mean for the item, the minimum and maximum values, and the variable label. A variation of the `codebook, compact` command includes one or more variable names in the command, for example, `codebook gender, compact`. This gives you the same information as the `codebook, compact` command, but only for the variable(s) you specify. The `compact` option cannot be combined with the other options.

Another alternative to see a brief summary of variables is to actually use the `describe` command. Enter the command in the Command window, or use the Data ▷ Describe data ▷ Describe data in memory or in a file menu, which will display the dialog box in figure 2.10.

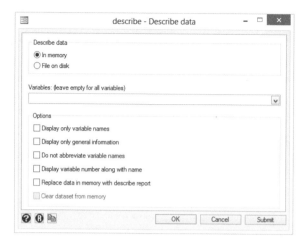

Figure 2.10. The `describe` dialog box

If you do not list any variables in the *Variables* box, all the variables will be included, as shown in the following output:

```
. describe
Contains data from C:\data\firstsurvey.dta
  obs:            20
  vars:            7                              15 Jan 2012 07:54
  size:          560                              (_dta has notes)

                 storage  display   value
variable name    type     format    label    variable label

id               float    %9.0g              Respondent's identification
gender           float    %9.0g     sex      Participant's gender
education        float    %9.0g              Years of education
sch_st           float    %9.0g     rating   Ratings of schools in your state
sch_com          float    %9.0g     rating   Ratings of schools in your
                                                community of origin
prison           float    %16.0g    harsh    Ratings of prison sentences
conserv          float    %17.0g    conserv  * Conservatism/liberalism
                                             * indicated variables have notes

Sorted by:
```

As we can see by examining the output, the path to the data file is shown (directory and filename), as are the number of observations, the number of variables, the size of the file, and the last date and time the file was written (called a *timestamp*). We can attach a label to a dataset, too, and if we had done so, it would have been shown above the timestamp. Below the information that summarizes the dataset is information on each variable: its name, its storage type (there are several types of numbers and strings), the format Stata uses when printing values for the variable, the value-label mapping associated with it, and the variable label. This is clearly not as complete a description of each variable as that provided by `codebook`, but it is often just what you want.

## Working with Excel files

Many people create data files in Excel that they want to open to Stata, or they have Stata files they want to save to Excel. Excel is much more limited than Stata in what it can record in a data file. For example, Excel cannot store both variable names and variable labels nor can it store value labels. If you have a variable named `race` that is coded `1` for Hispanic, `2` for non-Hispanic–whites, `3` for non-Hispanic–blacks, `4` for `Asian`–Pacific Islander, and `5` for other, you can see the problem. Stata will keep the value labels linked to the coded numerical value; Excel will not do this. However, Stata makes it easy to translate from a Stata to an Excel file given the limitations of what Excel can have in a file. Importing from Excel to Stata is also a straightforward process.

**Importing**:  An Excel file should have the first row be variable names and the data should be below that in a wide format, that is, one row for each observation. Using a file called `auto.xlsx`, in Excel the first few lines would look like the following:

| | A | B | C | D | E | F | G | H | I | J | K | L |
|---|---|---|---|---|---|---|---|---|---|---|---|---|
| 1 | make | price | mpg | rep78 | headroom | trunk | weight | length | turn | displacem | gear_ratio | foreign |
| 2 | AMC Conc | 4099 | 22 | 3 | 2.5 | 11 | 2930 | 186 | 40 | 121 | 3.58 | Domestic |
| 3 | AMC Pace | 4749 | 17 | 3 | 3 | 11 | 3350 | 173 | 40 | 258 | 2.53 | Domestic |
| 4 | AMC Spirit | 3799 | 22 | | 3 | 12 | 2640 | 168 | 35 | 121 | 3.08 | Domestic |

Notice that the first row is the variable names. Also, notice that missing values appear as dots (.) or as blanks; for example, there is no information for the `rep78` variable for the AMC Pacer. To import this file into Stata, we click on File ▷ Import ▷ Excel spreadsheet (*.xls, *.xlsx). In the dialog box, use the Browse... button to locate the Excel file to open. Selecting the file and clicking on OK displays the results of the Excel file in the *Preview* window. Also check the *Import first row as variable names* box and change the *Variable case* to `lower`. We can review this visually and then click on OK. If we have a Stata dataset already open, Stata will issue a warning message asking if you want to continue and lose any unsaved data. We will select Yes to complete the importing.

**Exporting**:  Changing a Stata to an Excel file reverses this process. Select File ▷ Export ▷ Data to Excel spreadsheet(*.xls,*.xlsx). Leave the *Variables* box empty. Type `autoexcel.xlsx` in the *Excel filename* box—include the path as needed; clicking the Save As... button will assist you in locating the right path. Click on the *Save variable names to first row in Excel file* box. Under the Advanced tab, I like to check the *Export missing values as* box and add a dot (.). That's it!

## 2.9   Summary

We covered the following topics in this chapter:

- How to create a codebook for a questionnaire

- How to develop a system for coding the data

- How to enter the data into a Stata dataset

- How to add variable names and variable labels to a dataset

- How to create value-label mappings and associate them with variables

## 2.10   Exercises

1. From the coding sheet shown in table 2.2, translate the numeric codes back into the answers given on the questionnaire by using the codebook in table 2.1.

2. Using the Variables Manager, attach a note that provides a descriptive label to the dataset you created in this chapter. What command is printed in the Results window by the dialog box? Then run the **notes** command from the Command window. See if you can use the Variables Manager to remove the note. What are the advantages of keeping a note for a dataset inside the dataset?

3. Using the Variables Manager, attach a note for each variable you created in this chapter. What command is printed in the Results window when you do this? Then run the **notes** command from the Command window. See if you can use the Variables Manager to remove the notes. What are the advantages of keeping notes for variables inside the dataset?

4. Run the command to create a compact codebook from the Command window. Interpret what the codebook shows for **gender**, **education**, and **conserv**.

5. Administer your own small questionnaire to students in your class. Enter the data into a Stata dataset, and label the variables and values. Be sure to have at least 15 people complete your questionnaire, but do not try to enter data for a very large group of people.

6. Create a codebook for your dataset from the previous exercise that includes notes and any possible problems Stata can identify about your dataset. What command appears in the Results window when you do this?

7. What is the value of using a two-step process for creating value labels? When is this process especially helpful?

# 3 Preparing data for analysis

## 3.1 Introduction

Most of the time spent on any research project is spent preparing for analysis. Exactly how much time will depend on the data. If you collect and enter your own data, most of the actual time you spend on the project will not be analyzing the data; it will be getting the data ready to analyze. If you are performing secondary analysis and are using well-prepared data, you may spend substantially less time preparing the data.

Stata has an entire reference manual on data management, the *Data-Management Reference Manual*. We will cover only a few of Stata's capabilities for managing data. Some of the data-management tasks required for a project you will be able to anticipate and plan, but others will just arise as you progress through a project, and you will need to deal with them on the spot. If your data are always going to be prepared for you and you will use Stata only for statistics and graphs, then you might skip this chapter.

## 3.2 Planning your work

The data we will be using are from the U.S. Department of Labor, Bureau of Labor Statistics, National Longitudinal Survey of Youth, 1997 (NLSY97). I selected a set of four items to measure how adolescents feel about their mothers and another set

49

of four parallel items to measure how adolescents feel about their fathers. I used an extraction program available from http://www.bls.gov/nls/nlsdata.htm and extracted a Stata dictionary that contains a series of commands for constructing the Stata dataset, which we will use along with the raw data in plain text. I created a dataset called `relate.dta`. In this chapter, you will learn how to work with this dataset to create two variables, Youth Perception of Mother and Youth Perception of Father.

### What is a Stata dictionary file?

A Stata dictionary file is a special file that contains Stata commands and raw data in plain text. We will not cover this type of file here, but we use this format to import the raw data into a Stata dataset. Selecting File ▷ Import ▷ Text data in fixed format with a dictionary opens a window with a blank box for the *Dictionary filename* and a blank box for the *Text dataset filename*. Because the raw data and the commands to create the Stata file are both in the dictionary file, we can simply enter the name of the dictionary file there and leave the *Text dataset filename* empty. The name of the file is `relate.dct`. If we did not have this file stored in the working directory, we could click on Browse... to help us search for it. The extension `.dct` is always used for a dictionary file. Clicking on OK imports the data into Stata. We can then save it using the name `relate.dta`. The dictionary provides the variable names and variable labels. Had the variable labels not been included in the dictionary, we would have created them prior to saving the dataset. If you need to create a dictionary file yourself, you can learn how to do that in the *Stata Data-Management Reference Manual*.

It is a good idea to make an outline of the steps we need to take to go from data collection to analysis. The outline should include what needs to be done to collect or obtain the data, enter (or read) and label the data, make any necessary changes to the data prior to analysis, create any composite variables, and create an analysis-ready version (or versions) of the dataset.

An outline serves two vital functions: it provides direction so we do not lose our way in what can often be a complicated process with many details and it provides us with a set of benchmarks for measuring progress. For a large project, do not underestimate the benefit to measuring progress because it is the key to successful project management.

Our project outline, which is for a simple project on which only one person is working, is shown in table 3.1.

Table 3.1. Sample project task outline

---

○ Consult NLSY97 documentation (see appendix 9 [PDF], page 23, available at
  https://www.nlsinfo.org/sites/nlsinfo.org/files/attachments/12125/app9pdf.pdf)
  to determine which variables are needed.

○ Download the data that are in the dictionary file and the codebook. (This was done
  for you as `relate.dct` and `relate.cdb`.)

○ Create a basic Stata dataset. (This was done for you as `relate.dta`.)

○ Create variable and value labels.

○ Generate tables for variables to compare against the codebook to check for errors.

○ Convert missing-value codes to Stata missing values.

○ Reverse code those variables that need it; verify.

○ Copy variables not reversed to named variables; verify.

○ Create the scale variable.

○ Save the analysis-ready copy of the dataset.

---

This is a pretty skeletal outline, but it is a fine start for a small project like this. The larger the project, the more detail needed for the outline. Now that we have the NLSY97-supplied codebook to look at, we can fill in some more details on our outline.

First, we will run a `describe` command on the dataset we created, `relate.dta`. Remember that you can download `relate.dta` and other datasets used in this book from the book's webpage. Alternatively, if you are using Stata 13 or later, you can open the file `relate.dta` by typing the command

```
. use http://www.stata-press.com/data/agis4/relate
```

Let's show the results for just the four items measuring the adolescents' perception of their mothers. From the documentation on the dataset, we can determine that these items are R3483600, R3483700, R3483800, and R3483900. First, we can run the command `describe R3483600 R3483700 R3483800 R3483900`, and then we can get more information by running the command `codebook, compact`. For the codebook, we do not list the variables because the dataset includes only a few variables. Thus we use the following command to produce a description of the data:

```
. describe R3483600 R3483700 R3483800 R3483900

              storage  display   value
variable name   type   format    label        variable label

R3483600        float  %9.0g                  MOTH PRAISES R DOING WELL 1999
R3483700        float  %9.0g                  MOTH CRITICIZE RS IDEAS 1999
R3483800        float  %9.0g                  MOTH HELPS R WITH WHAT IMPT TO R
                                                 1999
R3483900        float  %9.0g                  MOTH BLAMES R FOR PROBS 1999

. codebook, compact

Variable    Obs Unique      Mean  Min   Max  Label

R0000100   8984   8984   4504.302    1  9022  PUBID - YTH ID CODE 1997
R3483600   8984      9  -.5638914   -5     4  MOTH PRAISES R DOING WELL 1999
R3483700   8984      9  -1.425312   -5     4  MOTH CRITICIZE RS IDEAS 1999
R3483800   8984      9  -.5893811   -5     4  MOTH HELPS R WITH WHAT IMPT TO R
R3483900   8984      9  -1.769813   -5     4  MOTH BLAMES R FOR PROBS 1999
R3485200   8984      9  -1.535508   -5     4  FATH PRAISES R DO WELL 1999
R3485300   8984      9  -2.143143   -5     4  FATH CRITICIZE IDEAS 1999
R3485400   8984      9  -1.586487   -5     4  FATH HELPS R WITH WHAT IMPT TO R
R3485500   8984      9  -2.388023   -5     4  FATH BLAME R FOR PROBS 1999
R3828100   8984      8   15.02538   -5    20  SYMBOL!KEY!AGE 1999
R3828700   8984      3   .9319902   -5     2  SYMBOL!KEY!SEX 1999
```

These variable names are not helpful, so we might want to change them later. For example, we may rename R3483600 to `mompraise`. The variable labels make sense, but notice that there are no value labels listed under the column labeled value label in the results of the `describe` command. `codebook, compact` gives us a bit more information. We see there are 8,984 observations, we see a mean for each item, and we see the minimum (−5) and maximum value (4) reported for the items. Still, without value labels, we do not know what these values mean.

Let's read the codebook that we downloaded at the same time we downloaded this dataset. So you do not need to go through the process of getting the codebook from the NLS website, the part we use is called `relate.cdb`. You can open this file by pointing your web browser to http://www.stata-press.com/data/agis4/relate.cdb. Table 3.2 shows what the downloaded codebook looks like for the R3483600 variable.

Table 3.2. NLSY97 sample codebook entries

```
R34836.00    [YSAQ-022]                              Survey Year: 1999

             MOTHER PRAISES R FOR DOING WELL

How often does she praise you for doing well?

      118      0 NEVER
      235      1 RARELY
      917      2 SOMETIMES
     1546      3 USUALLY
     1701      4 ALWAYS
   -------
     4517

Refusal(-1)              20
Don't Know(-2)            3
TOTAL =========>       4540   VALID SKIP(-4)   3669   NON-INTERVIEW(-5)   775
```

We can see that a code of 0 means Never, a code of 1 means Rarely, etc. We need to make value labels so that the data will have nice value labels like these. Also notice that there are different missing values. The codebook uses -1 for Refusal, -2 for Don't know, -4 for Valid skip, and -5 for Noninterview. But there is no -3 code. Searching the documentation, we learn that this code was used when there was an interviewer mistake or an Invalid skip. We did not have any of these for this item, but we might for others. The Don't know responses are people who should have answered the question but instead they said that they did not know what they thought or that they had no opinion. By contrast, notice there are 3,669 valid skips representing people who were not asked the question for some reason. We need to check this out, and reading the documentation, we learn that these people were not asked because they were more than 14 years old, and only youth 12–14 were asked this series of questions. Finally, 775 youth were not interviewed. We need to check this out as well. These are 1999 data, the third year the data were collected, and the researchers were unable to locate these youth, or they refused to participate in the third year of data collection. In any event, we need to define missing values in a way that keeps track of these distinctions.

How does the Stata dataset look compared with the codebook that we downloaded? We can run the command codebook to see. Let's restrict it to just the variable R3483600 by entering the command codebook R3483600 in the Command window. The problem with the data is apparent when we look at the codebook: the absence of value labels makes our actual data uninterpretable.

```
. codebook R3483600
```

| R3483600 | MOTH PRAISES R DOING WELL 1999 |
|---|---|

```
              type:  numeric (float)
             range:  [-5,4]                    units:  1
     unique values:  9                    missing .:  0/8984
        tabulation:  Freq.  Value
                      775   -5
                     3669   -4
                        3   -2
                       20   -1
                      118    0
                      235    1
                      917    2
                     1546    3
                     1701    4
```

Before we add value labels to make our codebook look more like the codebook we downloaded, we need to replace the numeric missing values (−5, −4, −3, −2, and −1) with values that Stata recognizes as representing missing values. Stata recognizes up to 27 different missing values for each variable: . (a dot), .a, .b, ..., .z. Note that a missing value will be larger than any numeric value. When there is only one type of missing value, as discussed in chapter 2, we could change all numeric codes of −9 to . (a dot).

Because we have five different types of missing values, we will replace −5 with .a (read as dot-a), −4 with .b, −3 with .c, −2 with .d, and −1 with .e. Stata uses a command called mvdecode (missing-value decode) to do this. Because all items involve the same decoding replacements, we can do this with the following command:

```
. mvdecode _all, mv(-5=.a\-4=.b\-3=.c\-2=.d\-1=.e)
    R3483600: 4467 missing values generated
    R3483700: 4469 missing values generated
    R3483800: 4468 missing values generated
    R3483900: 4470 missing values generated
    R3485200: 5608 missing values generated
    R3485300: 5611 missing values generated
    R3485400: 5610 missing values generated
    R3485500: 5611 missing values generated
    R3828100: 775 missing values generated
    R3828700: 775 missing values generated
```

We use the _all that is just before the comma to tell Stata to do the missing-value decoding for all variables. If we just wanted to do it for the R3483600 and R3828700 variables, the mvdecode _all would be replaced by mvdecode R3483600 R3828700. Some datasets might use different missing values for different variables, which would require several mvdecode commands. For example, a value of −1 might be a legitimate answer for some items and for these items a value of 999 might be used for Refusal. Each replacement above is separated by a backslash (\). The results appear right below the mvdecode command and tell us how many missing values are generated for each variable. For example, `R3483600` has 4,467 missing values generated (775 replace the $-5$, 3,669 replace the $-4$, 3 replace the $-2$, and 20 replace the $-1$).

**Stata and capitalization**

Stata commands and options are all lowercase. Because Stata is case sensitive, if you typed a command, say, `Summarize` instead of `summarize`, or an option in uppercase, the word would be meaningless to Stata. For some reason, many national datasets put everything in capital letters. When we get an email message with a lot of capitalization, this does not make us happy; it makes us think the sender is yelling at us and it is hard to read. Other researchers mix the case in names using such names as `MyID` or `MOMfeels`. They might have an item named `MomFeels` and a scale named `MOMFEELS`. Switching case can be confusing, and if you make a mistake, say, you enter `MyId` instead of `MyID`, you will get an error message. Many experienced Stata users use lowercase for variable names and nicknames for value labels. If you do this consistently, you never have to worry about shifting from lowercase to uppercase. Many users capitalize the first word in the label assigned to a value, such as "Very strongly". This capitalization helps make output more readable, and you never have to type these labels again once the dataset is created.

Stata's `rename` command can change the case of any or all variables in a dataset. Suppose that we have five variables in a dataset—ID, `Age`, `WaistSize`, `Ed`, and `MOMFEELS`—and want to make all of these lowercase. We type `rename ID-MOMFEELS, lower` in the Command window and press Enter. Regardless of how the variable names used case originally, they will all now be lowercase. We could select Data ▷ Data utilities ▷ Rename groups of variables. Then we can click on *Change the case on groups of variable names*, enter our list of variables, click on *Lowercase variable names*, and then click on OK.

*lower case*

## 3.3   Creating value labels

Now let's add value labels, including labels for the five types of missing values. To do this, we will open the Variables Manager by selecting Data ▷ Variables Manager or by typing `varmanage`. Click on `R3483600` to highlight the variable. Click on Manage... next to *Value Label* on the Variable Properties pane. This opens the *Manage Value Labels* dialog box, where we click on Create Label. A new *Create Label* dialog box opens. In the *Create Label* dialog box, enter `often` for *Label name*; enter 0 for *Value*; enter `Never` for *Label*; and then click on Add. Repeat this step where *Value* is 1 and *Label* is `Rarely`, *Value* is 2 and *Label* is `Sometimes`, *Value* is 3 and *Label* is `Usually`, *Value* is 4 and *Label* is `Always`, *Value* is `.a` and *Label* is `Noninterview`, *Value* is `.b` and *Label* is `Valid skip`, *Value* is `.c` and *Label* is `Invalid skip`, *Value* is `.d` and *Label* is

`Don't know`, and *Value* is `.e` and *Label* is `Refused`.  At this point, our three windows appear as figure 3.1:

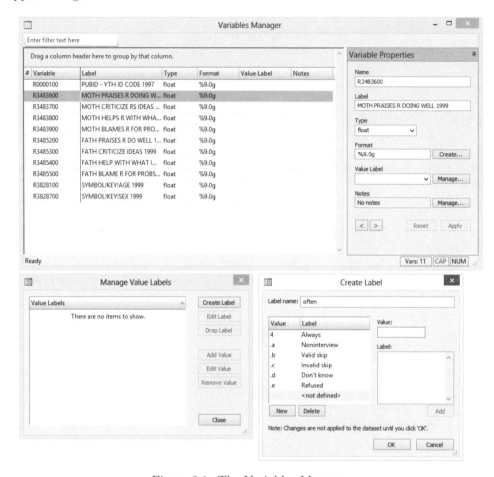

Figure 3.1. The Variables Manager

From here, we can click on **OK** to close the *Create Label* dialog box and then click on **Close** in the *Manage Value Labels* dialog box. We now need to assign our value label, `often`, to the appropriate variables. In the Variables Manager, we can click on each of the variables. By doing this, all properties of the variable are shown in the Variable Properties pane. We will click on `R3483600`. The *Value Label* in the Variable Properties pane is shown as a blank because we have not yet assigned the value label `often` to this variable. Click on the drop-down menu next to *Value Label* and select `often`, and then click on **Apply**. After we have repeated this for all eight variables, the Variables Manager appears as figure 3.2:

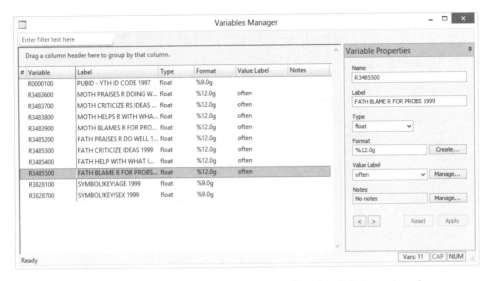

Figure 3.2. The Variables Manager with value labels assigned

Notice that each of the appropriate variables now shows that the value label nick-named `often` is assigned to it. At some point, you might want to remove a value label. For example, you might accidentally have applied the `often` value label to R3828100, the age of the observation. To drop the value label, you click on the variable, erase the label `often`, and then click on Apply.

The variable labels (remember, we can create or change variable names, variable labels, and value labels) are okay as they are, but let's change the case so that it does not look like somebody is screaming at us. We also might make them simpler. For instance, the label of `PUBID - YTH ID CODE 1997` for variable R0000100 could be improved. Because once entered, we never have to type a variable label again, we will use capitalization rules for a normal sentence. We will make a few minor word changes as well. In doing this, we want to keep the labels reasonably short.

In our Variables Manager, we can click on a variable, say, R3483600, and then in the *Label* field of the Variable Properties pane, change `MOTH PRAISES R DOING WELL 1999` to `Mother praises child for doing well`. Click on Apply and the label for this variable in the *Label* column is immediately changed. Using the Variables Manager, change the variable labels for the other seven variables as indicated in the commands below.

```
. label variable R3483700 "Mother criticizes child's ideas"
. label variable R3483800 "Mother helps child with what is important to child"
. label variable R3483900 "Mother blames child for problems"
. label variable R3485200 "Father praises child for doing well"
. label variable R3485300 "Father criticizes child's ideas"
. label variable R3485400 "Father helps child with what is important to child"
. label variable R3485500 "Father blames child for problems"
```

Do these changes help? Let's do a codebook for R3483600 and see how it compares with the codebook that appeared in table 3.2.

```
. codebook R3483600
```

| R3483600 | Mother praises child for doing well |
|---|---|

```
              type:  numeric (float)
             label:  often

             range:  [0,4]                          units:  1
     unique values:  5                          missing .:  0/8984
   unique mv codes:  4                         missing .*:  4467/8984

        tabulation:  Freq.   Numeric  Label
                       118         0  Never
                       235         1  Rarely
                       917         2  Sometimes
                      1546         3  Usually
                      1701         4  Always
                       775        .a  Noninterview
                      3669        .b  Valid skip
                         3        .d  Don't know
                        20        .e  Refused
```

## 3.4   Reverse-code variables

We have two items for the mother and two for the father that are stated negatively, so we should reverse code these. R3483700 refers to the mother criticizing the adolescent, and R3485300 refers to the father criticizing the adolescent. R3483900 and R3485500 refer to the mother and father blaming the adolescent for problems. For these items, an adolescent who reports that this always happens would mean that the adolescent had a low rating of his or her parent. We always want a higher score to signify more of the variable. A score of 0 for never blames the child on this pair of items should be the highest score on these items (4), and a response of always blames the child should be the worst response and would have the lowest score (0).

It is good to organize all the variables that need reverse coding into groups based on their coding scheme. It is important to create new variables instead of reversing the originals. This step ensures that we have the original information if any questions arise about what we have done. I recommend that you write things out before you start working, as I have done in table 3.3.

Table 3.3. Reverse-coding plan

| Old value | New value |
|:---:|:---:|
| 0 | 4 |
| 1 | 3 |
| 2 | 2 |
| 3 | 1 |
| 4 | 0 |

For a small dataset like the one we are using, writing it out may seem like overkill. When you are involved in a large-scale project, the experience you gain from working out how to organize these matters on small datasets will serve you well, particularly if you need to assign tasks to several people and try to keep them straight.

There are several ways to do this recoding, but let's use `recode` right now. Select Data ▷ Create or change data ▷ Other variable-transformation commands ▷ Recode categorical variable.

*Recode*

The `recode` command and its dialog box are straightforward. We provide the name of the variable to be recoded and then give at least one rule for how it should be recoded. Recoding rules are enclosed in parentheses, with an old value (or values) listed on the left, followed by an equal sign, and then the new value. The simplest kind of rule lists one value on either side of the equal sign. So for example, the first recoding rule we would use to reverse code our variables would be (0=4).

### More on recoding rules

More than one value can be listed on the left side of the equal sign, as in (1 2 3=0), which would recode values of 1, 2, and 3 as 0. Occasionally, you may want to collapse just one end of a scale, as with (5/max=5), which recodes everything from 5 up, but not including the missing values, as 5. You might use this to collapse the highest income categories into one category, for example. There is also a corresponding version to recode from the smallest value up to some value. For example, you might use (min/8=8) to recode the values for highest attained grade in school if you wanted everybody with fewer than nine years of education coded the same way.

When you use `recode`, you can choose to recode the existing variables or create new variables that have the new coding scheme. I strongly recommend creating new variables. The dialog boxes shown in figures 3.3 and 3.4 show how to do this for all four variables being recoded. Figure 3.3 shows the original variables on the Main tab.

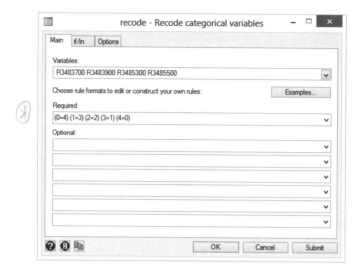

Figure 3.3. `recode`: specifying recode rules on the Main tab

Figure 3.4 shows the names of the new variables to be created on the Options tab.

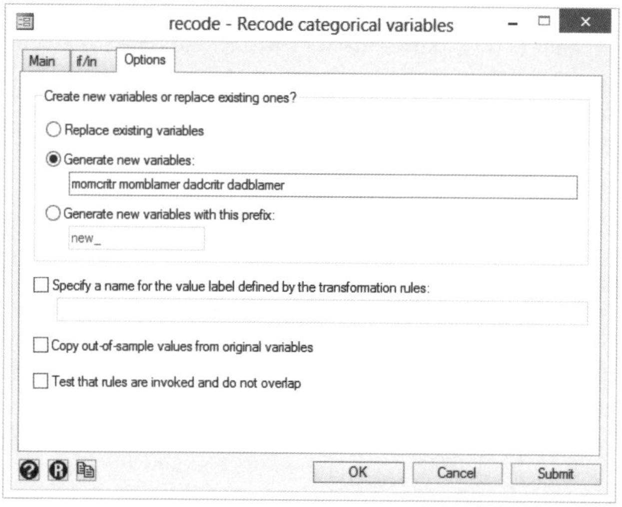

Figure 3.4. `recode`: specifying new variable names on the Options tab

The variables for mothers begin with `mom`, and the variables for fathers begin with `dad`. Adding an `r` at the end of each variable reminds you that they are reverse coded. Stata will pair up the names, starting with the first name in each list, and recode according to the rules stated. We must have the same number of variable names in each list. After you click on OK, the output from the selections shown in figures 3.3 and 3.4 is

```
. recode R3483700 R3483900 R3485300 R3485500 (0=4) (1=3) (2=2) (3=1) (4=0),
> generate(momcritr momblamer dadcritr dadblamer)
(3230 differences between R3483700 and momcritr)
(4037 differences between R3483900 and momblamer)
(2562 differences between R3485300 and dadcritr)
(3070 differences between R3485500 and dadblamer)
```

The output above shows how Stata presents output that is too long to fit on one line: it moves down a line and inserts a > character at the beginning of the new line. This differentiates command-continuation text from the output of the command. If you are trying to use the command in output as a model for typing a new command, you should not type the > character.

The `recode` command labels the variables it creates. The text is generic, so we may wish to replace it. For example, the `momcritr` variable was labeled with the text

```
RECODE of R3483700 (Mother criticizes child's ideas)
```

This text tells us the original variable name (and its label in parentheses), but it does not tell us how it was recoded. It may be worth the additional effort to create more explicit variable labels. We can do this with the Variables Manager. If you want to try entering the commands directly, they are

```
. label variable momcritr "Mother criticizes child's ideas, reverses R3483700"
. label variable momblamer "Mother blames child for problems, reverses R3483900"
. label variable dadcritr "Father criticizes child's ideas, reverses R3485300"
. label variable dadblamer "Father blames child for problems, reverses R3485500"
```

The `recode` command is useful. You can specify missing values either as original values or as new values, and you can create a value label as part of the `recode` command. You can also recode a range (say, from 1 to 5) easily, which is useful for recoding things like ages into age groups. If the resulting values are easily listed (most often integers, but they could be noninteger values), it is worth thinking about using the `recode` command. See the *Stata Data-Management Reference Manual* entry for `recode` or type `help recode` in Stata for more information.

Whenever we generate a new variable, we should always check to make sure we did what we wanted to do. We can quickly check by running a cross-tabulation (see chapter 6) of the old variable and the new variable. We will do this for R3483700 and `momcritr`. Although we will learn a lot more about the `tabulate` command later, for now we can just run the command to create a cross-tabulation in the Command window:

```
. tabulate momcritr R3483700
```

| Mother criticizes child's ideas, reverses R3483700 | Never | Rarely | Mother criticizes child's ideas Sometimes | Usually | Always | Total |
|---|---|---|---|---|---|---|
| 0 | 0 | 0 | 0 | 0 | 145 | 145 |
| 1 | 0 | 0 | 0 | 324 | 0 | 324 |
| 2 | 0 | 0 | 1,285 | 0 | 0 | 1,285 |
| 3 | 0 | 1,654 | 0 | 0 | 0 | 1,654 |
| 4 | 1,107 | 0 | 0 | 0 | 0 | 1,107 |
| Total | 1,107 | 1,654 | 1,285 | 324 | 145 | 4,515 |

Here the first variable after `tabulate`, that is, `momcritr`, appears as the row variable and the second variable, `R3483700`, appears as the column variable. We do not have the numeric values, 0, 1, 2, 3, and 4 for `R3483700`, because we have their labels, but we know from our codebook that "Never" is 0, "Rarely" is 1, "Sometimes" is 2, "Usually" is 3, and "Always" is 4. We see, for example, that the 1,107 people who had a 0 on `R3483700` now have a 4. What is missing? We need value labels for our new variable. We cannot assign the value label nicknamed `often` to `momcritr` because we have reversed the numbering. We need a new label where 4 is `Never`, 3 is `Rarely`, 2 is `Sometimes`, 1 is `Usually`, and 0 is `Always`.

We could use the Variables Manager to do this like we did to create the value label `often`. We would just give it another name, such as `often_r`, where the `r` stands for reverse order. You can use the Variables Manager, but you might want to type the command in the Command window:

```
. label define often_r 4 "Never" 3 "Rarely" 2 "Sometimes" 1 "Usually" 0 "Always"
> .a "Noninterview" .b "Valid skip" .c "Invalid skip" .d "Don't know"
> .e "Refusal"
```

As we have noted, when a line is too long to fit in the Command window, it just wraps around, but when it appears in the Results window, it has separate lines with `>` signifying that the lines are tied. In the Command window, you would not type the dot prompt (`.`) or the greater than signs (`>`).

Finally, we need to assign this nickname, `often_r`, to the new variables. We can do this with the Variables Manager or by using the following `label values` command:

```
. label values momcritr momblamer dadcritr dadblamer often_r
```

After the `label values` command, we list the variable names that share the same set of value labels, and at the end we have the nickname for that set of value labels, `often_r`.

# 3.! ...ating and modifying variables

...e several commands that are used in creating, replacing, and modifying vari-
...eful commands include **generate**, **egen** (short for extended generate), **rename**,
...evar. Suppose that we want to change the positive items so that their names
...re sense to us. A simple way to do this is to make the new variables be clones.
The **clonevar** *newname = oldname* command does this. **clonevar** has a couple of
advantages: the actual command is simple, it keeps the missing values coded the same
way as they were in the old variable, and it keeps the same value labels. We could run
the following commands from the Command window:

```
. clonevar mompraise = R3483600
. clonevar momhelp = R3483800
. clonevar dadpraise = R3485200
. clonevar dadhelp = R3485400
```

Let's do this for the other three variables in the dataset: identification number, age,
and gender.

```
. clonevar id = R0000100
. clonevar sex = R3828700
. clonevar age = R3828100
```

You can open the dialog box for **generate** by selecting Data ▷ Create or change
data ▷ Create new variable. The **generate** command does much the same thing as the
**clonevar** command, but it does not transfer value labels. It does transfer the missing-
value codes we used: .a, .b, .c, .d, and .e. If we wanted to run the **generate** command
directly without using the dialog box, we would type the following commands. (Do not
type these commands now because we already created the new variables by using the
**clonevar** command.)

```
. generate mompraise = R3483800
. generate momhelp = R3483800
. generate dadpraise = R3485200
. generate dadhelp = R3485400
. generate id = R0000100
. generate sex = R3828700
. generate age = R3828100
```

The **generate** command can also be used to create new variables by using an arithmetic expression. Table 3.4 shows the arithmetic symbols that can be used in these
expressions.

Table 3.4. Arithmetic symbols

| Symbol | Operation | Example |
|--------|-----------|---------|
| + | Addition | mscore + fscore + sibscore |
| − | Subtraction | balance − expenses − penalty |
| * | Multiplication | income * .75 |
| / | Division | expenses/income |
| ^ | Exponentiation $(x^2)$ | x^2 |

Attempts to do arithmetic with missing values will lead to missing values. So in the addition example in table 3.4, if `sibscore` were missing (say, it was a single-child household), the whole sum in the example would be set to missing for that observation. From the Command window, type `help generate` to see several more examples of what can be done with the `generate` command. If you ever have trouble finding a dialog box but you know the command name, you can open the help file first. The upper right corner of a help file opened in the Viewer window will show a **Dialog** menu. Select the command name from the drop-down menu, and the dialog box for that command will open.

For more complicated expressions, order of operations can be important, and you can use parentheses to control the order in which things are done. Parentheses contain expressions too, and those are calculated before the expressions outside of parentheses. Parentheses are never wrong. They might be unnecessary to get Stata to calculate the correct value, but they are not wrong. If you think they make an expression easier to read or understand, use as many as you need.

Fortunately, the rules are pretty simple. Stata reads expressions from left to right, and the order in which Stata calculates things inside expressions is to

1. Do everything in parentheses. If one set of parentheses contains another set, do the inside set first.

2. Exponentiate (raise to a power).

3. Multiply and divide.

4. Add and subtract.

Let's go step by step through an example:

```
. generate example = weight/.45*(5+1/age^2)
```

When Stata looks at the expression to the right of the equal sign, it notices the parentheses (priority #1) and looks inside them. There it sees that it first has to square `age` (priority #2), then divide 1 by the result (priority #3), and then add 5 to that result (priority #4). Once it is done with all the stuff in parentheses, it starts reading from left to right, so it divides `weight` by 0.45 and then multiplies the result by the final value it got from calculating what was inside parentheses. That final value is put into a variable called `example`.

Stata does not care about spaces in expressions, but they can help readability. So for example, instead of typing something like we just did, we can use some spaces to make the meaning clearer, as in

```
. generate example = weight/.45 * (5 + 1/age^2)
```

If we wanted to be even more explicit, it would not be wrong to type

```
. generate example = (weight/.45) * (5 + 1/(age^2))
```

Let's take another look at reverse coding. We already reverse-coded variables by using a set of explicit rules like `(0=4)`, but we could accomplish the same thing using arithmetic. Because this is a relatively simple problem, we will use it to introduce some of the things you may need to be concerned about with more complex problems.

Reversing a scale is swapping the ends of the scale around. The scale is 0 to 4, so we can swap the ends around by subtracting the current value from 4. If the original value is 4 and we subtract it from 4, we have made it a 0, which is the rule we specified with `recode`. If the original value is 0 and we subtract it from 4, we have made it a 4, which is the rule we specified with `recode`.

This scale starts at 0, so to reverse it, you just subtract each value from the largest value in the scale, in this case, 4. So if our scale were 0 to 6, we would subtract each value from 6; if it were 0 to 3, we would subtract from 3. If the scale started at 1 instead of 0, we would need to add 1 to the largest value before subtracting. So for a 1 to 5 scale, we would subtract from 6 ($6 - 1 = 5$ and $6 - 5 = 1$); for a 1 to 3 scale, we would subtract from 4 ($4 - 3 = 1$ and $4 - 1 = 3$).

What we have said so far is correct, as far as it goes, but we are not taking into account missing values or their codes. The missing values are coded in this dataset as $-1$ to $-5$, and if you subtract those from 4 along with the item responses that are not missing-value codes, we will end up with $4 - (-1) = 5$ to $4 - (-5) = 9$. So we must first convert the missing-value codes to Stata's missing-value code (`.`) and then do the arithmetic to reverse the scale.

Many researchers would code the values 1–5 rather than 0–4. To reverse a scale that starts at 1 and goes to 5, you need to subtract the current value from 1 more than the maximum value. Thus you would use $6 - variable$. If the variable's current value is 5, then $6 - 5 = 1$.

Let's try this example. We have already run the `mvdecode` command on our variables, so let's reverse code R3485300 and call it `facritr`. We will use the `generate` dialog box to create the variable. Select Data ▷ Create or change data ▷ Create new variable, type `facritr` for the *Variable name*, and type 4 - R3485300 in the *Specify a value or an expression* box. Figure 3.5 shows the completed dialog box.

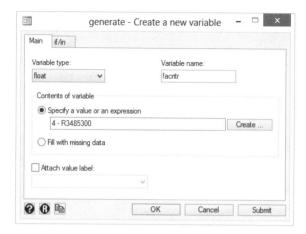

Figure 3.5. The `generate` dialog box

Here is the output of the `generate` command:

```
. generate facritr = 4 - R3485300
(5611 missing values generated)
```

Whenever you see that missing values are generated (there are 5,611 of them in this example!), it is a good idea to make sure you know why they are missing. These variables have only a small set of values they can take, so we can compare the original variable with the new variable in a table and see what got turned into what. Select Statistics ▷ Summaries, tables, and tests ▷ Frequency tables ▷ Two-way table with measures of association, which will bring up the dialog shown in figure 3.6.

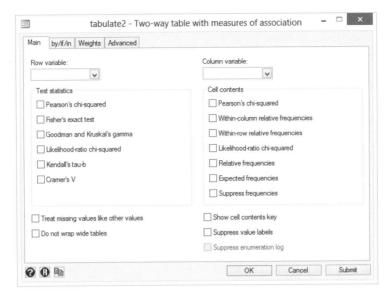

Figure 3.6. Two-way tabulation dialog box

Here we select `facritr` as the *Row variable* and `R3485300` as the *Column variable*. We also check two of the options boxes: Because we are interested in the actual values the variables take, select the option to suppress the value labels (*Suppress value labels*). We are also interested in the missing values, so we check the box to have them included in the table (*Treat missing values like other values*).

```
. tabulate facritr R3485300, miss nolabel
```

|  | Father criticizes child's ideas | | | | | |
|---|---|---|---|---|---|---|
| facritr | 0 | 1 | 2 | 3 | 4 | Total |
| 0 | 0 | 0 | 0 | 0 | 117 | 117 |
| 1 | 0 | 0 | 0 | 247 | 0 | 247 |
| 2 | 0 | 0 | 811 | 0 | 0 | 811 |
| 3 | 0 | 1,078 | 0 | 0 | 0 | 1,078 |
| 4 | 1,120 | 0 | 0 | 0 | 0 | 1,120 |
| . | 0 | 0 | 0 | 0 | 0 | 5,611 |
| Total | 1,120 | 1,078 | 811 | 247 | 117 | 8,984 |

|  | Father criticizes child's ideas | | | | |
|---|---|---|---|---|---|
| facritr | .a | .b | .d | .e | Total |
| 0 | 0 | 0 | 0 | 0 | 117 |
| 1 | 0 | 0 | 0 | 0 | 247 |
| 2 | 0 | 0 | 0 | 0 | 811 |
| 3 | 0 | 0 | 0 | 0 | 1,078 |
| 4 | 0 | 0 | 0 | 0 | 1,120 |
| . | 775 | 4,816 | 4 | 16 | 5,611 |
| Total | 775 | 4,816 | 4 | 16 | 8,984 |

From the table generated, we can see that everything happened as anticipated. Those adolescents who had a score of 0 on the original variable (the column with a 0 at the top) now have a score of 4 on the new variable. There are 1,120 of these observations. We need to check the other combinations, as well. Also the missing-value codes were all transferred to the new variable, but we lost the distinctions between the different reasons an answer is missing. The new variable, `facritr`, is a reverse coding of our old variable, `R3485300`.

## 3.6   Creating scales

We are finally ready to calculate our scales. We will construct two scale variables: one for the adolescent's perception of his or her mother and one for the adolescent's perception of his or her father. With four items, each scored from 0 to 4, the scores could range from 0 to 16 points. Higher scores indicate a more positive relationship with the parent.

At first glance, this assertion is straightforward. We just add the variables together:

```
. generate ymomrelate = mompraise + momcritr + momhelp + momblamer
. generate ydadrelate = dadpraise + dadcritr + dadhelp + dadblamer
```

Before we settle on this, we should understand what will happen for missing values, and we should check through the documentation to find out what Stata does about missing values. The first thing to do is to determine how many observations have one, two, three, or all four of the items answered, that is, not missing. This need introduces a new command, **egen** (short for extended generate). Selecting Data ▷ Create or change data ▷ Create new variable (extended) opens the **egen** dialog box, shown in figure 3.7.

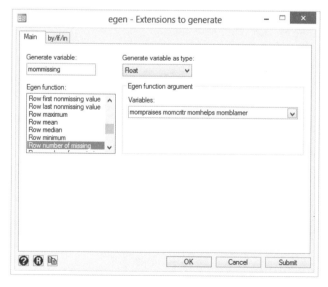

Figure 3.7. The Main tab for the **egen** dialog box

As you can see, the dialog box allows us to enter the name of the variable to be created, which we will call `mommissing` because this will tell us how many of the items about mothers that each adolescent missed answering. The next step is to select the function from the alphabetical list. In this case, we want to look among the row functions for one that will count missing values, which appears as *Row number of missing*. Last, we enter the items about the mother in the box provided. These selections produce

```
. egen float mommissing = rowmiss(mompraise momcritr momhelp momblamer)
```

### Beyond egen

We have seen how `generate` and `egen` cover a wide variety of functions. You can always type `help generate` or `help egen` to see a description of all the available functions. Sometimes this is not enough. Nicholas J. Cox, a Stata user, wrote in 2000, with subsequent revisions, a command called `egenmore`, which adds many more functions that are helpful for data management. You can type `ssc install egenmore, replace` in your Command window to install these added capabilities. The `replace` option will check for updates Cox has added if `egenmore` is already installed on your machine. If you now enter `help egenmore`, you will get a description of all the capabilities Cox has added.

If we look at a tabulation of the `mommissing` variable, the rows will be numbered 0–4, and each entry will indicate how many observations have that many missing values. For example, there are 4,510 observations for which all four items were answered (no missing values), 6 for which there are three items answered (one missing value), 2 for which there are two items answered (two missing values), and 4,466 for which there are no items answered. Our table does not include a row for 3 because there are no observations in our example that have three missing values.

```
. tab mommissing
```

| mommissing | Freq. | Percent | Cum. |
|---|---|---|---|
| 0 | 4,510 | 50.20 | 50.20 |
| 1 | 6 | 0.07 | 50.27 |
| 2 | 2 | 0.02 | 50.29 |
| 4 | 4,466 | 49.71 | 100.00 |
| Total | 8,984 | 100.00 | |

We can see that there are a total of eight observations for which there are some, but not complete, data. There should be little doubt about what to do for the 4,510 cases with complete information, as well as for the 4,466 with no information (all missing). The problem with using the sum of the items as our score on `ymomrelate` is that the eight observations with partial data will be dropped, that is, given a value of missing.

There are several solutions. We might decide to compute the mean of the items that are answered. We can do this with the `egen` dialog box; select Data ▷ Create or change data ▷ Create new variable (extended) to display the egen dialog box displayed in figure 3.7. This time, call the new variable mommeana and pick *Row mean* from the list of egen functions. This function generates a mean of the items that were answered, regardless of how many were answered, as long as at least one item was answered. If we tabulate mommeana, we have 4,518 observations, meaning that everybody who answered at least one item has a mean of the items they answered.

Another solution is to have some minimum number of items, say, 75% or 80% of them. We might include only people who answered at least three of the items. These people would have fewer than two missing items. We can go back to the dialog box for egen and rename the variable we are computing to mommeanb. The by/if/in tab allows us to stipulate the condition that mommissing < 2, as shown in figure 3.8.

Figure 3.8. The by/if/in tab for the egen dialog box

The Stata command this dialog box generates is

```
. egen float mommeanb = rowmean(mompraise momcritr momhelp momblamer)
> if mommissing < 2
```

This command creates a variable, mommeanb, that is the mean of the items for each observation that has at least three of the four items answered. Doing a tabulation on this shows that there are 4,516 observations with a valid score. This command allows us to keep the six cases that answered all but one of the items and drop the two cases that were missing more than one item.

**Deciding among different ways to do something**

> When you need to decide among different ways of performing a data-management
> task—which is what we have been doing this whole chapter—it is usually better to
> choose based on how easy it will be for someone to read and understand the record
> rather than based on the number of commands it takes or based on some notion of
> computer efficiency. For example, computing the row mean has a major advantage
> over computing the sum of items. The mean is on the same 0–4 scale that the
> items have, and this often makes more sense than a scale that has a different range.
> For example, if you had nine items that ranged from 1 for `strongly agree` to 5
> for `strongly disagree`, a mean of 4 would tell you that the person has averaged
> `agree` over the set of items. This might be easier to understand than a total or
> sum score of 36 on a scale that ranges from 9 (one on each item) to 45 (five on each
> item).

## 3.7   Saving some of your data

Occasionally, you will want to save only part of your data. You may wish to save only
some variables, or you may wish to save only some observations. Perhaps you have
created variables that you needed as part of a calculation but do not need to save, or
perhaps you wish to work only with the created scales and not the original items. You
might want to drop some observations based on some characteristic, such as gender or
parent's educational level or geographic area.

To drop a variable from your dataset, you can use the Variables Manager. Right-
click on the variable name. A drop-down menu appears in which you can select *Drop
Selected Variables*. When you do this, as a safety check, Stata asks you to confirm that
you want to drop the variable. Suppose that we want to drop all the original variables,
R0000100–R3828700, because we have given all of them a new name that is easier to
understand. We could click on R0000100 to highlight it, and then press Shift+Down
arrow to highlight the rest of the variables that we want to drop. Right-click on the         *DROP*
highlighted variables and select *Drop Selected Variables* to drop the set of variables.
The command produced by this example is

```
. drop R0000100 R3483600 R3483700 R3483800 R3483900 R3485200 R3485300 R3485400
> R3485500 R3828100 R3828700
```

Sometimes it is more efficient to indicate the variables you want to keep rather than
the variables you want to drop. In the Variables Manager, right-clicking gives us the
*Keep Selected Variables* choice. If we only wanted to keep a few variables, this might          *KEEP*
be a bit easier to do. We would highlight the ones we wanted to keep before clicking
*Keep Selected Variables*.

Sometimes we want to drop or keep selected observations rather than variables. For example, we might want to drop children who are under 12 or keep children if they are in an honors program. Similarly, we might want to keep people who are African American or drop people who are not African American. When we have an enormous dataset, we might want to keep the first 500 observations for preliminary analysis. We can use the same `drop` and `keep` commands we used for variables. Instead of using the Variables Manager, we can enter the commands directly, but rather than using `drop` *varlist*, we follow the syntax of `drop if` *exp* or `drop in` *range*. Here are a few examples:

```
. drop if age < 12
. drop if age > 17 & gender == 1
. keep if age >= 12 & age < 18 & gender == 1
. keep in 1/500
```

The first command, `drop if age < 12`, drops any observation who is under age 12. The second command drops any observation who is over age 17 and whose gender is coded as a 1. The double equal signs are read as "is", that is, `gender` is 1. If you wrote `gender = 1` here, it would create an error. The statement `gender = 1` would naturally tell Stata you want to assign a value of 1 on `gender` to everybody. We do not want to do this! We want to drop people if their `gender` "is" 1. Stata uses two equal signs, `==`, to say this statement. Notice the use of the ampersand, `&`, rather than typing `and`. If you used the word `and` instead of the symbol `&`, you would get an error message. The third example keeps people who are age 12 or over (`>=`) and are under age 18 (`<`) and have a `gender` that is coded as 1. The last example keeps only observations 1 to 500. The `/` sign can be read as "to".

## 3.8   Summary

This chapter has covered a lot of ground. You may need to refer to this chapter when you are creating your own datasets. We have covered how to label variables and create value labels for each possible response to an item. In explaining how to create a scale, we covered reverse-coding items that were stated negatively and creating and modifying items. We also covered how to create a scale and work with missing values, especially where some people who have missing values are included in the scale and some are excluded. Finally, we covered how to save parts of a file, whether the parts were selected items or selected observations.

In the next chapter, we will look at how to create a file containing Stata commands that can be run as a group. The book's webpage has such a command file for chapter 3 (`chapter3.do`). By recording all the commands into a program, we have a record of what we did, and we can rerun it or modify it in the future. We will also see how to record both output and commands in files.

## 3.9 Exercises

1. Open `relate.dta`. The R3828700 variable represents the gender of the adolescent and is coded as a 1 for males and 2 for females. There are 775 people who are missing because they dropped out of the study after a previous wave of data collection. These people have a missing value on R3828700 of -5. Run a tabulation on R3828700. Then modify the variable so that the code of -5 will be recognized by Stata as a missing value. Label the variable so that a 1 is male and a 2 is female. Finally, run another tabulation and compare this with your first tabulation.

2. Open `relate.dta`. Using the R3483600 variable, repeat the process you did for the first exercise. Go to the webpage for the book and examine the dictionary file `relate.dct` to see how this variable is coded. Modify the missing values so that -5 is .a, -4 is .b, -3 is .c, -2 is .d, and -1 is .e. Then label the values for the variable and run a tabulation.

3. Using the result of the second exercise, run the command `numlabel, add`. Repeat the tabulation of R3483600, and compare it with the tabulation for the second exercise. Next run the command `numlabel, remove`. Finally, repeat the tabulation, but add the `missing` option; that is, insert a comma and the word "missing" at the end of the command. Why is it good to include this option when you are screening your data?

4. `relate.dta` is data from the third year of a long-term study. Because of this, some of the youth who were adolescents the first year (1997) are more than 18 years old. Assign a missing value on R3828100 (age) for those with a code of -5. Drop observations that are 18 or older. Keep only R0000100, R3483600, R3483800, R3485200, R3485400, and R3828700. Save this as a dataset called `positive.dta`.

5. Using `positive.dta`, assign missing values to four of the variables: R3483600, R3483800, R3485200, and R3485400. Copy these four variables to four new variables called `mompraise`, `momhelp`, `dadpraise`, and `dadhelp`, respectively.

6. Create a scale called `parents` of how youth relate to their parents using the four items (R3483600, R3483800, R3485200, and R3485400). Do this separately for boys (if R3828700 == 1) and girls (if R3828700 == 2). Use the `rowmean()` function to create your scale. Do a tabulation of `parents`.

# 4 Working with commands, do-files, and results

## 4.1   Introduction

Throughout this book, I am illustrating how to use the menus and dialog boxes, but underneath the menus and dialogs is a set of commands. Learning to work with the commands lets you get the most out of Stata, and this is true of any statistical software. Even the official Stata documentation is organized by command name, as illustrated by the *Stata Base Reference Manual*, which has more than 2,500 pages explaining Stata commands.

With the logical organization of the menu system, you may wonder why you need to even think about the underlying commands. There are several reasons: Entering commands can be quicker than going through the menus. More importantly, you can put the commands into files that are called *do-files*, allowing for easy repetition of a series of commands. These do-files allow you to replicate your work, something you should always ensure you can do. When you collaborate with co-workers, they can use your do-file as a way to follow exactly what you did. It is hard enough to remember all the commands you create in a session, and if there is a delay between work sessions, it is impossible to remember all those commands. Even when you are using the menu system, it is useful to save the commands generated from the menus into a do-file, and Stata has a way to facilitate saving these commands; we will soon learn how to do this.

**What is a command?  What is a do-file?**

A command instructs Stata to do something, such as construct a graph, a frequency tabulation, or a table of correlations. A do-file is a collection of commands. "Do-file" is a good name because it is so descriptive: the series of commands in the file tells Stata what to "do". A simple do-file might open a dataset, summarize the variables, create a codebook, and then do a frequency tabulation of the categorical variables. A do-file can include all the commands you use to label variables and values; it can recode variables, average variables, and define how you treat missing values. Such a do-file might be only a few lines long, but complicated do-files can be thousands of lines long.

## 4.2   How Stata commands are constructed

Stata has many commands. Here are some of the commands covered in this book:

| | |
|---|---|
| list | List values of variables |
| summarize | Summary statistics |
| describe | Describe data in memory or in file |
| codebook | Describe data contents |
| tabulate | Tables of frequencies |
| generate | Create or change contents of variable |
| egen | Extensions to generate |
| correlate | Correlations (covariances) of variables or coefficients |
| ttest | Mean-comparison tests |
| anova | Analysis of variance and covariance |
| regress | Linear regression |
| logit | Logistic regression, reporting coefficients |
| factor | Factor analysis |
| alpha | Compute interitem correlations (covariances) and Cronbach's alpha |
| graph | The graph command |

Stata has a remarkably simple command structure. Stata commands are all lower-case. Virtually all Stata commands take the following form: *command varlist* if/in, *options*. The *command* is the name of the command, such as summarize, generate, or tabulate. The *varlist* is the list of variables used in the command. For many commands, listing no variables means that the command will be run on all variables. If we said summarize, Stata would summarize all the variables in the dataset. If we said summarize age education, Stata would summarize just the age and education variables. The variable list could include one variable or many variables. After the variable list come the if and in qualifiers regarding what will be included in the particular analysis. Suppose that we have a variable called male. A code of 1 means that the participant is a male, and a code of 0 means that the participant is female. We want to restrict the analysis to males. To restrict the analysis, we would say if male == 1. Here we use two equal signs, which is the Stata equivalent to the verb "is". So the command means "if male is coded with a value of 1". Why the two equal signs? The statement male = 1 literally means that the variable called male is a constant value of 1, but males are coded as 1 and females are coded as 0 on this variable. Sometimes we want to run a command on a subset of observations, and so we use the in qualifier. For example, we might have the command summarize age education in 1/200, which would summarize age and education in the first 200 observations.

Each command has a set of *options* that control what is done and how the results are presented. The options vary from command to command. One option for the summarize command is to obtain detailed results, summarizing the variables in more ways. If we wanted to do a detailed summary of scores on age and years of education for adult males, the command would be

```
. summarize age education if male == 1 & age > 17, detail
```

The command structure is fairly simple, which is helpful for us because it is absolutely rigid. This example used the ampersand (&), not the word "and". If we had entered the word "and", we would have received an error message. Here are more examples with if statements:

```
. summarize age education if sex == 0
. summarize age education if sex == 1 & age > 64
. summarize age sex if sex == 0 & age > 64 & education == 12
```

When you have missing values stored as . or .a, .b, etc., you need to be careful about using the if qualifier. Stata stores missing values internally as huge numbers that are bigger than any value in your dataset. If you had missing data coded as . or .a and entered the command summarize age if age > 64, you would include people who had missing values. The correct format would be

```
. summarize age if age > 64 & age < .
```

The < . qualifier at the end of the command is strange to read (less than dot) but necessary. Table 4.1 shows the relational operators available in Stata.

Table 4.1. Relational operators used by Stata

| Symbol | Meaning |
| --- | --- |
| == | is or is equal to |
| != or ~= | is not or is not equal to |
| > | is greater than |
| >= | is greater than or equal to |
| < | is less than |
| <= | is less than or equal to |

The in qualifier specifies that you will perform the analysis on a subset of cases based on their order in the dataset. If we had 10,000 participants in a national survey and we wanted to list the values in the dataset for age, education, and sex, this list would go on for screen after screen after screen, which would be a waste of time. We might want to list the data on age, education, and sex for just the first 20 observations by using in 1/20. The 1 is where Stata should start (called the first case), the "/" is read as "to", and the 20 is the last case listed. Thus in 1/20 tells Stata to do the command for the cases numbered from 1 to 20, or for the first 20 cases. The full command is

    . list age education sex in 1/20

Listing just a few cases is usually all you need to check for logical errors. Most Stata dialog boxes include an if/in tab for restricting data.

If your dataset contains few variables, it may be easier to just leave the Data Editor (Browse) open with the data while you are typing your commands instead of running a listing. Any changes made by your commands will appear immediately in the Data Editor (Browse). You can open the Data Editor (Browse) by typing the browse command or the browse, nolabel command in the Command window or by clicking on the toolbar icon that looks like a spreadsheet with a magnifying glass (see figure 2.3).

The final feature in a Stata command is a list of options. You must enter a comma before you enter the options. As you learn more about Stata, the options become increasingly important. If you do not list any options, Stata gives you what it considers to be basic results. Often these basic results are all you will want. The options let you ask for special results or formatting. For example, in a graph, you might want to add a title. In frequency tabulation, you might want to include cases that have missing values. One of the best reasons for using dialog boxes is that you can discover options that can help you tailor your results to your personal taste. Dialog boxes either include an Options tab or have the options as boxes that you can check on the Main tab. The most common mistake a beginner makes when typing commands directly in the Command window is leaving the comma out before specifying the options.

Here are a few Stata commands and the results they produce. You can enter these commands in the Command window to follow along.

```
. use http://www.stata-press.com/data/agis4/firstsurvey_chapter4

. summarize
      Variable |       Obs        Mean    Std. Dev.       Min        Max
---------------+--------------------------------------------------------
            id |        20        10.5     5.91608         1         20
        gender |        20         1.5    .5129892         1          2
     education |        20       14.45    2.946452         8         20
        sch_st |        18    3.444444    1.149026         2          5
       sch_com |        20         3.5    1.395481         1          5
---------------+--------------------------------------------------------
        prison |        17    3.176471    1.550617         1          5
        conserv |       19    2.947368    1.544657         1          5
```

This `summarize` command does not include a variable list, so Stata will summarize all variables in the dataset. It has no if/in restrictions and no options, so Stata summarizes all the variables, giving us the number of observations with no missing values, the mean, the standard deviation, and the minimum and maximum values. The statistics for the `id` variable are not useful, but it is easier to get these results for all variables than it is to list all the variables in a variable list, dropping only `id`.

We can add the `detail` option to our command to give more detailed information. Do this for just one variable.

```
. summarize education, detail
                       Years of education

              Percentiles      Smallest
 1%             8                 8
 5%             9.5              11
10%            11.5             12          Obs                  20
25%            12               12          Sum of Wgt.          20

50%            14.5                         Mean              14.45
                               Largest      Std. Dev.      2.946452
75%            16.5             17
90%            18               18          Variance       8.681579
95%            19               18          Skewness      -.1636124
99%            20               20          Kurtosis       2.522208
```

As expected, this method gives us more information. The 50% value is the median, which is 14.5. We also get the values corresponding to other percentiles, the variance, a measure of skewness, and a measure of kurtosis (we will discuss skewness and kurtosis later).

Next we will use the `list` command. Here are four commands you can enter, one at a time, to get three different listings:

```
. list gender education prison in 1/5
```

|     | gender | educat~n |        prison |
| --- | ------ | -------- | ------------- |
| 1.  | woman  | 15       | too long      |
| 2.  | man    | 12       | much too lenient |
| 3.  | man    | 16       | about right   |
| 4.  | man    | 8        | .             |
| 5.  | woman  | 12       | about right   |

```
. list gender education prison in 1/5, nolabel
```

|     | gender | educat~n | prison |
| --- | ------ | -------- | ------ |
| 1.  | 2      | 15       | 4      |
| 2.  | 1      | 12       | 1      |
| 3.  | 1      | 16       | 3      |
| 4.  | 1      | 8        | .      |
| 5.  | 2      | 12       | 3      |

```
. numlabel _all, add
. list gender education prison in 1/5
```

|     | gender   | educat~n |           prison |
| --- | -------- | -------- | ---------------- |
| 1.  | 2. woman | 15       | 4. too long      |
| 2.  | 1. man   | 12       | 1. much too lenient |
| 3.  | 1. man   | 16       | 3. about right   |
| 4.  | 1. man   | 8        | .                |
| 5.  | 2. woman | 12       | 3. about right   |

The first command shows the first five cases. Notice that the `education` variable appears at the top of its column as `educat~n`. We can use names with more than eight characters, but some Stata results will show only eight characters. For names with more than eight characters, Stata will keep the first six characters and the last character and insert the tilde (~) between them. Because we assigned value labels to `gender` and `prison`, the value labels are printed in the list. However, notice that the numerical values are omitted.

The second command adds the `nolabel` option, which gives us a listing with the numerical values we used for coding but not the labels. The next command, `numlabel _all, add`, adds a numeric value to each label for all variables (the `_all` tells Stata to apply this command to all variables). If we wanted to remove the values later, we would enter `numlabel _all, remove`. Finally, the last listing gives us both the values and the labels for each variable.

## 4.3   Creating a do-file

You may be asking yourself how you will ever learn to use all the options and qualifiers. One way is to read the Stata documentation, but it is often easier to use the menus and

record the command in a do-file. Stata has a simple text editor called the Do-file Editor
in which you can enter a series of commands. You can run all the commands in this
file or just some of them. You can then edit, save, and open the commands in a do-file
at a later date. Saving these do-files means not only that you can replicate what you
did and make any needed adjustments but also that you will develop templates you can
draw on when you want to do a similar analysis. To open the Do-file Editor window,
use Window ▷ Do-file Editor ▷ New Do-file Editor. You can also open the Do-file Editor
by clicking on the toolbar icon that looks like a notepad; see figure 4.1 for the icon in
Stata for Windows.

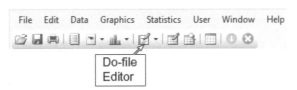

Figure 4.1. The Do-file Editor icon on the Stata menu

Stata allows you to have multiple do-files open at once, but we will use only one
do-file in this book. When you gain confidence using Stata, the ability to have several
do-files open simultaneously is a useful feature. You might use one do-file primarily
for data-management purposes; you might use another for general analysis of the data;
you might use a third for analyzing a subset of the data. You might even work on a
couple of different projects in one session. Using multiple do-files in a session is sort of
like juggling balls: some of us are better at it than others. For now, one Do-file Editor
window is all we need. When several do-files are open, clicking on the arrow to the right
of the icon shows the name of each open file. A blank do-file appears in figure 4.2.

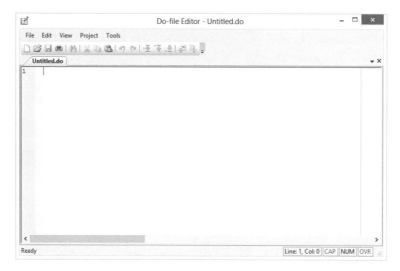

Figure 4.2. The Do-file Editor of Stata for Windows

When you click on another window, the Do-file Editor can be hidden by it. You can bring the Do-file Editor to the front again by clicking on it in the system toolbar or by using the Alt+Tab key combination to move through the open windows until the Do-file Editor is highlighted. You can avoid having the window disappear by arranging your desktop so that other windows do not overlap with the Do-file Editor window. If you have two monitors, it is convenient to have the main Stata window on one monitor and the Do-file Editor on the other monitor.

The Do-file Editor lacks special features, such as underlining or the special fonts you would get with a word processor such as MS Word. Word processors add a lot of hidden features that would be confusing to Stata. The Do-file Editor is a plain-text editor. It does, however, have some nice features for writing a do-file. As you can see in figure 4.3, the toolbar has many features that you would expect in an editor.

Figure 4.3. The Do-file Editor toolbar of Stata for Windows

The first 10 icons on the left are fairly standard. The first four icons let you open a new do-file, open an existing do-file, save your file, and print the file. The fifth icon allows you to perform a search, although most people prefer to use Ctrl+f to perform a search. The next three icons will cut, copy, and paste, although most people prefer to use Ctrl+x, Ctrl+c, and Ctrl+v (Cmnd+x, Cmnd+c, and Cmnd+v for Macs), respectively. The curved arrows provide undo and redo functions, like a word processor. The next three icons have to do with bookmarks, marking lines in a file you are editing so you can easily jump back to them later. The second from last icon will show content of the file in a Viewer window. Finally, the last icon runs either your entire do-file or the part of your do-file that you have selected. This is the key icon for running your do-file. You select lines of code in the same way you would in a word processor, by highlighting them. If you want to select a section of several lines, you do not need to highlight all the lines completely; you can highlight some of each line. Here is an example where we want to run two commands, describe and summarize. In figure 4.4, we have selected only part of the describe command but all the summarize command.

```
1        * selecting text.do
2        version 13
3        use firstsurvey_chapter4
4        describe
5        summarize
```

Figure 4.4. Highlighting in the Do-file Editor

The Do-file Editor automatically numbers your lines. Only the number 1 for the first line appeared in figure 4.2 because the rest of the file is blank. In long do-files, the line numbers help you find your way around. In the lower right, the Do-file Editor also gives the exact location of the cursor, such as "Line: 1, Col: 0" (again, see figure 4.2).

I recommend that you include the name of the file as a comment in the first line of the do-file. Placing an asterisk (*) at the beginning of the line marks the text as a comment that Stata prints but does not interpret. I include this line because my do-file will have the name of the file that created the output so that I know where to find the file if I need to change something later on. Let's type * my_first.do at the top of our blank do-file.

Another way of adding comments, especially long comments, to a file is by typing /* just before the comment and */ right after the comment. Anything between the /* and */ will be treated as a comment. Figure 4.5 shows what we will use in our example do-file. This type of long comment is helpful for organizing a long do-file into sections with a new comment explaining the purpose of each new section.

We will now add our first command, version 13. This command can be important because it will execute the commands as they were written in Stata 13. Stata is continually being improved, and updates may include changes to the commands that you are using. By including the version number you are using when you create the do-file, you ensure that everything will work as intended even when you are using future versions of Stata. This is called version control. It is how Stata maintains compatibility when new versions add new capabilities.

Notice that the comments appear in the Do-file Editor in a green font. This makes them easy to find in a long do-file. Stata commands appear in a bold blue font.

Next we need to open the dataset. In our do-file, we type

```
. use firstsurvey_chapter4.dta, clear
```

Stata automatically looks in our working directory for this file. If we already have a dataset open, the `clear` option will clear those data from Stata's memory. If we had a dataset open with changes made to the data and we did not include the `clear` option with our `use` command, Stata would issue a warning because it does not want us to accidentally clear an active dataset before we save our changes.

You do not have to store your data and do-files in the working directory. You can create a separate directory for each project, or you can use an external flash drive. To open a dataset that is not saved in the working directory, you must specify the full path in your command. Say that our dataset, `firstsurvey_chapter4.dta`, was saved to an external flash drive in a folder called `first project`. We would type

```
. use "E:\first project\firstsurvey_chapter4.dta", clear
```

The quotation marks are required if your folder name or filename contains embedded spaces like the space in `first project` above.

The folder path will vary for Mac and Unix. For example on a Mac if your file is in the `Documents` folder, you would type

```
. use "Users/username/Documents/firstsurvey_chapter4.dta", clear
```

After using Stata for some time, it is usually best to organize your folders in such a way that each project has a separate folder creating a good work flow. Scott Long has written a wonderful book called *The Workflow of Data Analysis Using Stata* (2009), which is available from Stata Press at http://www.stata-press.com/books/wdaus.html. You should read this book if you plan to be involved in a series of major projects using Stata.

The next two commands we need to type are `describe` and `summarize`, which will describe the dataset and then perform a summary of the variables, giving us the number of observations, mean, standard deviation, minimum value, and maximum value. Unless there are too many variables in your dataset, these are good commands to include at the beginning of your do-file.

If you are wondering how Stata knows when it is done with one command and ready to start another, the answer lies with the Enter key. Every time you press the Enter key, Stata assumes you are either done or starting a new command. Having a hard return between them makes `describe` and `summarize` two separate operations. All programs have some way to designate the end of a command. SAS, for example, ends each procedure with a semicolon. The semicolon in SAS is the same as the Enter key in Stata.

In general, one line equals one command in a do-file. This is a great way to distinguish one command from another as long as each command is short enough to fit on one line. One line equals one command. What happens when you have a long command that extends for more than one line? Stata needs a way to know that when you press the Enter key, you are not really done with the command. The solution is to put `///` at the end of a line. This tells Stata that the next line is a continuation of the previous one. We illustrate the use of /// in the `graph pie` commands shown in figure 4.5. (We will cover these commands in the next chapter. For now, just enter them into the Do-file Editor.)

If you are using dialog boxes to create your commands, you can click on the Copy button in the bottom left corner to copy the command and then press Ctrl+v in the Do-file Editor to paste the command into your do-file.

You must remember to save your do-file. You do this much like you would save a file in any other program; that is, click on File ▷ Save As.... Then you can type the name of the file; in our example, we are using `my_first.do`. You can also browse to find the project folder where you want to save the file, such as `E:\first project`. Until you have more experience, it is probably best to store your do-files and your data in the same folder.

Figure 4.5 shows our do-file, `my_first.do`, as it appears in the Do-file Editor. Notice that when we saved the do-file, the filename now appears at the top of the Editor.

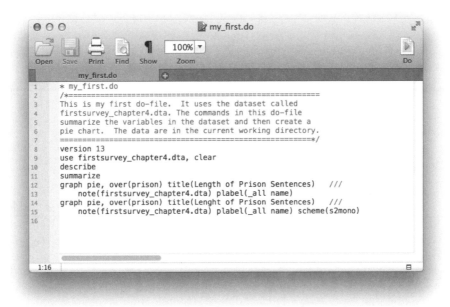

Figure 4.5. Commands in the Do-file Editor window of Stata for Mac

Are you convinced that you should keep a do-file? I hope so. Imagine that one month from now, you decide to create a pie chart for another variable, `conserv`, from `firstsurvey_chapter4.dta`. You can open the `my_first.do` file and replace `over(prison)` with `over(conserv)`. Click on the icon to run the do-file, and you are done.

As things get more complex, having a set of do-files that perform certain tasks becomes extremely valuable. One thing you might do is create an MS Excel spreadsheet to keep a record of the do-files you create. You might have three columns: the name and path of the do-file, the dataset it uses, and a brief statement of the purpose of each do-file. Eventually, you will have 30 or more do-files, and you can scan this listing to find one that is close to what you plan to do next to serve as a template. If you start doing several major projects, you probably should have a spreadsheet like this for each of the projects.

### Stata do-files for this book

The webpage for this book, http://www.stata-press.com/data/agis4/, has do-files for each chapter along with the datasets used. You can copy the do-files and datasets to your computer and reproduce the results in this book. These do-files may also be useful as templates when you do your own work because they can be modified as you need. You will have to change the paths in these files to the paths where you have stored your data unless you have stored the data in Stata's default directory, `C:\Data`. You could also change the default directory. If you stored the data files in `C:\Data\agis4`, you could enter the command `cd "C:\Data\agis4"` to make that the default directory. Then to open a file for Stata to use, you would simply type `use` *filename*, where *filename* is the name of the data file you want to use.

## 4.4   Copying your results to a word processor

Many people start using Stata to do their homework for a class in statistical methodology. In those cases, the datasets are often fully prepared, and you will not need to keep a record of how the data were created, how you labeled variables, how you recoded some items and dealt with missing values, or how you produced your results. In fact, many of the analyses will have short results. Here it may be simplest and best to save your results by highlighting the text that you want to save in the Results window, right-clicking on the highlighted text, and then selecting Copy from the menu (or using Ctrl+c after you have highlighted the text to copy). You can then paste the text into your favorite word processor. It is a good idea to include the commands that are in the Results window, because these give you a record of what you did. Except when there is no data manipulation, commands like these are no substitute for a do-file that includes everything you did in preparing the data.

When you copy results from Stata's Results window to your word processor, the format may look like a hopeless mess because Stata output is formatted using a fixed-width font; when you paste your results, things will likely not line up properly. The simplest solution to this alignment problem is to change the font and probably the font size, depending on your margins. I usually use the Courier or Courier New font at 9 point. Sometimes the lines may still wraparound, and you may need to widen the margins. Most Stata results will fit nicely if you have 1-inch margins and a 9- or 10-

point Courier font. With most word processors, you change the font of the lines that
contain the results by highlighting the lines you want to change and then selecting the
Courier font and the 9-point font size.

### Saving tabular output

If you are using the Windows version of Stata and have tabular output (say, the
results of a `summarize` command), you may want to select just that portion of the
text that appears as a table, right-click on it, and select Copy Table, Copy Table
as HTML, or Copy as Picture from the menu. You can then paste this text into
your word processor. The Copy Table as HTML option pastes it as a table like one
you would see on a webpage. However, if you are working with a specific style
format, such as the APA requirements for tables, you will need to reenter the table
in your word processor to meet those requirements. The Copy as Picture option
works nicely as long as you do not need to format the output to a particular
style. The Copy as Picture option takes a picture of the output you highlighted.
If you paste this picture into a word processor, it will look just like it did in the
Results window. You will then be able to resize it, but you will not be able to edit it.

If you are using the Mac version of Stata and select Edit ▷ Copy as Picture,
Stata will copy the table as both a TIFF image and as a PDF image to the
Clipboard. When you paste the file into another application, that application will
automatically determine which version of the file to take from the Clipboard.

As you progress in your class or as you start to use Stata in your own research, you
will find that copying and pasting is really not up to the task of creating a record of
your work. For more complex work, you will want to use log files, which we will now
discuss.

## 4.5   Logging your command file

Stata can write a copy of everything that is sent to the Results window into a file. The
file is called a *log file*. When you start logging, you can create a new file, or you can
add on to the end of an existing log. You can temporarily suspend logging at any time
and then restart it again. If you do not have a running Stata session, start one now,
and let's take a look at output logs.

We can open a log by selecting File ▷ Log ▷ Begin..., which will bring up the file
selector window. Navigate to the directory in which we want to keep our log, and then
enter a name (or select one from the list). By default, Stata will save this log using a
markup language called Stata Markup and Control Language (SMCL) that only Stata
can read in a Viewer or Results window. The log will have a *filename*.`smcl` name, where
.`smcl` is the extension. It looks nice in Stata but in another application, although it

is a text file, the SMCL tags will be displayed because the word processor or other text editors will not understand them.

The other file format is called "log", which is a simple text file that your word processor can read. We need to pick the option of having a log file rather than a SMCL file from the dialog box. At the bottom of the *Begin logging Stata output* dialog box, we can specify *Save as type*. From the pulldown menu, select *Log (*.log)* for the type. Like the Results window, the .log format uses a fixed-width format, and if you insert this file into a word processor, you need to make sure that the font is fixed width (for example, Courier) and the font size is small enough (for example, 9 point). Make sure to select a location where you can find this log file. We can insert this log into an open file in our word processor, or we can open the log file itself in our word processor. Because the extension will be .log, when we open it into a word processor, we need to make sure that the word processor is not just looking for certain other extensions. For example, in MS Word, you would need to browse for the log file where the type is *.* rather than *.docx.

If you are using MS Word as your word processor, you have two options when working with a .log file. You can open it as a new document in Word as long as you remember to change the file type from *.docx to *.* or show all. If you do this, the Word document will have a fixed font. You may need to change the font size, say, to 9 point, or make the margins wider. If you have an existing Word document, you need to use the Insert option rather than the Open option, and insert the text where you want it.

Open a new log file called results.log. Make sure to specify that it is a log file rather than a SMCL file. Then run a summarize command on the file. Now open the log by selecting File ▷ Log ▷ View..., which will show us the basic information about our file, where it is stored, and the date and time it was created, and will give the results of our summary in a Viewer. When we go back to our Command window, the Viewer will seem to disappear, but it will be on the taskbar at the bottom of our screen. We can click on it to open it again.

Run a few more commands, such as a tabulation and a graph. Now select the Viewer from the Windows toolbar, and the Viewer returns, but it goes only as far as the original summarize command. However, at the top of the Viewer window is a Refresh button. Clicking on this button will update the Viewer to include the tabulate and the graph commands. What happened to the graph? Unfortunately, the log does not include the graphic output in the log file.

Experienced users use log files a lot. For beginners, log files may not be necessary. Also, if you make a lot of mistakes and need to run each command several times before you get it just right, the log will have all the bad output that you do not want, along with the good output that you do want. You might want to pause the log file while you try out a command or do-file. Then when you have the command the way you want it, you can restart the log to minimize the bad output. To do this, select File ▷ Log ▷ Suspend or Resume, respectively.

A strength of using log files is that they provide a record of both your commands and your results. The commands precede each result, so you can immediately see what you did. The advantage of the do-file from the Do-file Editor is that the commands are all together in a compact form, so it is easier to review the programming you did. A major limitation of the log file is that it does not save graphic output, such as the pie chart we did in this chapter. Graphs need to be saved by right-clicking within the graph and then choosing an option, whether you want to save it (select Save As...) or copy it (select Copy) so you can paste it into your word processor.

*graphs*

## 4.6  Summary

Let's review what you have learned in this chapter:

- How to open a Do-file Editor and copy commands from the Results window

- How to use the dialog boxes in Stata to generate Stata commands and then copy these to the Do-file Editor

- How to string a series of commands together in the Do-file Editor and add comments to this file

- How to run an individual command and groups of commands from the Do-file Editor

- How to save your do-file and retrieve it for later analysis

- How to cut and paste between Stata and a word processor, like Word

- How to create and view a log file

This is a lot of new knowledge for you to absorb. Never feel bad if you need to review this material because you forgot a step along the way. Even as you gain experience with Stata, this chapter will be a handy resource.

This chapter focused on the mechanics of using the Do-file Editor, writing a simple do-file, and saving your results. As you go through the following chapters, you will learn how to write more complex do-files and do more complex data management.

The rest of the book will focus on performing graphic and statistical analysis, building on what we have done so far. Most people learning a statistics program want to learn how to do analyses rather than what we have done so far. Still, what we have done so far sets the groundwork for doing the analyses. Chapter 5 will go over graphic presentations and descriptive statistics.

## 4.7  Exercises

1. Open `firstsurvey_chapter4.dta` by selecting File ▷ Open.... Open a Do-file Editor, and copy into the Editor the command that opened the dataset. Open the dialog box to summarize the dataset, run the `summarize` command for all the variables, and copy this command from the Results window into the Do-file Editor. Save this do-file as `4-1.do` in a place where you can find it.

2. Open the do-file you created in the first exercise, and add appropriate comments. Save the new do-file under the name `4-2.do`.

3. Open the file `4-2.do`. Put your cursor in the Do-file Editor just below the command that opened the dataset and above the command that summarized the variables (you will need to insert a new line to do this). Type the `describe` command in the Do-file Editor. Add a command at the bottom of the file that gives you the median score on education. Save the new do-file under the name `4-3.do`.

4. Open `4-3.do`, run the `describe` command and the command that gave you the median score on education, highlight the results, paste the results into your word processor, and change the font so that it looks nice.

5. Open `4-3.do`. Open a log file with the log file type. Call the file `4results.log`. Run the entire `4-3.do` file, and exit Stata. Open a new session in your word processor, and open your log file into this session. Format it appropriately.

# 5 Descriptive statistics and graphs for one variable

## 5.1 Descriptive statistics and graphs

The most basic use of statistics is to provide descriptive statistics and graphs for individual variables. Advanced statistics and graph presentation can disentangle complex relationships between groups of variables, but for many purposes, simple descriptive statistics and graphs are exactly what is needed. Virtually every issue of a major city newspaper will have many descriptive statistics, and most issues will have one or more graphs. One article might report the percentage of teenagers who smoke cigarettes. Another article might report the average value of new homes. Each spring, there will be one or more articles estimating the average salary new college graduates will earn.

If you are working in a position related to public health, economics, or any of the social sciences, you will be a regular consumer of descriptive statistics, and many of you will be producers of these statistics. A parole office may need a graph showing trends in different types of offenses. A public health agency may need to demonstrate the need for more programs focused on sexually transmitted diseases: How much of a problem are sexually transmitted diseases? Is the problem getting worse or better?

Our society depends more and more on descriptive statistics. Policy makers are reluctant to make decisions without knowing the appropriate descriptive statistics. Social programs need to justify themselves to survive, much less to grow, and descriptive statistics and graphs are critical parts of this justification. In this chapter, you will learn how to produce these statistics and graphs using Stata.

## 5.2    Where is the center of a distribution?

Descriptive statistics are used to describe distributions. Three measures of central tendency describe the middle of the distribution: mean, median, and mode. The term "average" in statistics is typically defined as a synonym of the mean. Occasionally, this term is used to refer to the other measures of central tendency (the median and mode). When you read a newspaper article, it may say that the average family in a community has two children (this is probably the median, but it could be the mode) or that the average income in the community is $55,218 (also probably the median, but it could be the mean). The article might say that the average SAT score at your university is 1840 (probably the mean). It might say that the average person in a community has a high school diploma (this could be the mean, median, or mode). It is important to know when each central tendency measurement is appropriate.

The mode is the value or the category that occurs most often. We might say that the mode for political party in a parliament is the Labor Party. This would be the mode if there were more members of the parliament who were in the Labor Party than members of any other party. If we said the mode was 17 for age of high school seniors, this means that there are more high school seniors who are 17 than there are seniors who are any other age. This would be a reasonable measure of central tendency because most high school seniors are 17 years old. The mode represents the average in the sense of being the most typical value or category. If there is not a category or value that characterizes a distribution clearly, the mode is not descriptive of the central tendency of a distribution. For example, the heights of eighth-grade class members would not have a descriptive mode because each adolescent might be a different height, and there is no single height that is typical of eighth graders.

When you have unordered categorical variables, such as gender, marital status, or race/ethnicity, the mode is the only measure of central tendency. Even here, the mode is helpful only if one category is much more common than the others. If 79% of the adults in a community are married, saying that the modal marital status is married is a fair description of the typical member of the community. However, if 52% of adults in a community are female and 48% are male, it does not make much sense to say that the modal gender is female because there are nearly as many men as there are women.

The median is the value or the category that divides a distribution into two parts. Half the observations will have a higher value and half will have a lower value than the median. The median can be applied to categories that are ordered (political liberalism, religiosity, job satisfaction) or to quantitative variables (age, education, income). If we said that the median household income of a community is $55,218, we mean that half the households in that community have an income more than $55,218 and half the households have an income less than $55,218. When we used the summarize, detail command in chapter 4, we saw that Stata refers to the median as the 50th percentile.

The median is not influenced by extreme cases. If Bill Gates moved to this community, his multibillion dollar income would not influence the median. He would simply be in the half of the distribution that made more than the median income. Because of this property, the median is sometimes used with quantitative variables that are skewed (a distribution is skewed if it trails off in one direction or the other). Income trails off at the high end because relatively few people have huge incomes.

The median is occasionally used with variables that are ordered categories. When there are relatively few ordered categories, there may not be a category that has exactly half the cases above it and half below it. You might ask people about their marital satisfaction and give them response options of a) very dissatisfied, b) somewhat dissatisfied, c) neither satisfied nor dissatisfied, d) somewhat satisfied, and e) very satisfied. Because we usually code numbers rather than letters, we might code very dissatisfied as 1, somewhat dissatisfied as 2, neither satisfied nor dissatisfied as 3, somewhat satisfied as 4, and very satisfied as 5. The median satisfaction for men might be in the category we coded with a 4, somewhat satisfied. The median satisfaction for women might be 3, neither satisfied nor dissatisfied.

More often, researchers compute the mean for variables like this, and the mean for men might be 4.21 compared with 3.74 for women. These values indicate that men are, on average, a little above the somewhat-satisfied level and women are a little bit below the somewhat-satisfied level.

The mean is what lay people usually think of when they hear the word "average". It is the value every case would be, if every case had the same value. It is a fulcrum point that considers both the number of cases above and below it and how far they are above or below it. Although Bill Gates would scarcely change the median income of a community, his moving to a small town would raise the mean by a lot. Some people use $M$ (recommended by the American Psychological Association) to represent the mean, and others use $\overline{X}$ (recommended by most statisticians). The formula for the mean is

$$\overline{X} = \frac{\Sigma X}{n}$$

In plain English, this says the mean is the sum of all the values, $\Sigma X$ (pronounced sigma X or sum of X), divided by the number of observations, $n$. For example, if you had five college women who weighed 120, 110, 160, 140, and 210 pounds, respectively, the mean would be

$$\overline{X} = \frac{120 + 110 + 160 + 140 + 210}{5} = 148$$

From now on, we will use $M$ instead of $\overline{X}$ to represent the mean.

What measure of central tendency should you use? This decision depends on the level of measurement you have, how your variable is distributed, and what you are trying to show (see table 5.1).

Table 5.1. Level of measurement and choice of average

| Level of measurement | Mode | Median | Mean |
|---|---|---|---|
| Categorical, no order (nominal, e.g., gender) | Yes | No | No |
| Categorical, ordered (ordinal, e.g., social support) | Yes | Yes | Yes* |
| Quantitative (interval or ratio, e.g., age) | Yes | Yes | Yes |

*Many researchers use the mean when there are several ordered categories.

- When you have categories with no order (gender, religion), you can use only the mode. The mode for religion in Saudi Arabia, for example, is "Muslim". Unordered categorical variables are called *nominal-level variables*.

- When you have ordered categories (religiosity, marital satisfaction), the median is often recommended. Such variables are often labeled as ordinal measures. You might read that the median religiosity response in Chicago is "somewhat religious". Ordered categories can be ordered along some dimension, such as low to high or negative to positive. When there are several categories, many researchers treat them as quantitative variables and use the mean. If religiosity has seven ordinal categories from 1 for not religious at all to 7 for extremely religious, you might use the mean by treating these numbers from 1 to 7 as if they are an interval-level measure. You might say that the mean is 3.4, for example.

- When you have quantitative data (meaningful numbers), you can use the mean, median, or mode. Quantitative data are often called *interval-level variables*. You will usually use the mean. If, however, the variable is extremely skewed, you would use the median.

Suppose that we want an average value for the number of children in households that have at least one child. The distribution is highly skewed in a positive direction because it trails off on the positive tail (see figure 5.1).

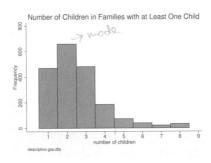

Figure 5.1. How many children do families have?

In this distribution, the mode is 2, the median (Mdn) is 2, and the mean ($M$) is 2.5. Notice how the small number of families with a lot of children drew the mean toward the tail but did not influence either the mode or the median. When a distribution is skewed, the mean will be bigger or smaller than the median, depending on the direction the distribution trails off.

There are two specialized averages you could use. The harmonic mean is useful when you want to average rates. Suppose that you go on a 60-mile bicycle ride where the first half of the course is a major hill climb and the second half of the course is a major descent. You might average just 5 miles per hour for the first half and then average 25 miles per hour for the second half of the ride. If we call your rate $a$, the harmonic mean ($H$) is

$$H = \frac{n}{\frac{1}{a_1} + \frac{1}{a_2} + \cdots + \frac{1}{a_k}}$$

$$= \frac{2}{\frac{1}{5} + \frac{1}{25}}$$

$$= 8.33$$

This harmonic mean of $H = 8.33$ miles per hour is a much better estimate of your average speed for the 60-mile ride than the arithmetic mean, which would be $(5+25)/2 = 15$ miles per hour.

The geometric mean is useful when you have a growth process where the growth is at a constant rate. This happens with population size or annual income, such as having your income grow at a rate of 3% per year. If you made 52,500 in 2000 and made 73,500 in 2010, what did you make in 2005? The arithmetic mean $(52500 + 73500)/2 = 63000$ exaggerates your income in 2005. The geometric mean ($G$) is

$$G = \sqrt[n]{a_1 \times a_2 \times \cdots \times a_n}$$

$$= \sqrt{52500 \times 73500}$$

$$= 62118.84$$

where $G = \$62{,}118.84$ is a much better estimate of your 2005 income.

The Stata command **ameans** *varlist* computes the arithmetic mean, the geometric mean, and the harmonic mean.

## 5.3   How dispersed is the distribution?

Besides describing the central tendency or average value in a distribution, descriptive statistics describe the variability or dispersion of observations. Are they concentrated in the middle? Do they trail off in one direction? Are they widely dispersed? Some suburbs are extremely homogeneous with rows of houses that are similar in style and value. These communities are highly concentrated around the average on a range of variables (income, education, ethnic background). Other communities are heterogeneous, and although they may have the same average values as the first community, they differ by having a mix of people who range widely on income, education, and ethnic background. This means that to understand a distribution, we need to know how it is distributed as well as its average value.

When there are only a few values or categories, we can use a frequency distribution (tabulation) to describe the variable. This distribution shows each value or category and how many people have that value or fall into that category. Stata calls this a tabulation. We can also use graphs to describe the dispersion of a distribution. When there are only a few categories being shown, the most common graphs to use are pie charts and bar charts.

When a variable is quantitative, we will usually want one number to represent the dispersion in the same way that we use one number to represent the central tendency. The standard deviation (SD) is used, especially with variables that have many possible values.

There are three parts of the SAT test: reading, writing, and math. Each of these parts will have a score between 200 and 800. If we just consider the reading and math sections of the SAT, the maximum possible score would be 1600. Let's assume that the total score on the math and reading sections has a mean of 1000 ($M = 1000$) and an SD of 100 (SD = 100). How can we interpret the mean and SD? Nearly all the students (about 95% of a normal distribution) are within two SDs of the mean and so they would have scores between 800 and 1200 on the combined math and reading sections.[1] This is the tall, but skinny, distribution in figure 5.2. If another school has the identical mean ($M = 1000$) but has an SD of 200, then nearly all the students had scores between 600 and 1400. The dispersion is much greater at the second school than at the first. We can see this in a graph of the two schools (figure 5.2). These two schools would have very different students even though the mean is identical. The smaller the SD, the more homogeneous the distribution is. From the graph, you can see how students at the first school are much more clustered around the mean ($M = 1000$) than are the students at the second school. The greater the SD, the more heterogeneous the distribution is.

---

1. The numbers used do not include the writing part of the SAT.

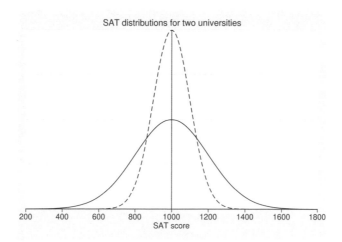

Figure 5.2. Distributions with same $M = 1000$ but SDs $= 100$ or 200

Both distributions in figure 5.2 are normal and have identical means of 1000 ($M = 1000$). The distribution that is tightly packed around the mean has an SD of 100 (SD $= 100$). The distribution that is more dispersed, the distribution that is low and wide in the figure, has an SD of 200 (SD $= 200$).

Skewness is known as the third moment of the distribution. A positive value indicates a positive skew, and I mentioned that income is often used as an example because there are relatively few people who have enormous incomes. A negative value indicates a negative skew. An example of a negatively skewed variable would be marital satisfaction. Surveys show that most married people are satisfied or very satisfied, few people are dissatisfied, and even fewer are very dissatisfied. Neither of the distributions in figure 5.2 is skewed. Both are symmetrical around their respective means.

Kurtosis is known as the fourth moment of the distribution. A distribution with high kurtosis tends to have a bigger peak value than a normal distribution. Correspondingly, a low kurtosis goes with a distribution that is too flat to be a normal distribution. The value of kurtosis for a normal distribution is 3. Some programs (both SAS and IBM SPSS Statistics, for example) subtract 3 from the kurtosis to center it on zero, and some statistics books may use this approach, but Stata uses the correct formula. If you are reporting the kurtosis in a field where either SAS or IBM SPSS Statistics is a widely used program, you need to know that for Stata a kurtosis of 3 indicates a normal distribution. In such fields, you might report the value of kurtosis minus 3 to be consistent with common practice in that field. When you report a value of kurtosis minus 3, a kurtosis with an absolute value greater than 10 is problematic.

## 5.4     Statistics and graphs—unordered categories

About all we can do to summarize a categorical variable that is unordered is to report the mode and show a frequency distribution or a graph (pie chart or bar chart). In this chapter, you will use a dataset, `descriptive_gss.dta`, that includes three categorical, unordered variables. `sex`, `marital`, and `polviews` are nominal-level variables. The variable `sex` is coded as `male` or `female`, `marital` is coded by marital status, and `polviews` is coded by political view. There is no order to `sex` in that being coded `male` or `female` does not make one higher or lower on `sex`. Similarly, there is no order to `marital` in that having a particular status (that is, never married, married, separated, divorced, or widowed) does not make one higher or lower on marital status. These are just different statuses.

We can use the `tabulate` command to get frequency distributions for `sex`, `marital`, and `polviews`. We could type `tabulate sex`, then `tabulate marital`, and then `tabulate polviews`. However, Stata has another command named `tab1` that will repeatedly issue the `tabulate` command on each of the specified variables: `tab1` `sex marital polviews`. This is so simple that you probably want to enter it directly, but if you want to use the dialog box, select Statistics ▷ Summaries, tables, and tests ▷ Frequency tables ▷ Multiple one-way tables; see figure 5.3. Be sure to select Multiple one-way tables rather than One-way tables.

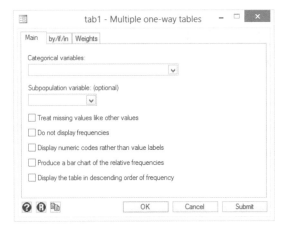

Figure 5.3. Dialog box for frequency tabulation

### Tabulating a series of variables and including missing values

There are three variations of the `tabulate` command. `tabulate`, followed by one variable name, produces a one-way table of frequency counts for the values of that variable. To do a tabulation of a variable, say, `educ`, the command is `tabulate educ`.

The `tab1` command, as mentioned above, takes a list of variable names and issues the `tabulate` command separately for each of them. Suppose that you want to do a tabulation on `educ`, `sex`, and `polviews`, and you want to do this using one command. You would use `tab1 educ sex polviews`. Sometimes you might want to have the tabulation show missing values. To do this, add the `missing` option. To do a tabulation of the three variables in one command and show the missing values, the command is `tab1 educ sex polviews, missing`.

`tabulate`, followed by two variable names, produces a two-way table of frequency counts for the combinations of values of those variables. The `tab2` command takes a list of variable names and issues the `tabulate` command separately for each possible pair of variable names.

Using the dialog box, we enter the variables `sex`, `marital`, and `polviews`. Clicking on OK produces the following result:

```
. tab1 sex marital polviews

-> tabulation of sex
```

| respondents sex | Freq. | Percent | Cum. |
|---|---|---|---|
| male | 1,228 | 44.41 | 44.41 |
| female | 1,537 | 55.59 | 100.00 |
| Total | 2,765 | 100.00 | |

```
-> tabulation of marital
```

| marital status | Freq. | Percent | Cum. |
|---|---|---|---|
| married | 1,269 | 45.90 | 45.90 |
| widowed | 247 | 8.93 | 54.83 |
| divorced | 445 | 16.09 | 70.92 |
| separated | 96 | 3.47 | 74.39 |
| never married | 708 | 25.61 | 100.00 |
| Total | 2,765 | 100.00 | |

```
-> tabulation of polviews
```

| think of self as liberal or conservative | Freq. | Percent | Cum. |
|---|---|---|---|
| extremely liberal | 47 | 3.53 | 3.53 |
| liberal | 143 | 10.74 | 14.27 |
| slightly liberal | 159 | 11.95 | 26.22 |
| moderate | 522 | 39.22 | 65.44 |
| slghtly conservative | 209 | 15.70 | 81.14 |
| conservative | 210 | 15.78 | 96.92 |
| extrmly conservative | 41 | 3.08 | 100.00 |
| Total | 1,331 | 100.00 | |

The first tabulations for sex and marital tell us a lot. Some 55.6% of our sample of 2,765 adults comprises women, and 44.4% comprises men. We have 1,537 women and 1,228 men. For the marital variable, 45.9% (1,269) of adults are married. This is a clear mode because this marital status is so much more frequent than any of the other statuses. By contrast, the mode for sex is not as predominant a category.

A Stata user, Ben Jann, in 2007, with subsequent revisions wrote a command called fre, which provides more details than tab1. To install the fre command, type search fre. A new Viewer window will open with a list of all packages with the keyword "fre". You can scroll through these to find the command fre, or you can press Ctrl+f on Windows or Cmnd+f on Mac, type fre in the resulting box, and then press Enter until you find the command fre. Click on the link, and in the new window, click on the click here to install link, which will install the fre ado-file and help file. Because you know the name of the command, you could install the command more simply by typing ssc install fre in the Command window. This will automatically install the command.

Enter the command fre sex marital polviews. This gives you the result we had before, but this is more useful when there are missing values. In this dataset, there are 1,434 people who were not asked—or at least did not report—their political views. The fre command provides the percentage of the total sample who selected each category (18.88% picked moderate). We usually do not want to use the percentages in this column because we are usually interested in the percentage of the valid responses. The valid responses are those 1,331 survey participants who answered the item. The column headed with "Valid" shows us that 39.22% of those valid responses picked the moderate category. The last column on the right is labeled "Cum.", which is the cumulative percentage for the valid responses. We see that 65.44% of the participants picked moderate or more liberal responses, but only 14.27% picked the liberal or the extremely liberal categories. One of the nicest features of the fre command is the way it provides both the value label and the numeric value in one table. Being able to see the label and value together is quite useful when you are recoding or creating a scale. Comparing the fre results with the tab1 results above, we see that the variable label is easier to read with the fre command.

```
. fre sex marital polviews
```

sex ⎯ respondents sex

|       |          | Freq.  | Percent | Valid  | Cum.   |
|-------|----------|--------|---------|--------|--------|
| Valid | 1 male   | 1228   | 44.41   | 44.41  | 44.41  |
|       | 2 female | 1537   | 55.59   | 55.59  | 100.00 |
|       | Total    | 2765   | 100.00  | 100.00 |        |

marital ⎯ marital status

|       |                 | Freq.  | Percent | Valid  | Cum.   |
|-------|-----------------|--------|---------|--------|--------|
| Valid | 1 married       | 1269   | 45.90   | 45.90  | 45.90  |
|       | 2 widowed       | 247    | 8.93    | 8.93   | 54.83  |
|       | 3 divorced      | 445    | 16.09   | 16.09  | 70.92  |
|       | 4 separated     | 96     | 3.47    | 3.47   | 74.39  |
|       | 5 never married | 708    | 25.61   | 25.61  | 100.00 |
|       | Total           | 2765   | 100.00  | 100.00 |        |

polviews ⎯ think of self as liberal or conservative

|         |                        | Freq.  | Percent | Valid  | Cum.   |
|---------|------------------------|--------|---------|--------|--------|
| Valid   | 1 extremely liberal    | 47     | 1.70    | 3.53   | 3.53   |
|         | 2 liberal              | 143    | 5.17    | 10.74  | 14.27  |
|         | 3 slightly liberal     | 159    | 5.75    | 11.95  | 26.22  |
|         | 4 moderate             | 522    | 18.88   | 39.22  | 65.44  |
|         | 5 slghtly conservative | 209    | 7.56    | 15.70  | 81.14  |
|         | 6 conservative         | 210    | 7.59    | 15.78  | 96.92  |
|         | 7 extrmly conservative | 41     | 1.48    | 3.08   | 100.00 |
|         | Total                  | 1331   | 48.14   | 100.00 |        |
| Missing | .                      | 1434   | 51.86   |        |        |
| Total   |                        | 2765   | 100.00  |        |        |

## Obtaining both numbers and value labels

Before doing the tabulations, you might want to type the command `numlabel _all, add`. After you enter this command, whenever you type the `tabulate` command, Stata reports both the numbers you use for coding the data (1, 2, 3, 4, and 5) and the value labels (married, widowed, divorced, separated, and never married). Later, if you do not want to include both of these, you can drop the numerical values by using the command `numlabel _all, remove`. Practice using these commands as an exercise on your own.

The tables with both numbers and value labels may not look great, so you may want two tables for each variable, with one showing the value labels without the numeric codes and the other showing the numeric codes without the value labels. The default gives you the value labels. On the dialog box, there is an option to *Display numeric codes rather than value labels*. This option produces the numeric values without the value labels. If you want both the numeric values and the value labels using official Stata commands, you need to run either the `numlabel` command or run the `tab1` command twice—once with *Display numeric codes rather than value labels* checked and once with it not checked. It is probably simpler just to run Ben Jann's `fre` command once you have installed it.

In chapter 4, we created a pie chart. Here we will create a pie chart for marital status. Select Graphics ▷ Pie chart and look at the Main tab. If this dialog still has information entered from a previous pie chart, you should click on the Ⓡ (reset) icon in the lower left of the view screen to clear the dialog box. Type `marital` as the *Category variable*. This uses the categories we want to show as pieces of the pie. Leave the *Variable: (optional)* box blank. Under the Titles tab, enter a nice title in the *Title* box and the name of the dataset we used as a *Note*. Under the Options tab, click on *Order by this variable* and type `marital`. Also check *Exclude observations with missing values (casewise deletion)* because we do not want these, if there are any, to appear as a separate piece of the pie. The dialog box for the Options tab is shown in figure 5.4.

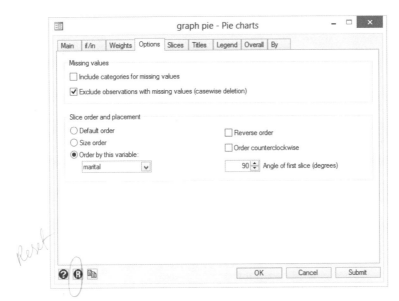

Figure 5.4. The **Options** tab for pie charts (by category)

The initial pie chart on the left in figure 5.5 provides a visual display of the distribution of marital statuses in the United States. The size of each piece of the pie is proportional to the percentage of the people in that status. This pie chart shows that the most common status of adults is married.

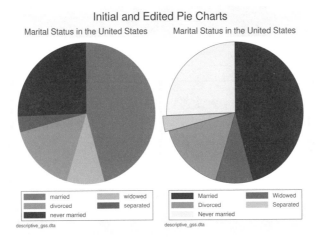

Figure 5.5. Pie charts of marital status in the United States

It is possible to improve the default pie chart. The default pie chart is a bit hard to read because it assumes you want each slice a different color, and this will not work well when printing in black and white. Because many publications require black and white printing, we should edit the pie chart. From our dialog box for the pie chart, we could click on the Overall tab, from which we could select a monochrome scheme from the drop-down *Scheme* list. However, there are several other ways we can improve this pie chart, so let's open the Graph Editor.

We can open the Graph Editor by right-clicking on the pie chart and selecting Start Graph Editor, clicking on the icon above the graph that has a bar chart with a pencil, or in Windows, we can select File ▷ Start Graph Editor. This expands the window that has the graph and adds a panel on the side of the chart with things we might want to change (see figure 5.6). On the left is the pie chart we will edit, and on the right are the names of the parts of the pie chart.

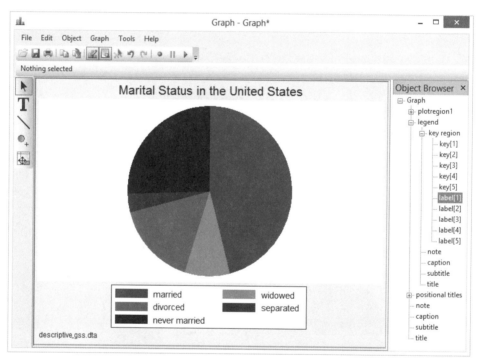

Figure 5.6. The Graph Editor

Notice that the labels in the legend of the initial pie chart are not capitalized. Click on the plus sign by *legend* and the plus sign by *key region*. Double-click on *label[1]* and change the *Text* from `married` to `Married`. Do the same for each of the other labels. We could also do this by double-clicking on the label `married`.

Next click on the plus sign by *plotregion1*, and then double-click on *pieslices[1]*. Here we will pick *Black* as the *Color* and *100%* as the *Fill intensity*. For *pieslices[2]*, pick *Black* and *70%*. For *pieslices[3]*, pick *Black* and *50%*. For *pieslices[4]*, pick *Black* and *30%*, check *Explode slice*, and make sure the *Distance* is *Medium*. Finally, for *pieslices[5]*, pick *Black* and *10%*. You can experiment with other options. The pie chart on the right in figure 5.5 is what we have created. The exploded slice, `Separated`, would be useful if you wanted to emphasize the size of this group.

We have only scratched the surface of what you can do with the powerful Graph Editor. For example, you can click somewhere on the figure, then click on the T (Add Text Tool) to the left of the graph, and a dialog box opens so you can add text. You could then click on the \ (slash or Add Line Tool) just below the T, and draw a line from the text to the piece of the pie it describes. As an exercise, you might add the text "Less than half married" with a line pointing to the piece of the pie for married.

Within Stata's Graph Editor, you can record the changes you make to a graph by using the Graph Recorder. When the Graph Editor is open, there are symbols like those you might see on a recorder or a video player (at the top right of the Editor). Clicking on the red circle starts the Recorder, and clicking on the pair of vertical bars pauses a recording. When you are done making your changes, click on the red circle again before you save the graph or exit the Graph Editor, and the Graph Recorder will prompt you to name the recording. Suppose that we call the recording `myscheme` and save it. The next time we do a similar graph and want to make the same changes, we can click on the arrow to the right of the pair of vertical bars, and it will give us a list of recordings we have saved. We can pick `myscheme` to apply the changes we recorded and saved in `myscheme` to our current graph.

A bar chart is more attractive than a pie chart for many applications. Instead of selecting Bar Chart from the Graphics menu, select Histogram. Here we are creating a bar chart rather than a histogram, but this is the best way to produce a high-quality bar chart using Stata.

On the Main tab, type `marital` in the *Variable* box. Click on the button next to *Data are discrete*. In the section labeled *Y axis*, click on the button next to *Percent*. The trick to making this a bar chart is to click on *Bar properties* in the lower left of the Main tab. This opens another dialog box where the default is to have no gap between the bars. Change this to a gap of 10, which sets the gap between bars to 10 percent of the width of a bar. Click on Accept. If you switch to the Titles tab, you can enter a title, such as `Marital Status in the United States`. Next switch to the X axis tab and click on *Major tick/label properties*. This opens another dialog box where you select the Labels tab and check the box for *Use value labels*. Sometimes the value labels are too wide to fit under each bar. You may need to create new value labels that are shorter. If they are just a little bit too wide, you can change the angle. Click on *Angle* and select *45 degrees* from the drop-down menu. Click on Accept. Finally, switch back to the Main tab. In the lower right corner of the dialog box, click on *Add height labels to bars*. Because we are reporting percentages, this option will show the percentage in each marital status at the top of each bar. The Main tab is shown in figure 5.7.

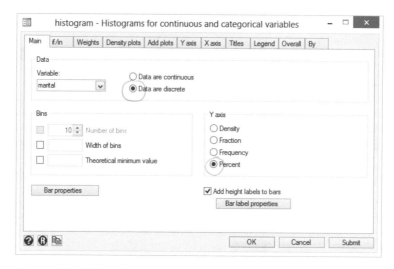

Figure 5.7. Using the `histogram` dialog box to make a bar chart

Figure 5.8 shows the resulting bar chart, which has the percentage in each status at the top of each bar. Married is the most common status, but never married is second. This dataset includes people who are 18 and older, and it is likely that many of those in the never-married status are between 18 and 30.

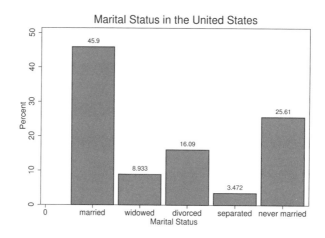

Figure 5.8. Bar chart of marital status of U.S. adults

Sometimes you may have a larger number of categories. When this happens, Stata's default will show only a limited number of value labels, so some of the bars will be unlabeled. If you want to label all of them, you need to go to the X axis tab and click on *Major tick/label properties* to open the dialog box we opened before. On the Rule tab, click on *Suggest # of ticks* and enter the number of bars in the box by *Ticks*. Another

issue can arise if you have value labels that are no longer used. For the `polviews`
variable, we could use seven ticks because there are seven categories for `polviews`.
However, our missing category includes people who had originally been coded as 0 `na`
for not applicable. They were not asked this question for some reason. If you create a
bar chart for `polviews`, you will need to change the value label or make a new set of
value labels that does not have this option.

## 5.5   Statistics and graphs—ordered categories and variables

When our categories are ordered, we can use the median to measure the central tendency. When there are only a couple of categories, however, the median does not
work well. Here is an example where there are several categories for the `polviews`
variable, which asks people their political views on a seven-point scale from extremely
liberal to extremely conservative. We might want to report the median or mean and
the SD. We have done a `summarize` and a `tabulate` before using the dialog system,
so this time we will just enter the commands directly: `tab1 polviews` and `summarize
polviews, detail`. Before you run these commands, make sure that you run the com-
mand `numlabel _all, add` so that both the numbers and the value labels are shown.

```
. numlabel _all, add

. tab1 polviews

-> tabulation of polviews
```

| think of self as liberal or conservative | Freq. | Percent | Cum. |
|---|---|---|---|
| 1. extremely liberal | 47 | 3.53 | 3.53 |
| 2. liberal | 143 | 10.74 | 14.27 |
| 3. slightly liberal | 159 | 11.95 | 26.22 |
| 4. moderate | 522 | 39.22 | 65.44 |
| 5. slghtly conservative | 209 | 15.70 | 81.14 |
| 6. conservative | 210 | 15.78 | 96.92 |
| 7. extrmly conservative | 41 | 3.08 | 100.00 |
| Total | 1,331 | 100.00 | |

```
. summarize polviews, detail
```

think of self as liberal or conservative

| | Percentiles | Smallest | | |
|---|---|---|---|---|
| 1% | 1 | 1 | | |
| 5% | 2 | 1 | | |
| 10% | 2 | 1 | Obs | 1331 |
| 25% | 3 | 1 | Sum of Wgt. | 1331 |
| 50% | 4 | | Mean | 4.124718 |
| | | Largest | Std. Dev. | 1.385016 |
| 75% | 5 | 7 | | |
| 90% | 6 | 7 | Variance | 1.918268 |
| 95% | 6 | 7 | Skewness | -.1509408 |
| 99% | 7 | 7 | Kurtosis | 2.693351 |

The frequency distribution produced by `tab1` is probably the most useful way to describe the distribution of an ordered categorical variable. We can see that it is fairly symmetrically distributed around the mode of moderate, with somewhat more people describing themselves as conservative rather than liberal.

Although the tabulation gives us a good description of the distribution, we often will not have the space in a report to show this level of detail. The median is provided by the `summarize` command, which shows that the 50th percentile occurs at the value of 4, so the Mdn is 4, corresponding to a political moderate. Even though these are ordinal categories, many researchers would report the mean. The mean assumes that the quantitative values, 1–7, are interval-level measures. However, the mean ($M = 4.12$) is usually a good measure of central tendency. The mean reflects the distribution somewhat more accurately here than does the median because the mean shows that the average response is a bit more to the conservative end than to the liberal. We know that the mean is a bit more conservative because the higher numeric score on `polviews` corresponds to more conservative views. You should be able to see this from reading the frequency distribution carefully. Although this variable is clearly ordinal, many researchers treat variables like this as if they were interval and rely on the mean as a measure of central tendency. If you are in doubt, it may be a good idea to report both the median and the mean.

Here is how you can create a histogram showing the distribution of political views. Type the command

```
. histogram polviews, discrete percent
> title(Political Views in the United States) subtitle(Adult Population)
> note(General Social Survey 2002) xtitle(Political Conservatism) scheme(s1mono)
```

Remember, this is the way Stata writes it in the Results window. You do not need to type the . or the >. In a do-file, you would actually need to use the ///, preceded by a space at the end of each but the last line, so Stata will know the three lines are all one command. You would write this command in the Do-file Editor as

```
histogram polviews, discrete percent ///
  title(Political Views in the United States) subtitle(Adult Population) ///
  note(General Social Survey 2002) xtitle(Political Conservatism) scheme(s1mono)
```

Like most graph commands, this is an example of a complicated command that can be easily produced using the dialog box. Commands for making graphs can get complicated, so it is usually best to use the dialog box with graphs. Select Graphics ▷ Histogram.

I will not show the resulting dialogs. On the Main tab, select `polviews` from the *Variable* list, and click on the button by *Data are discrete*. Go to the X axis tab, and enter the title you want to appear on the $x$ axis (the $x$ axis is the horizontal axis of the graph). Go back to the Main tab and click on the radio button by *Percent*. If you want your graph to appear without colors, you can pick a monochrome scheme. On the Overall tab, select `s1 monochrome` from the *Scheme* list. Finally, go to the Titles tab;

enter the title, subtitle, and notes that you want to appear on the graph. Click on OK. The example graph appears in figure 5.9.

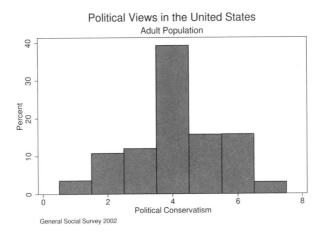

Figure 5.9. Histogram of political views of U.S. adults

This histogram allows the reader to quickly get a good sense of the distribution. In 2002, moderate was the overwhelming choice of adults in the United States. The bars on the right (conservative) are a bit higher than the bars on the left, which indicates a tendency for people to be conservative ($M = 4.12$, Mdn = 4, Mode = 4, SD = 1.39).

Some researchers, when making a graph to show the distribution of an ordinal variable, are reluctant to have the bars for each value touch. Also some like to have the labels posted on the $x$ axis, rather than the coded values. These options and more can be customized in the dialog box.

## 5.6  Statistics and graphs—quantitative variables

We will study three variables: `age`, `educ`, and `wwwhr` (hours spent on the World Wide Web). Two types of useful graphs are the already familiar histogram and a new graph called the *box plot*. We will usually use the mean or median to measure the central tendency for quantitative variables. The SD is the most widely used measure of dispersion, but a statistic called the *interquartile range* is used by the box plots that are presented below.

Let's start with `wwwhr`, hours spent in the last week on the World Wide Web. These data were collected in 2002, and by now, the hours have probably increased a lot.

Computing descriptive statistics for quantitative variables is easy. Let's skip the dialog box and just enter the command:

```
. summarize wwwhr, detail
```

                              www hours per week

|  | Percentiles | Smallest |  |  |  |
|---|---|---|---|---|---|
| 1% | 0 | 0 |  |  |  |
| 5% | 0 | 0 |  |  |  |
| 10% | 0 | 0 | Obs | | 1574 |
| 25% | 1 | 0 | Sum of Wgt. | | 1574 |
| 50% | 3 |  | Mean | | 5.907878 |
|  |  | Largest | Std. Dev. | | 8.866734 |
| 75% | 7 | 60 |  |  |  |
| 90% | 15 | 64 | Variance | | 78.61897 |
| 95% | 21 | 100 | Skewness | | 3.997908 |
| 99% | 40 | 112 | Kurtosis | | 30.39248 |

*(handwritten: median)* *(handwritten: = Average)* *(handwritten: SD)*

This output says that the average person spent a mean of 5.91 hours on the World Wide Web in the week before the survey was taken. The median is 3 hours. Because the mean is greater than the median when a distribution is positively skewed, we can assume that the distribution is positively skewed (trails off on the right side). A positively skewed distribution makes sense because the value for hours on the World Wide Web cannot be less than zero, but we all know a few people who spend many hours on the web. The SD is 8.87 hours, which tells us that the time on the web varies widely. For a normally distributed variable, about two-thirds of the cases are within one SD of the mean ($-2.96$ hours and 14.77 hours) and 95% will be within two SDs of the mean ($-11.83$ hours and 23.64 hours). Clearly, this does not make any sense because you cannot use the World Wide Web fewer than zero hours a week. Still, this information suggests that there is a lot of variation in how much time people spend on the web.

The skewness is 4.00, which means that the distribution has a positive skew (greater than zero), and the kurtosis is 30.39, which is huge compared with 3.0 for a normal distribution. Remember that a kurtosis greater than 10 is problematic; a kurtosis over 20 is very serious. This result suggests that there is a big clump of cases concentrated in one part of the distribution. Can you guess where this concentration was in 2002?

*(handwritten margin note: kurtosis эксцесс островершинность)*

Stata can test for normality based on skewness and kurtosis. For most applications, this test is of limited utility. It is extremely sensitive to small departures from normality when you have a large sample, and it is insensitive to large departures when you have a small sample. The problem is that, when we do inferential statistics, the lack of normality is much more problematic with small samples (where the test lacks power) than it is with large samples (where the test usually finds a significant departure from normality, even for a small departure).

To run the test for normality based on skewness and kurtosis, we can use the dialog box by selecting Statistics ▷ Summaries, tables, and tests ▷ Distributional plots and tests ▷ Skewness and kurtosis normality test. Once the dialog box is open, enter the variable wwwhr and click on OK. Unlike with the complex graph commands, with statistical tests it is often easier to enter the command in the Command window (unless you cannot remember it). The command is simply sktest wwwhr.

```
. sktest wwwhr
                    Skewness/Kurtosis tests for Normality
                                                          ——— joint ———
    Variable │   Obs   Pr(Skewness)   Pr(Kurtosis)  adj chi2(2)   Prob>chi2
    ─────────┼──────────────────────────────────────────────────────────────
       wwwhr │ 1.6e+03    0.0000         0.0000          .           0.0000
```

These results show that, based on skewness, the probability that `wwwhr` is normal is 0.000 and, based on kurtosis, the probability that `wwwhr` is normal is also 0.000. Anytime either probability is less than 0.05, we say that there is a statistically significant lack of normality. Testing for normality based on skewness and kurtosis jointly, Stata reports a probability of 0.000, which reaffirms our concern. It is best to report this as $Pr < 0.001$ rather than as $Pr = 0.000$. This test computes a statistic called chi-squared ($\chi^2$), and it is so big that Stata cannot print it in the available space; instead, Stata inserts a ".". The results also report the number of observations as `1.6e+03`. This format is used with large numbers. The `e+03` means to move the decimal place three places to the right, so the number of observations is 1,600. The actual number of valid responses is 1,574, so you can see that Stata is rounding to the nearest hundred.

We need to be quite thoughtful when using `sktest`. When we have a large sample, this command is quite powerful, meaning that it will show even a small departure from normality to be statistically significant. However, when we have a large sample, the assumption of normality is less crucial than it is with a small sample. With a small sample, the `sktest` command may fail to show a substantial departure from normality to be significant because the test has very little power for a small sample. Unfortunately, the violation of the assumption of normality is most important when the sample size is small. So it is a catch-22: `sktest` may show an unimportant violation to be significant for a large sample but fail to show an important violation to be significant for a small sample. This is why we need to look at the actual size of the skewness and kurtosis as a measure, as well as a histogram, and not depend just on the significance test.

When we are describing several variables in a report, space constraints usually limit us to reporting the mean, median, and standard deviation. You can read these numbers along with the measure of skewness and kurtosis and have a reasonable notion of what each of the distributions looks like. However, it is possible to describe `wwwhr` nicely with a few graphs. First, we will create a histogram by using the dialog box described in section 5.5. Or for a basic histogram (shown in figure 5.10), we could enter

```
. histogram wwwhr, frequency
```

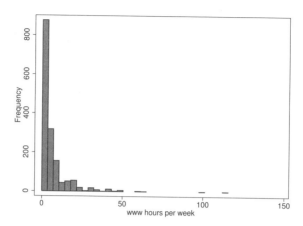

Figure 5.10. Histogram of time spent on the World Wide Web

This simple command does not include all the nice labeling features you can get using the dialog box, but it gives us a quick view of the distribution. This graph includes a few outliers (observations with extreme scores) who surf the World Wide Web more than 25 hours a week. Providing space in the histogram for the handful of people using the World Wide Web between 25 hours and 150 hours takes up most of the graph, and we do not get enough detail for the smaller number of hours that characterizes most of our web users.

We will get around this problem by creating a histogram for a subset of people who use the web fewer than 25 hours a week, and we will do a separate histogram for women and for men. You can get these histograms using the dialog box by inserting the restriction `wwwhr < 25` in the *If: (expression)* box under the if/in tab and by clicking on *Draw subgraphs for unique values of variables* and then inserting the `sex` variable under the By tab. Here is the command we could enter directly:

```
. histogram wwwhr if wwwhr < 25, frequency by(sex)
```

Notice that the `frequency by(sex)` part of the command appears after the comma. The new histograms appear in figure 5.11.

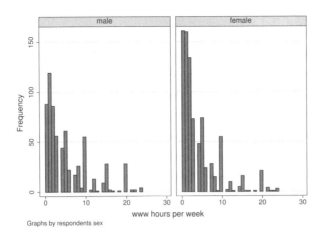

Figure 5.11. Histogram of time spent on the World Wide Web (fewer than 25 hours a week, by gender)

By using the interface, we could improve the format of figure 5.11 by adding titles, and we might want to report the results by using percentages rather than frequencies. We also could experiment with different widths of the bars. Still, figure 5.11 shows that the distribution is far from normal, as the measures of skewness and kurtosis suggested. By creating the histogram separately for women and men, we can see that at the time these data were collected in 2002, far more women were in the lowest interval. Although we do not have more recent data, it would be interesting to compare these data from 2002 with a current histogram. Both of these distributions are surely quite different today, and the gender differences of 2002 may no longer be present.

We could also open the Graph Editor by right-clicking on the graph we just created. It is often easier to make changes using the Graph Editor than to work with the command. The advantage of working with the command is that we have a record of what we did. We might want to replace `male` and `female` with `Men` and `Women`. To do this in the Graph Editor, just click on the headers and change the text in the appropriate boxes. You probably can think of additional changes that would make the graph nicer.

When you want to compare your distribution on a variable with how a normal distribution would be, you can click on an option for Stata to draw how a normal distribution would look right on top of this histogram. We will not show an illustration of this, but all you need to do is open the Density plots tab for the `histogram` dialog box and check the box that says *Add normal-density plot*. This is left for you to do on your own. There is another option on the dialog box for adding a kernel density plot. This is an estimate of the most likely population distribution for a continuous variable that would account for this sample distribution. This will smooth out some of the bars that are extremely high or low because of variation from one sample to the next.

To get the descriptive statistics for men and women separately (but not restricted to those using the web fewer than 25 hours a week), we need a new command:

```
. by sex, sort: summarize wwwhr
```

We can do this from the **summarize** dialog box by checking *Repeat command by groups* and entering **sex** under the by/if/in tab. This command will sort the dataset by **sex** and then summarize separately for women and men.

Another way to obtain a statistical summary of the **wwwhr** variable is to use the **tabstat** command, which gives us a nicer display than what we obtained with the **summarize** command. Select Statistics ▷ Summaries, tables, and tests ▷ Other tables ▷ Compact table of summary statistics to open the dialog box.

Under the Main tab, type **wwwhr** under *Variables*. Check the box next to *Group statistics by variable* and type **sex**. Now pick the statistics we want Stata to summarize. Check the box in front of each row and pick the statistic. The **tabstat** command gives us far more options than did the **summarize** command. The dialog box in figure 5.12 shows that we asked for the mean, median, SD, interquartile range, skewness, kurtosis, and coefficient of variation.

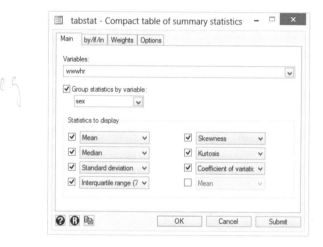

Figure 5.12. The Main tab for the **tabstat** dialog box

Under the Options tab, go to the box for *Use as columns* and select *Statistics*, which will greatly enhance the ease of reading the display. Next we could go to the by/if/in tab and enter **wwwhr < 25** under *If: (expression)*, but we will not do that here. Here is the resulting command:

```
. tabstat wwwhr, statistics(mean median sd iqr skewness kurtosis cv) by(sex)
> columns(statistics)
```

Summary for variables: wwwhr
      by categories of: sex (respondents sex)

| sex | mean | p50 | sd | iqr | skewness | kurtosis | cv |
|---|---|---|---|---|---|---|---|
| male | 7.106892 | 4 | 9.98914 | 9 | 3.608189 | 25.2577 | 1.405557 |
| female | 4.920046 | 2 | 7.688655 | 4 | 4.409274 | 36.78389 | 1.56272 |
| Total | 5.907878 | 3 | 8.866734 | 6 | 3.997908 | 30.39248 | 1.500832 |

The table produced by the `tabstat` command summarizes the statistics we requested that it include, showing the statistics for males and females, and the total for males and females combined. Stata calls the median `p50` because the median represents the value corresponding to the 50th percentile. If you copied this table to a Word file, you might want to change the label to "median" to benefit readers who do not really know what the median is. If you highlight the `tabstat` output in the Results window and copy it as a picture to a Word document, you will not be able to make this change in Word. However, if you choose one of the other copy options, you will be able to make the change.

In addition to skewness and kurtosis, we selected two additional statistics I have not yet introduced. The coefficient of relative variation (CV) is simply the SD divided by the mean (that is, $CV = SD/M$). This statistic is sometimes used to compare SDs for variables that are measured on different scales, such as income measured in dollars and education measured in years. The interquartile range is the difference between the value of the 75th percentile and the value of the 25th percentile. This range covers the middle 50% of the observations.

Men, on average, spent far more time using the World Wide Web in 2002 than did women. Because the means are bigger than the medians, we can assume that the distributions are positively skewed (as was evident in the histograms we did). Men are a bit more variable than women because their SD is somewhat greater. Both distributions are skewed and have heavy kurtosis. The CV is 1.41 for men and 1.56 for women. Women have slightly greater variance relative to their mean than men do (based on comparing the CV values), even though the actual SD is bigger for men. Finally, the interquartile range of 9 for men is more than double the interquartile range of 4 for women. Thus the middle 50% of men are more dispersed than the middle 50% of women. Comparing the CVs suggests the opposite finding to comparing interquartile ranges. Because the scale (hours of using the World Wide Web) is the same, we would not rely on the CV.

A horizontal or vertical box plot is an alternative way of showing the distribution of a quantitative variable such as `wwwhr`. Select Graphics ▷ Box plot. Here we will use four of the tabs: Main, Categories, if/in, and Titles. Under the Main tab, check the radio button by *Horizontal* to make the box plot horizontal, and enter the name of our variable, `wwwhr`. Under the Categories tab, check *Group 1* and enter the *grouping variable*, that is, `sex`. This will create separate box plots for women and for men. We could have additional grouping variables, but these plots can get complicated. If we wanted a box plot that included both women and men, we would leave this tab blank. The Categories tab is similar to the by/if/in tab that we used for the **tabstat** command. Under the if/in tab, we need to make a command so that the plots are shown only for those who spend fewer than 25 hours a week on the web. In the *If: (expression)* box, we type **wwwhr < 25**. Finally, under the Titles tab, enter the title and any subtitles or notes we want to appear on the chart. The command generated from the dialog box is

```
. graph hbox wwwhr if wwwhr < 25, over(sex)
> title(Hours Spent on the World Wide Web) subtitle(By Gender)
> note(descriptive_gss.dta)
```

and the resulting graph appears in figure 5.13.

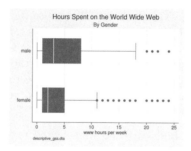

Figure 5.13. Box plot of time spent on the World Wide Web (fewer than 25 hours a week, by gender)

Histograms may be easier to explain to a lay audience than box plots. For a nontechnical group, histograms are usually a better choice. Many statisticians like box plots better because they show more information about the distribution. The white vertical line in the dark-gray boxes is the median. For women, you can see that the median is about 2 hours per week, and for men, about 4 hours per week.

The left and right sides of the dark-gray box are the 25th and 75th percentiles, respectively. Within this dark-gray box area are half of the people. This box is much wider for men than it is for women, showing how men are more variable than women. Lines extend from the edge of the dark-gray box 1.5 box lengths, or until they reach the largest or smallest cases. Beyond this, there are some dots representing outliers, or extreme values.

## 5.7  Summary

After four chapters about how to set up files and manage them, I hope you enjoyed getting to a substantive chapter showing you some of the output produced with Stata. We are just beginning to see the power of Stata, but you can already summarize variables and create several types of attractive graphs. This chapter covered the following topics:

- How to compute measures of central tendency (averages), including the mean, median, and mode

- When to use the different measures of central tendency based on level of measurement, distribution characteristics, and your purposes

- How to describe the dispersion of a distribution by using

  – Statistics (standard deviations)
  – Tables (frequency distributions)
  – Graphs (pie charts, bar charts, histograms, and box plots)

- How to use Stata to give you these results for nominal-, ordinal-, and interval-level variables

- How to use Stata's Graph Editor

The graphs we have introduced in this chapter show just a few of the graph capabilities offered by Stata. We will cover a few more types of graphs in later chapters, but if you are interested in producing high-quality graphs, see *A Visual Guide to Stata Graphics, Third Edition* by Michael Mitchell (2012), which is available from Stata Press.

This is just the start of the useful output you can produce using Stata. Statistics books tend to get harder and harder as you move toward more complicated procedures. I cannot help that, but Stata is just the opposite. Managing data and doing graphs are the two hardest tasks for statistical programs because Stata is designed primarily to do statistical analysis. In the next chapter, we will examine how to use graphs and statistics when we are examining the relationship between two or more variables.

## 5.8   Exercises

1. Open `descriptive_gss.dta` and do a detailed summary of the variable `hrs1` (hours worked last week). Also create a histogram of the variable. Interpret the mean and median. Looking at the histogram, explain why the skewness value is close to zero. What does the value of kurtosis tell us? Looking at the histogram, explain why the kurtosis is a positive value.

2. Open `descriptive_gss.dta` and do a detailed summary of the variable `satjob7` (job satisfaction). Type the command `numlabel satjob7, add`, and then do a tabulation of `satjob7`. Interpret the mean and median values. Why would some researchers report the median? Why would other researchers report the mean?

3. Open `descriptive_gss.dta` and do a tabulation of `deckids` (who makes decisions about how to bring up children). Do this using the by/if/in tab to select by `sex`. Create and interpret a bar chart by using the `histogram` dialog box. Why would it make no sense to report the mean, median, or standard deviation for `deckids`? Use Stata's Graph Editor to make the bar chart look as nice as you can.

4. Open `descriptive_gss.dta` and do a tabulation of `strsswrk` (job is rarely stress-ful) and a detailed summary. Do this using the by/if/in tab to select by `sex`. Create and interpret a histogram, using the By tab to do this for males and females. In the Main tab, be sure to select the option *Data are discrete*. Carefully label your histogram to show value labels and the percentage in each response category. Each histogram should be similar to figure 5.9. Interpret the median and mean for men and women.

5. Open `descriptive_gss.dta` and do a tabulation of `trustpeo`, `wantbest`, `advantge`, and `goodlife`. Use the `tabstat` command to produce a table that summa-rizes descriptive statistics for this set of variables by gender. Include the median, mean, standard deviation, and count for each variable. Interpret the means by using the variable labels you get with the tabulation command.

6. Open `descriptive_gss.dta` and do a tabulation of `polviews`. Create a bar chart for this variable showing the percentage of people who are in each of the seven categories. Next create a chart that has labels at a 45-degree angle for each of the bars. Finally, change the chart by using the X axis tab's *Minor tick/label properties* and *Major tick/label properties* (using the *Custom* option) so that the histogram does not have a null category and all the categories are labeled. Compare this final figure with figure 5.9.

7. Open `descriptive_gss.dta`. In figure 5.13, we created a box plot for the hours women and men spend surfing the World Wide Web. First, create a similar box plot for `hrs1` (hours worked last week). Now add a second grouping variable, `marital`. Using this graph, give a detailed interpretation of how marital status and gender are related to hours a person works for pay.

8. Open `descriptive_gss.dta`. Execute the following commands for `educ` and com-pare the two sets of results. What problem are you illustrating here?

```
. summarize educ, detail
. sktest educ
. histogram educ
. preserve
. sample 10, count
. summarize educ, detail
. sktest educ
. histogram educ
. restore
```

9. Open `descriptive_gss.dta`. Construct a graph showing two histograms, one for women and one for men for the `educ` variable. Use the Graph Editor to improve the appearance of the graph by adding an overall title and making other changes you can think of. Is this graph helpful for comparing the educational achievement of women and men? How so?

10. Repeat all parts of exercise 9, but construct a box plot instead of histograms.

11. Repeat exercise 10 using the `tabstat` command, and ask for the mean, median, standard deviation, skewness, kurtosis, and interquartile range. Interpret each of these statistics to compare the women and men, and explain how the differences are reflected (or not) in the graphs done in exercises 9 and 10.

# 6 Statistics and graphs for two categorical variables

## 6.1   Relationship between categorical variables

Chapter 5 focused on describing single variables. Even there, it was impossible to resist some comparisons, and we ended by examining the relationship between gender and hours per week spent using the web. Some research simply requires a description of the variables, one at a time. You do a survey for your agency and make up a table with the means and standard deviations for all the quantitative variables. You might include frequency distributions and bar charts for each key categorical variable. This information is sometimes the extent of statistical research your reader will want. However, the more you work on your survey, the more you will start wondering about possible relationships.

- Do women who are drug dependent use different drugs from those used by drug-dependent men?

- Are women more likely to be liberal than men?

- Is there a relationship between religiosity and support for increased spending on public health?

You know you are "getting it" as a researcher when it is hard for you to look at a set of questions without wondering about possible relationships. Understanding these relationships is often crucial to making policy decisions. If 70% of the nonmanagement employees at a retail chain are women, but only 20% of the management employees are women, there is a relationship between gender and management status that disadvantages women.

In this chapter, you will learn how to describe relationships between categorical variables. How do you define these relationships? What are some pitfalls that lead to misinterpretations? In this chapter, the statistical sophistication you will need increases, but there is one guiding principle to remember: the best statistics are the simplest statistics you can use—as long as they are not too simple to reflect the inherent complexity of what you are describing.

## 6.2   Cross-tabulation

Cross-tabulation is a technical term for a table that has rows representing one categorical variable and columns representing another. These tables are sometimes called contingency tables because the category a person is in on one of the variables is contingent on the category the person is in on the other variable. For example, the category people are in on whether they support a particular public health care reform may be contingent on their gender. If you have one variable that depends on the other, you usually put the dependent variable as the column variable and the independent variable as the row variable. This layout is certainly not necessary, and several statistics books do just the opposite. That is, they put the dependent variable as the row variable and the independent variable as the column variable.

Let's start with a basic cross-tabulation of whether a person says abortion is okay for any reason and their gender. Say you decide that whether a person accepts abortion for any reason is more likely if the person is a woman because a woman has more at stake when she is pregnant than does her partner. Therefore, whether a person accepts abortion will be the dependent variable, and gender will be the independent variable.

We will use gss2006_chapter6.dta, which contains selected variables from the 2006 General Social Survey, and we will use the cross-tabulation command, tabulate, with two categorical variables, sex and abany. To open the dialog box for tabulate, select Statistics ▷ Summaries, tables, and tests ▷ Frequency tables ▷ Two-way table with measures of association. This dialog box is shown in figure 6.1.

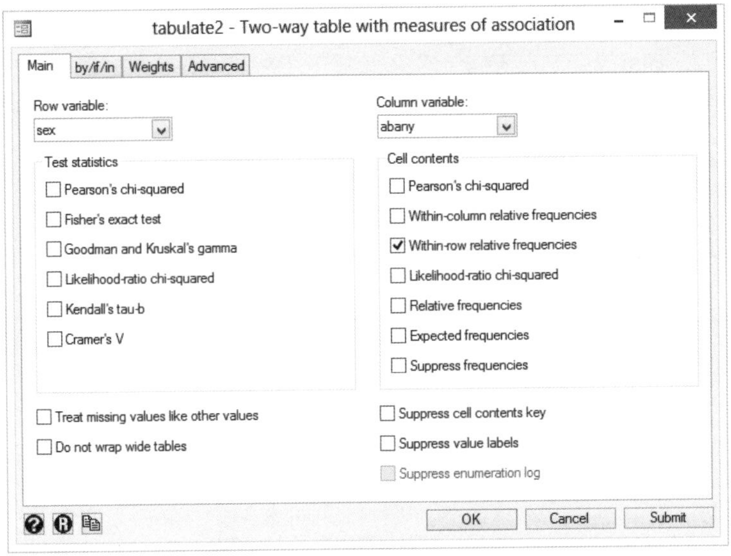

Figure 6.1. The Main tab for creating a cross-tabulation

If we had three variables and wanted to see all possible two-by-two tables (vara with varb, vara with varc, and varb with varc), we could have selected instead Statistics ▷ Summaries, tables, and tests ▷ Frequency tables ▷ All possible two-way tables. With our current data, we continue with the dialog box in figure 6.1.

Select sex, the independent variable, as the *Row variable* and abany, the dependent variable, as the *Column variable*. We are assuming that abany is the variable that depends on sex. Also check the box on the right side under *Cell contents* for the *Within-row relative frequencies* option. This option tells Stata to compute the percentages so that each row adds up to 100%. Here are the resulting command and results:

```
. tabulate sex abany, row
```

| Key |
|-----|
| frequency |
| row percentage |

| | ABORTION IF WOMAN WANTS FOR ANY REASON | | |
|---|---|---|---|
| Gender | YES | NO | Total |
| MALE | 350 | 478 | 828 |
| | 42.27 | 57.73 | 100.00 |
| FEMALE | 434 | 677 | 1,111 |
| | 39.06 | 60.94 | 100.00 |
| Total | 784 | 1,155 | 1,939 |
| | 40.43 | 59.57 | 100.00 |

## Independent and dependent variables

Many beginning researchers get these terms confused. The easiest way to remember which is which is that the dependent variable "depends" on the independent variable. In this example, whether a person accepts abortion for any reason depends on whether the person is a man or a woman. By contrast, it would make no sense to say that whether a person is a man or a woman depends on whether they accept abortion for any reason.

Many researchers call the dependent variable an "outcome" and the independent variable the "predictor". In this example, sex is the predictor because it predicts the outcome, abany.

It may be hard or impossible to always know that one variable is independent and the other variable is dependent. This is because the variables may influence each other. Imagine that one variable is your belief that eating meat is a health risk (there could be four categories: strongly agree, agree, disagree, or strongly disagree). Then imagine that the other variable is whether you eat meat or not. It might seem easy to say that your belief is the independent variable and your behavior is the dependent variable. That is, whether you eat meat or not depends on whether you believe doing so is a health risk. But try to think of the influence going the other way. A person may stop eating meat because of a commitment to animal rights issues. Not eating meat over several years leads to look for other justifications, and they develop a belief that eating meat is a health risk. For them, the belief depends on their prior behavior.

There is no easy solution when we have trouble deciding which is the independent variable. Sometimes, we can depend on time ordering. Whichever came first is the independent variable. Other times, we are simply forced to say that the variables are associated without identifying which one is the independent variable and which is the dependent variable.

The independent variable sex forms the rows with labels of male and female. The dependent variable abany, accepting abortion under any circumstance, appears as the columns labeled yes and no. The column on the far right gives us the total for each row. Notice that there are 828 males, 350 of whom find abortion for any reason to be acceptable, compared with 1,111 females, 434 of whom say abortion is acceptable for any reason. These frequencies are the top number in each cell of the table.

The frequencies at the top of each cell are hard to interpret because each row and each column has a different number of observations. One way to interpret a table is to use the percentage, which takes into account the number of observations within each category of the independent variable (predictor). The percentages appear just below the frequencies in each cell. Notice that the percentages add up to 100% for each row.  Overall, 40.43% of the people said "yes", abortion is acceptable for any reason, and 59.57% said "no".  However, men were relatively more likely (42.27%) than women (39.06%) to report that abortion is okay, regardless of the reason.  We get these percentages because we told Stata to give us the *Within-row relative frequencies*, which in the command is the `row` option.

Thus men are more likely to report accepting abortion under any circumstance.  We compute percentages on the rows of the independent variable and make comparisons up and down the columns of the dependent variable.  Thus we say that 42.27% of the men compared with 39.06% of the women accept abortion under any circumstance.  This is a small difference, but interestingly, it is in the opposite direction from what we expected.

## 6.3   Chi-squared test

The difference between women and men seems small, but could we have obtained this difference by chance?  Or is the difference statistically significant?  Remember, when you have a large sample like this one, a difference may be statistically significant even if it is small.

If we had just a handful of women and men in our sample, there would be a good chance of observing this much difference just by chance.  With such a large sample, even a small difference like this might be statistically significant.  We use a chi-squared $(\chi^2)$ statistic to test the likelihood that our results occurred by chance.  If it is extremely unlikely to get this much difference between men and women in a sample of this size by chance, you can be confident that there was a real difference between women and men, but you still need to look at the percentages to decide whether the statistically significant difference is substantial enough to be important.

The chi-squared test compares the frequency in each cell with what you would expect the frequency to be if there were no relationship.  The expected frequency for a cell depends on how many people are in the row and how many are in the column.  For example, if we asked a small high school group if they accept abortion for any reason, we might have only 10 males and 10 females.  Then we would expect far fewer people in each cell than in this example, where we have 828 men and 1,111 women.

In the cross-tabulation, there were many options on the dialog box (see figure 6.1). To obtain the chi-squared statistic, check the box on the left side for *Pearson's chi-squared*.  Also check the box for *Expected frequencies* that appears in the right column on the dialog box.  The resulting table has three numbers in each cell.  The first number in each cell is the frequency, the second number is the expected frequency if there were no relationship, and the third number is the percentage of the row total.  We would not usually ask for the expected frequency, but now you know it is one of Stata's capabilities.  The resulting command now has three options: `chi2 expected row`. Here is the command and the output it produces:

```
. tabulate sex abany, chi2 expected row
```

| Key |
|---|
| frequency |
| expected frequency |
| row percentage |

| | ABORTION IF WOMAN WANTS FOR ANY REASON | | |
|---|---|---|---|
| Gender | YES | NO | Total |
| MALE | 350 | 478 | 828 |
| | 334.8 | 493.2 | 828.0 |
| | 42.27 | 57.73 | 100.00 |
| FEMALE | 434 | 677 | 1,111 |
| | 449.2 | 661.8 | 1,111.0 |
| | 39.06 | 60.94 | 100.00 |
| Total | 784 | 1,155 | 1,939 |
| | 784.0 | 1,155.0 | 1,939.0 |
| | 40.43 | 59.57 | 100.00 |

Pearson chi2(1) =    2.0254   Pr = 0.155

_Probability_

In the top left cell of the table, we can see that we have 350 men who accept abortion for any reason, but we would expect to have only 334.8 men here by chance. By contrast, we have 434 women who accept abortion for any reason, but we would expect to have 449.2. Thus we have $350 - 334.8 = 15.2$ more men accepting abortion than we would expect by chance and $434 - 449.2 = -15.2$ fewer women than we would expect. Stata uses a function of this information to compute chi-squared.

At the bottom of the table, Stata reports `Pearson chi2(1) = 2.0254` and `Pr = 0.155`, which would be written as $\chi^2(1, N = 1939) = 2.0254$; $p$ not significant. Here we have one degree of freedom. The sample size of $N = 1939$ appears in the lower right part of the table. We usually round the chi-squared value to two decimal places, so 2.0254 becomes 2.03. Stata reports an estimate of the probability to three decimal places. We can report this, or we can use a convention found in most statistics books of reporting the probability as less than 0.05, less than 0.01, or less than 0.001. Because $p = 0.155$ is greater than 0.05, we say $p$ not significant. What would happen if the probability were $p = 0.0004$? Stata would round this to $p = 0.000$. We would not report $p = 0.000$ but instead would report $p < 0.001$.

_probability_

To summarize what we have done in this section, we can say that men are more likely to report accepting abortion for any reason than are women. In the sample of 1,939 people, 42.3% of the men say that they accept abortion for any reason compared with just 39.1% of the women. This relationship between gender and acceptance of abortion is not statistically significant.

## 6.3.1   Degrees of freedom

Because I assume that you have a statistics book explaining the necessary formulas, I have not gone into detail. Stata will compute the chi-squared, the number of degrees of freedom, and the probability of getting your observed result by chance.

You can determine the number of degrees of freedom yourself. The degrees of freedom refers to how many pieces of independent information you have. In a $2 \times 2$ table, like the one we have been analyzing, the value of any given cell can be any number between 0 and the smaller of the number of observations in the row and the number of observations in the column. For example, the upper left cell (350) could be anything between 0 and 784. Let's use the observed value of 350 for the upper left cell. Now how many other cells are free to vary? By subtraction, you can determine that 434 people must be in the female/yes cell because $784 - 350 = 434$. Similarly, 478 observations must be in the male/no cell ($828 - 350 = 478$), and 677 observations must be in the female/no cell ($1111 - 434 = 677$). Thus, with four cells, only one of these is free, and we can say that the table has 1 degree of freedom. We can generalize this to larger tables where degrees of freedom $= (R - 1)(C - 1)$, where $R$ is the number of rows and $C$ is the number of columns. If we had a $3 \times 3$ table instead of a $2 \times 2$ table, we would have $(3-1)(3-1) = 4$ degrees of freedom.

## 6.3.2   Probability tables

Many experienced Stata users have made their own commands that might be helpful to you. Philip Ender made a series of commands that display probability tables for various tests. The `search` command finds user-contributed ado-files and lets you install them on your machine. Typing the command `search chitable`[1] produces the results shown in figure 6.2.

---

1. You cannot install this user-written command by typing `ssc install chitable`, because the `ssc` command is limited to installing and uninstalling user-written commands that are located within the Statistical Software Components (SSC) archive or often called the Boston College Archive at http://www.repec.org. Although the SSC archive is by far the dominant depository of user-written Stata commands, it does not include all user-written commands.

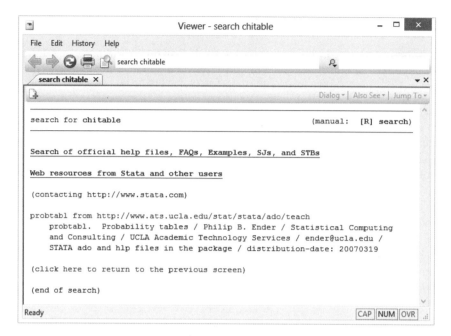

Figure 6.2. Results of `search chitable`

From here, click on the blue web link

probtabl from http://www.ats.ucla.edu/stat/stata/ado/teach

This link takes you to another screen where you can click on the blue link labeled `click here to install`. Once you have done this, anytime you want to see a chi-squared table, you merely type the command `chitable`. This installation also gives you other probability tables that we will use elsewhere in this book, including $t$-test tables (`ttable`) and $F$-test tables (`ftable`). Simply entering `chitable` is a lot more convenient than having to look up a probability in a textbook. Try it now.

```
. chitable
       Critical Values of Chi-square
  df    .50    .25    .10    .05    .025    .01    .001
  1    0.45   1.32   2.71   3.84   5.02   6.63   10.83
  2    1.39   2.77   4.61   5.99   7.38   9.21   13.82
  3    2.37   4.11   6.25   7.81   9.35  11.34   16.27
  4    3.36   5.39   7.78   9.49  11.14  13.28   18.47
  5    4.35   6.63   9.24  11.07  12.83  15.09   20.52
  6    5.35   7.84  10.64  12.59  14.45  16.81   22.46
  7    6.35   9.04  12.02  14.07  16.01  18.48   24.32
  8    7.34  10.22  13.36  15.51  17.53  20.09   26.12
  9    8.34  11.39  14.68  16.92  19.02  21.67   27.88
 10    9.34  12.55  15.99  18.31  20.48  23.21   29.59
 11   10.34  13.70  17.28  19.68  21.92  24.72   31.26
 12   11.34  14.85  18.55  21.03  23.34  26.22   32.91
 13   12.34  15.98  19.81  22.36  24.74  27.69   34.53
 14   13.34  17.12  21.06  23.68  26.12  29.14   36.12
 15   14.34  18.25  22.31  25.00  27.49  30.58   37.70
 16   15.34  19.37  23.54  26.30  28.85  32.00   39.25
 17   16.34  20.49  24.77  27.59  30.19  33.41   40.79
 18   17.34  21.60  25.99  28.87  31.53  34.81   42.31
 19   18.34  22.72  27.20  30.14  32.85  36.19   43.82
 20   19.34  23.83  28.41  31.41  34.17  37.57   45.31
 21   20.34  24.93  29.62  32.67  35.48  38.93   46.80
 22   21.34  26.04  30.81  33.92  36.78  40.29   48.27
 23   22.34  27.14  32.01  35.17  38.08  41.64   49.73
 24   23.34  28.24  33.20  36.42  39.36  42.98   51.18
 25   24.34  29.34  34.38  37.65  40.65  44.31   52.62
 26   25.34  30.43  35.56  38.89  41.92  45.64   54.05
 27   26.34  31.53  36.74  40.11  43.19  46.96   55.48
 28   27.34  32.62  37.92  41.34  44.46  48.28   56.89
 29   28.34  33.71  39.09  42.56  45.72  49.59   58.30
 30   29.34  34.80  40.26  43.77  46.98  50.89   59.70
 35   34.34  40.22  46.06  49.80  53.20  57.34   66.62
 40   39.34  45.62  51.81  55.76  59.34  63.69   73.40
 45   44.34  50.98  57.51  61.66  65.41  69.96   80.08
 50   49.33  56.33  63.17  67.50  71.42  76.15   86.66
 55   54.33  61.66  68.80  73.31  77.38  82.29   93.17
 60   59.33  66.98  74.40  79.08  83.30  88.38   99.61
 65   64.33  72.28  79.97  84.82  89.18  94.42  105.99
 70   69.33  77.58  85.53  90.53  95.02 100.43  112.32
 75   74.33  82.86  91.06  96.22 100.84 106.39  118.60
      (output omitted)
100   99.33 109.14 118.50 124.34 129.56 135.81  149.45
```

In a `chitable`, the first row shows the significance levels. The first column shows the degrees of freedom. You can see that with 1 degree of freedom, you need a chi-squared value of 3.84 to be significant at the 0.05 level. If you had 4 degrees of freedom, you would need a chi-squared of 9.49 to be significant at the 0.05 level. You might try the other commands for the other tables. If you have ever had to search for one of these tables in a textbook, you will appreciate the convenience of these commands.

**Reporting chi-squared results**

How do we report the significance level of chi-squared? How do we report that chi-squared varies somewhat from one field to another? A safe way to report the significance is as $p < 0.05$, $p < 0.01$, or $p < 0.001$. Suppose that we had a chi-squared of 10.05 with 1 degree of freedom and $p = 0.002$. This is less than 0.01 but not less than 0.001. We would say that $\chi^2(1) = 10.05$, $p < 0.01$. Notice that we put the degrees of freedom in parentheses. Some disciplines would also like you to list the sample size, for example, $\chi^2(1, N = 435) = 10.05$, $p < 0.01$. Still other disciplines would rather you report the probability value, for example, $p = 0.002$.

Reporting the probability value has the advantage of providing more-detailed information. For example, one result might have a $p = 0.052$ and another might have a $p = 0.451$. The first of these is almost statistically significant and is an unlikely result if the null hypothesis of no relationship is correct. The second of these is just about what you would expect to get by chance flipping a coin. Saying that both of these have the same classification of $p$ not significant conceals the clear difference between these two results.

## 6.4   Percentages and measures of association

We have already discussed the use of percentages. These are often the easiest and best way to describe a relationship between two variables. In our last example, the percentage of men who said abortion was okay for any reason was slightly greater than, although not statistically significantly greater than, the percentage of women who said abortion was okay for any reason. Percentages often tell us what we want to know. There are other ways of describing an association called measures of association; they try to summarize the strength of a relationship with a number.

   The value of chi-squared depends on two things. First, the stronger the association between the variables, the bigger chi-squared will be. Second, because we have more confidence in our results when we have larger samples, then the more cases we have, the bigger chi-squared will be. In fact, for a given relationship expressed in percentages, chi-squared is a function of sample size. If you had the same relationship as in our example but, instead of having 1,939 observations, you had 19,390 (10 times as many), then chi-squared would be 20.254, also 10 times as big. There would still be 1 degree of freedom, but here the results would be statistically significant, $p < 0.001$. With large samples, researchers sometimes misinterpret a statistically significant chi-squared value as indicating a strong relationship. With a large sample, even a weak relationship can be statistically significant. This makes sense because with a large sample we have the power to detect even small effects.

One way to minimize this potential misinterpretation is to divide chi-squared by the maximum value it could be for a table of a particular shape and number of observations. This is simple in the case of $2 \times 2$ tables, such as the one we are using. The maximum value of chi-squared for a $2 \times 2$ table is the sample size, $N$. Thus in our example, if the relationship were as strong as possible, chi-squared would be 1,939. Our chi-squared of 2.03 is tiny in comparison. The coefficient $\phi$ (phi) is defined as the positive or negative square root of the quantity chi-squared divided by $N$:

$$\phi = \sqrt{\frac{\chi^2}{N}}$$

Stata uses a different formula for calculating the value of $\phi$. Stata's formula produces a positive or negative value directly, depending on the arrangement of rows and columns; see the box on page 132. A $\phi$ with an absolute value from 0.0 to 0.19 is considered weak, from 0.20 to 0.49 is considered moderate, and from 0.50 and above is considered strong. In our example, $\phi = 0.03$, which we would thus describe as a weak relationship, meaning that the strength of the relationship is weak regardless of whether it is statistically significant. This is an important distinction between the strength of the relationship (some call this substantive or practical significance) and statistical significance. So long as the same distribution based on percentages describes a table, $\phi$ will have the same value whether we have 194 observations, 1,939 observations, or 19,390 observations.

Because both Cramér's $V$ and $\phi$ are the square root of chi-squared divided by its maximum possible value and because $\phi$ can be thought of as a special case of $V$, Stata simply has an option to compute Cramér's $V$. However, if you have a $2 \times 2$ table, you should call this measure of association $\phi$ to avoid confusion and recognize that it may be either a positive or negative value. On the dialog box for doing the cross-tabulation with measures of association, simply check *Cramer's V* under the list of *Test statistics*. This results in the command `tabulate sex abany, chi2 row V`. If you type this command directly into the Command window, remember to capitalize the V; this is a rare example where you must use an uppercase letter in Stata.

The ability to have a positive or negative $\phi$ does not extend to larger tables where Cramér's $V$ is the appropriate measure of association. The maximum value chi-squared can obtain for a larger table is $N$ times the smaller of $R - 1$ or $C - 1$, where $R$ is the number of rows and $C$ is the number of columns in the table. For these tables, we report the positive square root of the ratio of chi-squared to its maximum positive value and call it $V$:

$$V = \sqrt{\frac{\chi^2}{N \times \min(R - 1, C - 1)}}$$

**Why can $\phi$ be negative?**

Stata uses a special formula for calculating $\phi$, and this formula gives us the positive or negative sign for $\phi$ directly. The formulas that Stata uses is

$$\phi = \frac{n_{11}n_{22} - n_{12}n_{21}}{\sqrt{n_{1.}n_{2.}n_{.1}n_{.2}}}$$

Let's look at how this formula applies to a sample table. In the following table, the first subscript is the row and the second subscript is the column. Hence, $n_{11}$ is the number of cases in row 1, column 1, and $n_{21}$ is the number of cases in row 2, column 1. A dot is used to refer to all cases in both rows or both columns. Thus $n_{1.}$ is the number of people who are in row 1 summed over both columns; it is the row total. Similarly, $n_{.1}$ is the column total for the first column. Here is how this looks in a $2 \times 2$ table:

|       | column |        |        |
|-------|--------|--------|--------|
| row   | 1      | 2      | Total  |
| 1     | $n_{11}$ | $n_{12}$ | $n_{1.}$ |
| 2     | $n_{21}$ | $n_{22}$ | $n_{2.}$ |
| Total | $n_{.1}$ | $n_{.2}$ | $n_{..}$ |

Applying the above formula to a simple table, we obtain $\phi = -0.3333$ (reported in the table as Cramér's $V$):

|       | column |     |       |
|-------|--------|-----|-------|
| row   | 1      | 2   | Total |
| 1     | 10     | 20  | 30    |
| 2     | 20     | 10  | 30    |
| Total | 30     | 30  | 60    |

```
Pearson chi2(1) =   6.6667      Pr = 0.010
   Cramér's V =  -0.3333
```

However, if we rearrange the rows and columns, we obtain $\phi = 0.3333$. Here are the rearranged table and results:

|       | column |     |       |
|-------|--------|-----|-------|
| row   | 1      | 2   | Total |
| 1     | 20     | 10  | 30    |
| 2     | 10     | 20  | 30    |
| Total | 30     | 30  | 60    |

```
Pearson chi2(1) =   6.6667      Pr = 0.010
   Cramér's V =   0.3333
```

For most $2 \times 2$ tables, we do not care about how the rows and columns are arranged and so use only the positive value of $V$ and $\phi$.

## 6.5    Odds ratios when dependent variable has two categories

Odds ratios (ORs) are useful when the dependent variable has just two categories. We define ORs by using the independent variable. What are the odds that a man will say that abortion is okay for any reason? What are the odds a woman will say that abortion is okay for any reason? Let's look at our cross-tabulation, where I have added letter labels for each cell:

```
. tabulate sex abany

               │  ABORTION IF WOMAN
               │ WANTS FOR ANY REASON
     Gender    │     YES           NO  │    Total
───────────────┼───────────────────────┼─────────
        MALE   │   350 (a)      478 (b) │      828
      FEMALE   │   434 (c)      677 (d) │    1,111
───────────────┼───────────────────────┼─────────
       Total   │     784          1,155 │    1,939
```

We know that 350 men say abortion is okay for any reason and 478 do not, so the odds are $a/b = 350/478 = 0.73$. What about women? The odds for a woman supporting abortion for any reason are $c/d = 434/677 = 0.64$. The real question is the difference between men and women. We could say simply that the odds of a man supporting abortion are higher, 0.73, than a woman supporting abortion, 0.64. Alternatively, we can calculate an OR. This is the odds of a man supporting abortion divided by the odds of a woman supporting abortion. It is call the OR because it is the ratio of two odds.

$$\text{OR} = \frac{a/b}{c/d} = \frac{350/478}{434/677} = 1.14$$

The OR for men compared with women is (approximately) $0.73/0.64 = 1.14$. Thus the odds of a man supporting abortion for any reason are greater than the odds of a woman supporting abortion for any reason. How much greater? When an OR is greater than 1, we calculate $100 \times (\text{OR} - 1.00) = (1.14 - 1.00) = 14\%$. The odds of a man supporting abortion are 14% greater than the odds of a woman supporting abortion.

We could reverse this and calculate the OR with the odds for women in the numerator:

$$\text{OR} = \frac{c/d}{a/b} = \frac{434/677}{350/478} = 0.88$$

The OR for women compared with men is (approximately) $0.64/0.73 = 0.88$. When the OR is less than 1, we calculate $100 \times (1.00 - \text{OR}) = 100 \times (1 - 0.88) = 12\%$. The odds of a woman supporting abortion are 12% lower than the odds of a man supporting abortion.

Sometimes an OR can be calculated for a $2 \times 2$ table and provide much more useful information than a measure of association such as $\phi$. Here is an illustration of how $\phi$ can be very misleading. Utts (2014) reports a table showing the relationship between men taking a daily dose of aspirin and having a heart attack. This was originally reported by the Steering Committee of the Physicians' Health Study Research Group and involved a 5-year trial that followed 22,071 male physicians between 40 and 84 years of age who were randomly assigned to either take an aspirin every day or take a placebo pill. We will use `chapter6_aspirin.dta` for our example. Our command and the results are

```
. tabulate aspirin heartattack, chi2 row V
```

```
  ┌─────────────────┐
  │ Key             │
  ├─────────────────┤
  │     frequency   │
  │  row percentage │
  └─────────────────┘
```

|            |          | Heart Attack |          |
| Condition  | No Attack | Attack      | Total    |
|------------|-----------|-------------|----------|
| Placebo    | 10,845    | 189         | 11,034   |
|            | 98.29     | 1.71        | 100.00   |
| Aspirin    | 10,933    | 104         | 11,037   |
|            | 99.06     | 0.94        | 100.00   |
| Total      | 21,778    | 293         | 22,071   |
|            | 98.67     | 1.33        | 100.00   |

```
          Pearson chi2(1) =  25.0139    Pr = 0.000
              Cramér's V =  -0.0337
```

The results are inconsistent. Notice that the Pearson chi2(1) = 25.01, $p < 0.001$, so the results are highly significant. A problem with this finding is that even a small effect will be significant when we have a large sample, and 22,071 men is definitely a large sample. This is one reason why it is always good to look into other measures of effect size when doing a chi-squared test of significance. We see that Cramér's $V$ (or $\phi$) is just $-0.03$, which suggests a weak relationship. Stata and other statistics software packages report a value for $\phi$ that can be the positive or negative square root of chi-squared over $N$ and thus have a potential range of $-1.0$ to $+1.0$. In our example, we coded the aspirin group as 1 and the control group as 0, and those who had a heart attack as 1 and those who did not as 0. With this coding, the negative $\phi$ indicates that being in the aspirin group (coded 1) is negatively related to being in the heart attack group (coded 1). If we had coded the variables differently, the sign would have been positive.

ORs are also quite useful when interpreting the effect size shown in the above cross-tabulation, but they lead to a very different conclusion. The odds of a man in the aspirin group having a heart attack are $104/10933 = 0.0095$. The odds of a man in the placebo group having a heart attack are $189/10845 = 0.0174$. The OR of a man in the aspirin group having a heart attack compared with a man in the placebo group is the ratio OR $= (a/b)/(c/d) = 0.0095/0.0174 = 0.546$. Because the OR is less than 1.00, we subtract it from 1.00, $100 \times (1.0 - 0.546) = 45.4\%$. Thus we can say that the odds of a man having a heart attack in a 5-year period are reduced by 45.4% if he takes an aspirin every day. This is an important finding that has influenced medical care even though the measure of association is only $-0.034$. The $\phi$ coefficient can be very misleading, especially when one outcome (having a heart attack) is a rare event. Millions of people have been advised to take a daily small aspirin by their physician because of the OR $= 0.546$ and not because the $\phi = -0.034$.

What else can we use to aid our interpretation of this table? The percentage in each cell can be quite useful. The table shows that 1.71% of the men who took the placebo had a heart attack within 5 years, but only 0.94% of the men who took aspirin every day had a heart attack within 5 years. Comparing these percentages, we see that $1.71/0.94 = 1.81$. Thus men who were in the placebo group were 1.81 times as likely to have a heart attack than men in the aspirin group. Alternatively, we could say that $0.94/1.71 = 0.54$, or men taking aspirin were about half as likely to have a heart attack than men taking a placebo.

What is the best way to describe the results? The results of this study indicate that adult males taking a daily dosage of aspirin are significantly less likely to have a heart attack than men who take a placebo pill. Although a heart attack is a rare event for either the treatment or the control group, occurring in only 1.33% of the men, those taking the placebo were 1.81 times as likely to have an attack. The OR = 0.55 indicates that the odds of a man taking aspirin daily having a heart attack are 45% lower than if he took a placebo.

## 6.6   Ordered categorical variables

The example we have covered involves two unordered categorical variables (nominal level). Sometimes the categorical variables have an underlying order. For example, we might be interested in the relationship between health (`health`) and happiness (`happy`) from `gss2006_chapter6.dta`, which contains selected variables from the 2006 General Social Survey. Are people who are healthier also happier? When the question is asked this way, `health` is the independent variable because being happy is said to depend on your health.

$$\text{health} \rightarrow \text{happy}$$

Another researcher might reverse this premise and argue that the happier a person is then the more likely he or she is to rate everything, including his or her own health, as better than people who are unhappy. If this is your argument, then happiness is the independent variable and health depends on how happy you are.

$$\text{health} \leftarrow \text{happy}$$

A third researcher may simply say that the two variables are related without claiming the direction of the relationship. We say that happiness and health are reciprocally related. In other words, the happier you are, the more positive you will report your health to be, and the healthier you are, the happier you will report being.

$$\text{health} \leftrightarrow \text{happy}$$

The example above uses a double-headed arrow, meaning that happiness leads to better perceived health and better perceived health leads to happiness. Because both variables depend on each other, there is no single variable that we can call independent or dependent. This probably makes sense. We have all known people who are happy and see the world through rose-colored lenses, where they rate nearly everything positively.

They are likely to rate their health as positive. We have also known people for whom their health has a big influence on their happiness. If they have a health condition that varies, then when they are feeling relatively well they will report being happy, and when they are feeling bad they will report being unhappy.

If there is no clear independent or dependent variable, it is usually best to make the row variable the one with the most categories. Running the command codebook happy health shows that health has four categories (excellent, good, fair, and poor) and happy has just three categories (very happy, pretty happy, and not too happy). So we will make health the row variable and happy the column variable.

Select Statistics ▷ Summaries, tables, and tests ▷ Frequency tables ▷ Two-way table with measures of association. The resulting dialog box is the same as the one in figure 6.1. We need to enter the names of the variables and the options we want. Enter health for the row variable and happy for the column variable. We pick both *Within-column relative frequencies* and *Within-row relative frequencies*. Also check *Pearson's chi-squared*, *Goodman and Kruskal's gamma*, *Kendall's tau-b*, and *Cramer's V* to obtain these statistics. Here are the command and results that are produced:

```
. tabulate health happy, chi2 column gamma row taub V
```

| Key |
|---|
| frequency |
| row percentage |
| column percentage |

| CONDITION OF HEALTH | GENERAL HAPPINESS VERY HAPP | PRETTY HA | NOT TOO H | Total |
|---|---|---|---|---|
| EXCELLENT | 271 | 247 | 33 | 551 |
|  | 49.18 | 44.83 | 5.99 | 100.00 |
|  | 42.74 | 22.50 | 12.50 | 27.61 |
| GOOD | 261 | 567 | 103 | 931 |
|  | 28.03 | 60.90 | 11.06 | 100.00 |
|  | 41.17 | 51.64 | 39.02 | 46.64 |
| FAIR | 82 | 231 | 92 | 405 |
|  | 20.25 | 57.04 | 22.72 | 100.00 |
|  | 12.93 | 21.04 | 34.85 | 20.29 |
| POOR | 20 | 53 | 36 | 109 |
|  | 18.35 | 48.62 | 33.03 | 100.00 |
|  | 3.15 | 4.83 | 13.64 | 5.46 |
| Total | 634 | 1,098 | 264 | 1,996 |
|  | 31.76 | 55.01 | 13.23 | 100.00 |
|  | 100.00 | 100.00 | 100.00 | 100.00 |

```
          Pearson chi2(6) = 182.1737   Pr = 0.000
              Cramér's V =   0.2136
                  gamma =   0.3917   ASE = 0.030
          Kendall's tau-b =   0.2492   ASE = 0.020
```

Stata makes some compromises in making this table. If a value label is too big to fit, Stata simply truncates the label. You might need to do a codebook on your variables to make sure you have the labels correct. In this table, the value label "very happy" appears as `VERY HAPP`, and "not too happy" appears as `NOT TOO H`. If you were preparing this table for a report or publication, you would want to edit it so that it has proper labels.

Just like with unordered categories, we use chi-squared to test the significance of the relationship. The relationship between perceived health and happiness is statistically significant: $\chi^2(6, N = 1996) = 182.17$, $p < 0.001$.

The percentages are also useful. Here we will pick one variable, `health`, arbitrarily as the independent variable. The box just above the table tells us that the row percentages are the second number in each cell. Only 18.35% of those with poor health said they are very happy, compared with 49.18% of those in excellent health. Similarly, only 5.99% of those in excellent health said they were not too happy, but 33.03% of those in poor health said they were not too happy.

If you decide to treat happiness as the independent variable, you would say that 42.74% of those who were very happy reported being in excellent health, compared with just 12.50% of those who were not too happy. Analyzing the percentages provides far richer information about the relationship than a measure of association can provide. However, researchers often want the simplicity of having a number that summarizes the strength of the association, and this is exactly why we have measures of association.

Cramér's $V$ can be used as a measure of association, but it does not use the ordered nature of the variables. Both gamma ($\gamma$) and tau-b ($\tau_b$) are measures of association for ordinal data. Both of these involve the notion of concordance. If one person is happier than another, we would expect that person to report being in better health. We call this "concordance". If a person has worse health, we expect this person to be less happy. This is what we mean when we say that health and happiness are positively related. Gamma and tau-b differ in how they treat people who are tied on one or the other variable, but both measures are bigger when there is a predominance of concordant pairs. Because of the way it is computed, gamma tends to be bigger than tau-b, and tau-b is closer to what you would get if you treated the variables as interval level and computed a correlation coefficient. Values of tau-b less than 0.2 signify a weak relationship. Values between 0.2 and 0.49 indicate a moderate relationship. Values of 0.5 and higher indicate a strong relationship. Stata tells us that tau-b is 0.25, so we can say the relationship is moderate. After we study the percentages, they seem consistent with this judgment provided by tau-b.

The chi-squared test is an appropriate test for the significance of Cramér's $V$. Stata does not provide a significance level of gamma or tau-b, but it does provide an asymptotic standard error (ASE) for each. The asymptotic standard error for tau-b is ASE = 0.020. If you divide tau-b by this estimated standard error, you get the $z$-test value. For our example, $z = 0.249/0.020 = 12.45$. $|z| \geq 1.96$ is significant at the $p < 0.05$ level, $|z| \geq 2.60$ is significant at the 0.01 level, and $|z| \geq 3.32$ is significant at

*ordinal data*

the 0.001 level. Considering these levels, we can say that our tau-b $= 0.25$, $z = 12.45$, $p < 0.001$, meaning that there is a moderate relationship and that it is statistically significant. A note of caution: Because these are asymptotic standard errors, they are only good estimates when you have a large sample.

## 6.7    Interactive tables

If you are reading a report, you might find a cross-tabulation where the authors did not compute percentages, did not compute chi-squared, or did not compute any measures of association. Here is an example of a table showing the cross-tabulation of `sex` and `abany` (abortion is okay for any reason). We have been using data from the 2006 General Social Survey, but this table is from the 2002 General Social Survey. Let's say that you come across this table and want to study its contents to see if there is a statistically significant relationship and how strong the association is. The author may have presented this table to show that men are less supportive of abortion than women, and you want to make sure the author has interpreted the table correctly.

```
. use http://www.stata-press.com/data/agis4/gss2002_chapter6
. tabulate sex abany
```

| respondent sex | abortion if woman wants for any reason | | Total |
|---|---|---|---|
| | yes | no | |
| male | 215 | 269 | 484 |
| female | 172 | 244 | 416 |
| Total | 387 | 513 | 900 |

If you do not have access to the author's actual data, you need to enter the table into Stata. You can enter the raw numbers in the cells of the table, and Stata will compute percentages, chi-squared, and measures of association.

Select Statistics ▷ Summaries, tables, and tests ▷ Frequency tables ▷ Table calculator to open a dialog box in which you enter the frequencies in each cell. The format for doing this is rigid. Enter each row of cells, separating the rows with a backslash, \. Enter only the cell values and not the totals in the margins of the table. This entry format is illustrated in figure 6.3. You need to check the table Stata analyzes to make sure that you entered it correctly. Check the boxes to request chi-squared and within-row relative frequencies. Because this is not ordinal data, the only measure of association you request is Cramér's $V$.

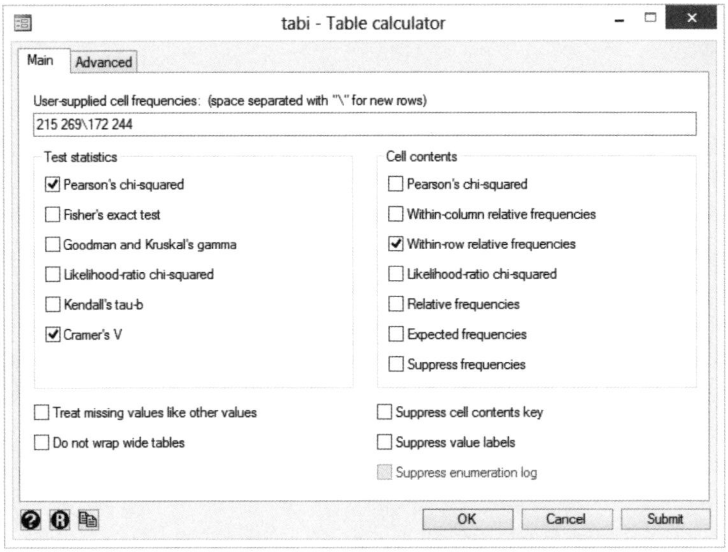

Figure 6.3. Entering data for a table

If you click on **Submit** in this dialog box, you obtain the same results as if you had entered all the data in a big dataset.

```
. tabi 215 269\172 244, chi2 row V
```

```
  Key

    frequency
  row percentage
```

```
              |         col
       row    |     1          2  |    Total
   -----------+------------------- +----------
         1    |    215        269  |      484
              |  44.42      55.58  |   100.00
   -----------+------------------- +----------
         2    |    172        244  |      416
              |  41.35      58.65  |   100.00
   -----------+------------------- +----------
      Total   |    387        513  |      900
              |  43.00      57.00  |   100.00

          Pearson chi2(1) =   0.8633   Pr = 0.353
            Cramér's V =   0.0310
```

These results indicate that the author overstated his findings. The results are not statistically significant; $p = 0.353$ and Cramér's $V$ is 0.03. A slightly higher percentage of men in the 2002 sample said that abortion was acceptable for any reason (44.42%) compared with women (41.35%), but this difference is not statistically significant.

## 6.8    Tables—linking categorical and quantitative variables

Many research questions can be answered by tables that mix a categorical variable and a quantitative variable. We may want to compare income for people from different racial groups. A study may need to compare the length of incarceration for men and for women sentenced for the same crime. A drug company may want to compare the time it takes for its drug to take effect with the time it takes another drug.

Here we switch back to 2006 General Social Survey, `gss2006_chapter6.dta`. We will use `hrs1`, hours a person worked last week, as our quantitative variable and compare men with women. Our hypothesis is that men spend more time working for pay each week than do women. We will use a formal test of significance in the next chapter, but here we will just show a table of the relationship. Select Statistics ▷ Summaries, tables, and tests ▷ Other tables ▷ Flexible table of summary statistics to get to the dialog box in figure 6.4.

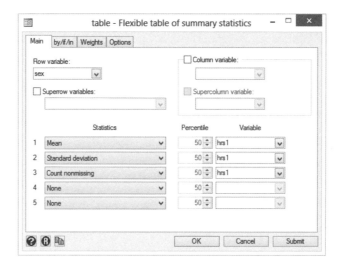

Figure 6.4. Summarizing a quantitative variable by categories of a categorical variable

The dialog box looks fairly complicated. Look closely at where we enter the variable names and statistics we want Stata to compute. The *Row variable* is the categorical variable, `sex`. Under *Statistics*, select the following from the menus: *Mean, Standard deviation*, and *Count nonmissing*. On the far right, there are places to enter variables. We want `hrs1` for the mean, standard deviation, and count. Click on Submit so that we can return to this dialog box at a later time. The resulting command is

```
. table sex, contents(mean hrs1 sd hrs1 count hrs1)
```

| Gender | mean(hrs1) | sd(hrs1) | N(hrs1) |
|--------|-----------|----------|---------|
| MALE   | 44.8767   | 14.27598 | 1,387   |
| FEMALE | 39.2034   | 13.60452 | 1,352   |

This table shows that the week before the survey, men spent on average over 5 more hours working for pay than did women: 44.88 hours for men versus 39.20 for women.

Now suppose that you want to see the effect of marital status and gender on hours spent working for pay. Go back to the dialog box, click on the checkbox next to *Column variable*, and add `marital` as the variable. Go to the Options tab and select *Add row totals*. The command for this is

```
. table sex marital, contents(mean hrs1 sd hrs1 count hrs1) row
```

I omitted the results here.

These summary tables are useful, but to communicate to a lay audience, a graph often works best. We can create a bar chart showing the mean number of hours worked by gender and by marital status. We created one type of bar chart in chapter 5, where we used the dialog box for a histogram to create the bar chart. Now we are doing a bar chart for a summary statistic, mean hours worked, over a pair of categorical variables, `sex` and `marital`. To do this, access the dialog box for a bar chart by selecting Graphics ▷ Bar chart. Under the Main tab, make sure that the first statistic is checked, and, by default, this is the mean. To the right of this, under *Variables*, type `hrs1` to make the graph show the mean hours worked. Next click on the Categories tab. Check *Group 1* and type `sex` as the *Grouping variable*. Check *Group 2* and type `marital` as the *Grouping variable*. There is a button labeled *Properties* to the right of where you entered `sex`. Click on this button and change the *Angle* to *45 degrees*. Click on Accept.

It would be nice to have the actual mean values for hours worked at the top of each bar, so switch to the Bars tab, and click on the radio button for *Label with bar height*. If we stop here, the numerical labels will have too many decimal places to look nice. We can fix these by making the numerical labels have a fixed format. Click on the *Properties* button in the section labeled *Bar labels*, and type the value `%9.1f` in the *Format* box. The *Bar label properties* dialog box is shown in figure 6.5.

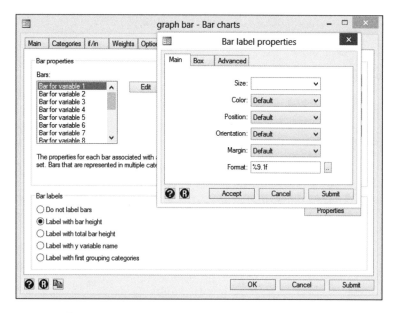

Figure 6.5. The *Bar label properties* dialog box

This effort produces a nice bar chart, but there are some things we can still improve. The marital status labels are too wide and run into each other. The label of the hours worked for pay could also be better. Open the Graph Editor and change the font of the labels of marital status and gender to size small. Then add the title `Hours Worked Last Week` and the subtitle `By Sex and Marital Status`. Your final bar chart is shown in figure 6.6.

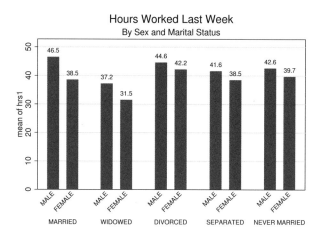

Figure 6.6. Bar graph summarizing a quantitative variable by categories of a categorical variable

# 6.9    Power analysis when using a chi-squared test of significance

Power analysis is discussed in more detail in chapter 7, but I will introduce it briefly here. Power is the ability to do what you want to do. In statistics, power means how likely you are to get a statistically significant result for your chi-squared test if the relationship in the population is strong enough for you to consider it important. One characteristic of chi-squared tests that sometimes leads to misinterpretation is that for a given relationship in terms of percentages, the value of chi-squared varies directly with the sample size. We can illustrate this variation by taking a 10% sample of `gss2006_chapter6.dta` and doing a cross-tabulation of health and gender. There are two steps to taking a 10% sample. The first step is to set an arbitrary value for the seed. The value of the seed controls where the random process starts the generation of our 10% subsample. By specifying a specific value, we ensure that we will get the same 10% sample if we repeat this process at a later time. The second step is to take the random sample. Here are the two commands needed for these steps:

```
. set seed 219
. sample 10
```

I have already done these steps and saved the results as a new dataset, `gss2006_chapter6_10percent.dta`, which we will now use to do our tabulation.

```
. tab sex health, lrchi2 row
```

| Key |
| --- |
| *frequency* |
| *row percentage* |

|  |  | CONDITION OF HEALTH |  |  |  |
| --- | --- | --- | --- | --- | --- |
| Gender | EXCELLENT | GOOD | FAIR | POOR | Total |
| MALE | 53 | 75 | 32 | 6 | 166 |
|  | 31.93 | 45.18 | 19.28 | 3.61 | 100.00 |
| FEMALE | 51 | 84 | 41 | 15 | 191 |
|  | 26.70 | 43.98 | 21.47 | 7.85 | 100.00 |
| Total | 104 | 159 | 73 | 21 | 357 |
|  | 29.13 | 44.54 | 20.45 | 5.88 | 100.00 |

```
           likelihood-ratio chi2(3) =   3.8933   Pr = 0.273
```

These results are not significant, chi-squared(3) = 3.893, ns. We used what is called the likelihood-ratio chi-squared with the `lrchi2` option in the command. This variation on how chi-squared is estimated is very similar to what we would have obtained using the `chi2` option, but we need to use the likelihood-ratio chi-squared to do our power analysis. If we look at the percentages, it seems that relatively fewer women, 26.7%, say their health is excellent than do men, 31.9%. Also women are more likely to rate their health as fair or poor. Let's assume these percentages are the true values in the population. If that were the case, how much power do we have to get a significant chi-squared with this sample size of $N = 357$?

Philip Ender at UCLA wrote a useful command, chi2power, that can help us. You can install this command by typing search chi2power and following the instructions. The chi2power command is run immediately after the tab command that included the option to get a likelihood-ratio chi-squared, lrchi2. Here is the command and the results:

```
. chi2power, startf(1) endf(10) incr(1)
alpha = .05
sample size factor = 1.00 power = 0.3496 for n = 357
sample size factor = 2.00 power = 0.6410 for n = 714
sample size factor = 3.00 power = 0.8293 for n = 1071
sample size factor = 4.00 power = 0.9272 for n = 1428
sample size factor = 5.00 power = 0.9715 for n = 1785
sample size factor = 6.00 power = 0.9896 for n = 2142
sample size factor = 7.00 power = 0.9964 for n = 2499
sample size factor = 8.00 power = 0.9988 for n = 2856
sample size factor = 9.00 power = 0.9996 for n = 3213
sample size factor = 10.00 power = 0.9999 for n = 3570
```

In the chi2power command, the first option, startf(1), asks Stata for the power of the current sample size, that is, $N = 357$. The endf(10) option asks Stata to do a power analysis where the percentage in each cell is the same but where there are up to 10 times as many observations (3,570). The incr(1) option asks Stata to do this for increments of a factor of 1. That is, 1 times the starting sample size, 2 times the starting sample size, up to 10 times the starting sample size. We assume we are using an alpha equal to 0.05.

The results show that for the sample-size factor of 1.0 of our current sample size of $N = 357$, we have a power of 0.350. This means that if the percentage distribution we observed were true in the population, we would obtain a statistically significant chi-squared just $100 \times 0.350 = 35.0\%$ of the time. If these percentages represent what we consider an important or interesting gender difference (women rating their health worse than men), then we would be doomed to failure with a sample of 357. If we had a sample that was four times bigger, making $N = 1,428$, we would have power of 0.927. Meaning that 92.7% of the time, we would get a sample of 1,428 people in which the chi-squared was significant, but the other 7.3% of the time, we would get a chi-squared that was not statistically significant.

Statisticians usually say that the power should be greater than 0.80, and in many fields, they say it should be greater than 0.90. The results of the chi2power command show that we would need 1,071 observations to have a power greater than 0.80 and 1,428 observations to have a power greater than 0.90.

Why is this important? If you are requesting a grant to support collecting data to test your hypothesis that women rate their health worse than men do, the funders want to be reasonably confident that you will find significant results when there is a substantively significant relationship. If you proposed to collect data on 500 people, you would have inadequate power to get a statistically significant result. The funders would not want to fund you because you would be asking for too little. By contrast, this sort of power analysis may show that you are collecting data from more people than necessary to have good power. Asking for support to collect data on 2,000 observations would be unnecessary and an inefficient investment of the funders' money. This type of power analysis is difficult to implement because the researcher needs to specify what result would be the smallest result that would be substantively important. It is often very difficult to do this, especially with larger tables.

## 6.10  Summary

Cross-tabulations are extremely useful ways of presenting data, whether the data are categorical variables or a combination of categorical and quantitative variables. This chapter has covered a lot of material. We have learned

- How to examine the relationship between categorical variables

- How to develop cross-tabulations

- How to perform a chi-squared test of significance

- How to use percentages and compute appropriate measures of association for unordered and ordered cross-tabulations

- What the odds ratio is

- How to use an interactive table calculator

- How to extend tables to link categorical variables with quantitative variables

- How a bar graph can show how means on a quantitative variable differ across groups on a categorical variable

Stata has many capabilities for dealing with categorical variables that we have not covered. Still, think about what you have learned in this chapter. Policies are made and changed because of highly skilled presentations. Imagine your ability to make a presentation to a research group or to a policy group. Now you can "show them the numbers" and do so effectively. Much of what you have learned would have been impossible without statistical software. The next chapter will continue to show how Stata helps us analyze the relationship between pairs of variables. It will focus on quantitative outcome variables and categorical predictors and move beyond the point biserial correlation.

## 6.11   Exercises

1. Open `gss2006_chapter6.dta`, do a `codebook` on `pornlaw`, and then do a cross-tabulation of this with `sex`. Which variable is the independent variable? Which variable will you put on the row, and how will you do the percentages? If you were preparing a document, how would you change the labels of the response options for `pornlaw`?

2. Based on the first exercise, what is the chi-squared value? How would you report the chi-squared and the level of significance? What is the value of Cramér's $V$, and how strong of an association does this show? Finally, interpret the percentages to answer the question of how much women and men differ in their attitude about legalizing pornography.

3. Open `gss2006_chapter6.dta` and do a cross-tabulation of `pres00` (whom you voted for in 2000) and `pres04` (whom you voted for in 2004). From the dialog box, check the option to include missing values. Do a `codebook` on both variables, and then use the by/if/in tab in the `tabulate` dialog box to repeat the table just for those who voted for Gore or Bush in 2000 and for Kerry or Bush in 2004. Treat the 2000 vote as the independent variable. Is there a significant relationship between how people voted in 2000 and 2004? Interpret the percentages and phi, as well as the statistical significance.

4. Open `gss2006_chapter6.dta` and do a cross-tabulation of `polviews` and `premarsx`. Treating `polviews` as the independent variable, compute percentages on the rows. Because these are ordinal variables, compute gamma and tau-b. Is there a significant relationship between political views and conservatism? Interpret the relationship using gamma and tau-b. Interpret the relationship using the percentages.

5. In exercise 4, Stata reports that there is 18 degrees of freedom. How does it get this number?

6. Open `gss2002_chapter6.dta`. Create a table showing the mean hours worked in the last week (`hrs1`) for each level of political views (`polviews`). In your table, include the standard deviation for hours and the frequency of observations. What do the means suggest about people who are extreme in their views (in either direction)?

7. Based on exercise 6, create a bar chart showing the relationship between hours worked in the last week and political views.

8. You are given the following dataset:

| RESPONDENT S SEX | FEELINGS ABOUT PORNOGRAPHY LAWS | | | Total |
|---|---|---|---|---|
| | ILLEGAL T | ILLEGAL U | LEGAL | |
| MALE | 234 | 568 | 38 | 840 |
| FEMALE | 541 | 557 | 29 | 1,127 |
| Total | 775 | 1,125 | 67 | 1,967 |

Calculate the chi-squared test of significance, percentages, and Cramér's $V$ using the `tabi` command. Interpret Cramér's $V$.

9. Examine the relationship between gender and self-reported health condition using `gss2006_chapter6.dta`. Recode the `health` variable so `excellent` and `good` are combined into a new category labeled `satisfactory`, and `fair` and `poor` are combined into a new category labeled `unsatisfactory`. Call the new variable `health2` and label the variable "Is your health satisfactory?" Conduct a chi-squared analysis along with Cramér's $V$ ($\phi$). Calculate the odds ratio. Write a short paragraph on how you would report these analyses.

# 7 Tests for one or two means

## 7.1   Introduction to tests for one or two means

Imagine that you are in the public relations department for a small liberal arts college. The mean score for the math and reading sections on the SAT at your college is $M = 1180$.[1] Suppose that the mean for all small liberal arts colleges in the United States is $\mu = 1080$. We use $M$ to refer to the mean of a sample and $\mu$, pronounced "mu", to refer to the mean of a population. For our purposes, we are treating the students at your college as the sample and all students at small liberal arts colleges in the United States as the population. Say that our supervisor asked if we could demonstrate that our college is more selective than its peer institutions. Could this difference, our $M = 1180$, and the mean for all colleges, $\mu = 1080$, occur just by chance?

---

1. The scores used do not include the writing part of the SAT.

Suppose that last year, our Boys and Girls Club had 30% of the children drop out of programs before the programs were completed. We have a new program that we believe does a better job of minimizing the dropout rate. However, 25% of the children still drop out before the new program is completed. Because a smaller percentage of children drop out, this is better, but is this enough difference to be statistically significant?

These two examples illustrate a one-sample test where we have one sample (students at our college or the children in the new Boys and Girls Club program), and we want to compare our sample mean with a population value (the mean SAT score for all liberal arts colleges or the mean dropout rate for all programs at our Boys and Girls Club).

At other times, we may have two groups or samples. For example, we have a program that is designed to improve reading readiness among preschool children. We might have 100 children in our preschool and randomly assign half of them to this new program and the other half to a traditional approach. Those in the new program are in the treatment group (because they get the new program), and those in the old program are in the control group (because they did not get the new program). If the new program is effective, the children in the treatment group will score higher on a reading-readiness scale than the children in the control group. Let's say that, on a 0–100 point scale, the 50 children randomly assigned to the control group have a mean of $M = 71.3$ and a standard deviation of $SD = 10.4$. Those in the treatment group, who were exposed to the new program, have $M = 82.5$ and $SD = 9.9$. It sounds like the program helped. The children in the treatment group have a mean that is more than one standard deviation higher than that of the students in the control group. This sounds like a big improvement, but is this statistically significant? Here we have a problem for a two-group or two-sample $t$ test.

We have a 20-item scale that measures the importance of voting. Each item is scored from 1, signifying that voting is not important, to 5, signifying that voting is very important. Thus the possible range for the total score on our 20-item scale is from 20 (all items answered with a 1, indicating that voting is not important) to 100 (all items answered with a 5, signifying that voting is very important). We hypothesize that minority groups will have a lower mean than will whites. Can we justify this hypothesis? Perhaps the lower mean results because minority groups have experienced a history of their votes not counting or because of the perception that their vote does not matter in the final outcome. To do an experiment with a treatment and control group, we would need to randomly assign people to minority status or majority status. Because we cannot do this, we can use a random sample of 200 people from the community. If this sample is random for the community, it is a random sample for any subgroup. Let's say that 70 minority community members are in our sample, and they have a mean importance of voting score of $M = 68.9$. The 130 white community members in our sample have a mean importance of voting score of $M = 83.5$. This is what we hypothesized would happen. A two-sample $t$ test will tell us if this difference is statistically significant.

Sometimes researchers use a two-sample $t$ test even when they do not have random assignment to a treatment or control group or when they do not have random sampling. You might want to know if a highly developed recreation program at a full-care retirement center leads to lower depression among center members. Your community might have two full-care retirement centers, one with a highly developed recreation program and the other without such a program. We could do a two-sample test to see whether the residents in the center with the highly developed recreation program had a lower mean score on depression.

Can you see the problems with the above example? First, without randomization to the two centers, we do not know whether the recreation program makes the difference or if people who are more prone to be depressed selected the center that does not offer much recreation. A second problem is that the programs being compared may be different in many respects other than our focus. A center that can afford a highly developed recreation program may have, for example, better food, more spacious accommodations, and better-trained staff. Even if we have randomization, we want the groups to differ only on the issue we are testing. More-advanced procedures, such as analysis of covariance and multiple regression, can help control for other differences between the groups being compared, but a two-sample $t$ test is limited in this regard without randomization or random sampling.

### Random sample and randomization

It is easy to confuse random sample and randomization. A random sample refers to how we select the people in our study. Once we have our sample, randomization is sometimes used to randomly assign our sample to groups. Some examples will better illustrate the distinction between the terms:

**Random sample without randomization**. We obtain a list of all housing units that have a water connection from our city's utility department. We randomly sample 500 of these houses.

**Randomization without a random sample**. We solicit volunteers for an experiment. After 100 people volunteer, we randomly assign 50 of them to the treatment group and 50 to the control group. We randomized the assignment, but we did not start with a random sample. Because we did not start with a random sample, we could have a *sample selection bias* problem because volunteers are often much more motivated to do well than the rest of the population.

**Randomization with a random sample**. Obtaining a list of all students enrolled in our university, we randomly sample 100 of these students. Next we randomly assign 50 of them to the treatment group and 50 to the control group.

# 7.2   Randomization

How can we select the people we want to randomly assign to the treatment group and the control group? There are two types of random selection. One is done with replacement, and one is done without replacement. When we sample with replacement, each observation has the same chance of being selected. However, the same observation may be selected more than once. Suppose that you have 100 people and want to do random sampling with replacement. You randomly pick one case, say, Mary. She has a 1/100 probability of being selected. If you put her back in the pool (replacement) before selecting your second case, whomever you select will have the same 1/100 chance of being selected. However, you may sample Mary again doing it this way.

Many statistical tests assume sampling with replacement, but in practice, we rarely do this. We usually sample without replacement. This means that the first person would have a 1/100 chance of being selected, but the second person would have a 1/99 chance of being selected. This method violates the assumption that each observation has the same chance of being selected. Sampling without replacement is the most practical way to do it. If we have 100 people and want 50 of them in the control group and 50 in the treatment group, we can sample 50 people without replacement (this gives us 50 different people) and put them in the control group. Then we put the other 50 people in the treatment group. If we had sampled with replacement, the 50 people we sampled might actually be only 40 different people, with a few of them selected twice or even three times.

Sampling without replacement is a simple task in Stata. If you have $N = 20$ observations, you would number these from 1 to 20. Then you would enter the command `sample 10, count` to select 10 observations randomly but without replacement. These 10 people would go into your treatment group, and the other 10 would go into your control group. Here is the set of commands:

```
clear
set obs 20
gen id = _n
list
set seed 220
sample 10, count
list
```

The first command clears any data we have in memory. You should save anything you were doing before entering this command. The second command, `set obs 20`, tells Stata that you want to have 20 observations in your dataset. The next command, `gen id = _n`, will generate a new variable called `id`, which will be numbered from 1 to the number of observations in your dataset, in this case, 20. The underscore n (`_n`) is a reserved name in Stata that refers to the observation number. This command generates a new variable, `id`, which is the same as the observation number, 1, 2, 3, ..., 20. You can see this in the first listing below. We then use the command `set seed 220`, where the number 220 is arbitrary. This is not necessary, but by doing this, we will be able to reproduce our results. When you set the seed at a particular value, every time you

run this set of commands, you will get the same result. Without setting the seed at a
specific value, we would get a different result every time we ran the set of commands.
Finally, we take the sample using the command `sample 10, count` and show a listing
of the 10 cases that Stata selected at random without replacement. These cases would
go into the control group, and the other 10 would go into the treatment group. Thus
observations numbered 8, 19, 9, 13, 18, 12, 6, 11, 17, and 3 (as shown below with the
`list` command) go into the control group, and the other 10 observations go into the
treatment group. By doing this, any differences between the groups that are not due to
the treatment are random differences.

```
. clear
. set obs 20
obs was 0, now 20
. gen id = _n
. list

         +----+
         | id |
         |----|
  1.     |  1 |
  2.     |  2 |
  3.     |  3 |
  4.     |  4 |
  5.     |  5 |
         |----|
  6.     |  6 |
  7.     |  7 |
  8.     |  8 |
  9.     |  9 |
 10.     | 10 |
         |----|
 11.     | 11 |
 12.     | 12 |
 13.     | 13 |
 14.     | 14 |
 15.     | 15 |
         |----|
 16.     | 16 |
 17.     | 17 |
 18.     | 18 |
 19.     | 19 |
 20.     | 20 |
         +----+

. set seed 220
. sample 10, count
(10 observations deleted)
```

```
. list
```

|      | id |
|------|----|
| 1.   | 8  |
| 2.   | 19 |
| 3.   | 9  |
| 4.   | 13 |
| 5.   | 18 |
| 6.   | 12 |
| 7.   | 6  |
| 8.   | 11 |
| 9.   | 17 |
| 10.  | 3  |

## 7.3   Random sampling

The process of selecting a random sample is similar. Suppose that you had 20,000 students in your university. Assume that each student has a unique identification number. You would have a Stata dataset with these numbers. Assume too that you need a sample of $N = 500$ students. You would simply set the seed at some arbitrary value, such as 3 with the command `set seed 3`, and then run the command `sample 500, count`. You could then list these ID numbers to see your sample.

If we have a target sample of 500, we can expect that several of our initial list of $N = 500$ either will not be located or will refuse to participate. Therefore, we should draw a second sample of, say, $N = 300$, and go down this list as we need replacements.

## 7.4   Hypotheses

Before we can run a $z$ test or a $t$ test, we need to have two hypotheses: a null hypothesis ($H_0$) and an alternative hypothesis ($H_a$). I will mention these briefly because they are covered in all statistics texts.

Although one-tailed tests are appropriate if we can categorically exclude negative findings (results in the opposite direction of our hypothesis), we will rarely be this confident in the direction of the results. Most statisticians report two-tailed tests routinely and make note that the direction of results is what they expected. Other statisticians rely on a one-tailed test when they have a clear prediction of the direction. For example, if they have a new drug that should reduce the level of cholesterol, they might use a one-tailed test because they have a clear directional hypothesis. If, however, this new drug increased the level of cholesterol, then they would fail to reject the null hypothesis without having to do a test of significance. A two-tailed test is always more conservative than a one-tailed test, so the tendency to rely on two-tailed tests can be viewed as a conservative approach to statistical significance.

Stata reports both one- and two-tailed significance for all the tests covered in this chapter. For more-advanced procedures, such as those based on regression, Stata reports only two-tailed significance. It is easy to convert two-tailed significance to one-tailed significance. If something has a probability of 0.1 using a two-tailed test, its probability would be 0.05 using a one-tailed test—we simply cut the two-tailed probability in half.

## 7.5   One-sample test of a proportion

Let's use items from the 2002 General Social Survey dataset gss2002_chapter7.dta to test how adults feel about school prayer. An existing variable, prayer, is coded 1 if the person favors school prayer and 2 if the person does not. We must recode the variable prayer into a new variable, schpray, so that a person has a score of 1 if he or she supports school prayer and a score of 0 if he or she does not. We can use Stata's recode command to accomplish this. We will do a cross-tabulation as well to make sure we did not make a mistake:

```
. recode prayer (1 = 1 Approve) (2 = 0 Disapprove), gen(schpray)
. tabulate schpray prayer, missing
```

Stata will not let you accidentally write over an existing variable, so if you try to create a variable that already exists, Stata will issue an error message. When we give schpray a score of 1 for those adults who favor school prayer and a score of 0 for those who oppose it, Stata can compute and test proportions correctly. If we ran a proportions test on the prayer variable, we would get an error message saying that prayer is not a 0/1 variable.

Although we can speculate that most people oppose school prayer, that is, have a score of 0 on schpray, let's take a conservative track and use a two-tailed hypothesis:

Alternative hypothesis $H_a$: proportion $\neq 0.5$

Null hypothesis $H_0$: proportion $= 0.5$

The null hypothesis, proportion $= 0.5$, uses a value of 0.5 because this represents what the proportion would be if there were no preference for or against school prayer.

Open the appropriate dialog box by selecting Statistics ▷ Summaries, tables, and tests ▷ Classical tests of hypotheses ▷ Proportion test. On the Main tab, select *One-sample*, if it is not already selected, in the *Tests of proportions* section. Enter the schpray variable and the hypothesized proportion for the null hypothesis, 0.5, and click on OK. The resulting command and results are

```
. prtest schpray == 0.5
One-sample test of proportion                    schpray: Number of obs =      859
```

| Variable | Mean | Std. Err. | | [95% Conf. Interval] |
|---|---|---|---|---|
| schpray | .3969732 | .0166937 | .3642542 | .4296923 |

```
   p = proportion(schpray)                                      z =  -6.0392
Ho: p = 0.5

     Ha: p < 0.5                Ha: p != 0.5                 Ha: p > 0.5
  Pr(Z < z) = 0.0000        Pr(|Z| > |z|) = 0.0000        Pr(Z > z) = 1.0000
```

Let's go over these results and see what we need. All the results in this chapter use this general format for output, so it is worth the effort to review it closely. The first line is the Stata command. The second line indicates that this is a one-sample test of a proportion, the variable is schpray, and we have $N = 859$ observations. The table gives a mean, $M = 0.397$, and standard error, 0.017. Because we recoded schpray as a dummy variable, where a 1 signifies that the person supports school prayer and a 0 signifies that the person does not, the mean has a special interpretation. The mean is simply the proportion of people coded 1. (This is true only of 0/1 variables and would not work if we had used the original variable, prayer.) Thus 0.397, or 39.7%, of the people in the survey say that they support school prayer. The 95% confidence interval tells us that we can be 95% confident that the interval 0.364 to 0.430 includes adults who support school prayer. Using percentages, we could say that we are 95% confident that the interval of 36.4% to 43.0% contains the true percentage of adults who support school prayer.

Below the table, we can see the null hypothesis that the proportion is exactly 0.5, that is, Ho: p = 0.5. Also just below the table but to the far right is the computed $z$ test, $z = -6.039$.

Below the null hypothesis are three results that depend on how we stated the alternative hypothesis. Here we used a two-tailed alternative hypothesis, and this result appears in the middle group. Stata's Ha: p != 0.5 is equivalent to stating the alternative hypothesis, $H_a$: proportion $\neq$ 0.5. Because Stata results are plain text, Stata uses != in place of $\neq$. Stata programmers use an exclamation mark (programmers often call this a "bang") to mean "not", so != literally means "not equal". Below the alternative hypothesis is the probability that the null hypothesis is true: Pr(|Z| > |z|) = 0.0000. The $p$-value = 0.0000 does not mean that there is no probability that the null hypothesis is true, just that the probability is all zeros to four decimal places. If $p$-value = 0.00002, this is rounded to 0.0000 in Stata output. A $p$-value = 0.0000 is usually reported as a significance level of $p < 0.001$. This finding is highly statistically significant and allows us to reject the null hypothesis. Fewer than 1 time in 1,000 would we obtain these results by chance if the null hypothesis were true.

We would write our results as follows: 39.7% of the sample support school prayer. With a $z = -6.039$, we can reject the null hypothesis that proportion is 0.5 at the $p < 0.001$ level. The 39.7% is in the direction expected, namely, that fewer than half of the adult population supported school prayer in 2002.

On the left side of the output is a one-tailed alternative hypothesis, `Ha: p < 0.5`. If another researcher thought that only a minority supported school prayer and could rule out the possibility that the support was in the opposite direction, then the researcher would have this as a one-tailed alternative hypothesis, and it would be statistically significant with a $p < 0.001$.

On the right side of the output is a one-tailed alternative hypothesis, `Ha: p > 0.5`. This means the researcher thought that most adults favored school prayer, and the researcher (incorrectly) ruled out the possibility that the support could be the other way. The results show that we cannot reject the null hypothesis. When we observe only 39.7% of our sample supporting school prayer, we clearly have not received a significant result that most adults supported school prayer in 2002.

### Distinguishing between two $p$-values

In testing proportions, we have two different $p$-values. Do not let this confuse you. One $p$-value is the proportion of the sample coded 1. In this example, this is the proportion that supports school prayer, 0.397 or 39.7%.

The second $p$-value is the level of significance. In this example, $p < 0.001$ refers to the probability that we would obtain this result by chance if the null hypothesis were true. Because we would get our result fewer than 1 time in 1,000 by chance, $p < 0.001$, we can reject the null hypothesis. Many researchers will reject a null hypothesis whenever the probability of the results is less than 0.05, that is, $p < 0.05$.

### Proportions and percentages

Often we report proportions as percentages because many readers are more comfortable trying to understand percentages than proportions. Stata, however, requires us to use proportions. As you may recall, the conversion is simple. We divide a percentage by 100 to get a proportion, so 78.9% corresponds to a proportion of 0.789. Similarly, we multiply a proportion by 100 to get a percentage, so a proportion of 0.258 corresponds to a percentage of 25.8%.

## 7.6   Two-sample test of a proportion

Sometimes a researcher wants to compare a proportion across two samples. For example, you might have an experiment testing a new drug (`wide.dta`). You randomly assign 40 study participants so that 20 are in a treatment group receiving the drug and 20 are in a control group receiving a sugar pill. You record whether the person is cured by assigning a 1 to those who were cured and a 0 to those who were not. Here are the data:

```
. list
```

|      | treat | control |
|------|-------|---------|
| 1.   | 1     | 1       |
| 2.   | 0     | 0       |
| 3.   | 1     | 0       |
| 4.   | 1     | 0       |
| 5.   | 1     | 0       |
| 6.   | 1     | 1       |
| 7.   | 1     | 1       |
| 8.   | 0     | 0       |
| 9.   | 1     | 0       |
| 10.  | 1     | 0       |
| 11.  | 1     | 0       |
| 12.  | 1     | 1       |
| 13.  | 1     | 1       |
| 14.  | 0     | 1       |
| 15.  | 1     | 1       |
| 16.  | 1     | 0       |
| 17.  | 0     | 0       |
| 18.  | 1     | 0       |
| 19.  | 0     | 0       |
| 20.  | 1     | 0       |

In the treatment group, we have 15 of the 20 people, or 0.75, cured; that is, they have a score of 1. In the control group, just 7 of the 20 people, or 0.35, are cured. Before proceeding, we need a null and an alternative hypothesis. The null hypothesis is that the two groups have the same proportion cured. The alternative hypothesis is that the proportions are unequal, and therefore, the difference between them will not equal zero: $p_{(\text{treat})} - p_{(\text{control})} \neq 0$. Your statistics book may state the alternative hypothesis as $p_{(\text{treat})} \neq p_{(\text{control})}$. The two ways of stating the alternative hypothesis are equivalent. They are both two-tailed tests because we are saying that the proportions cured in the two groups are not equal. We could argue for a one-tailed test that the proportion in the treatment group is higher, but this means that we need to rule out the possibility that it could be lower. The null hypothesis is that the two proportions are equal; hence, there is no difference between them: $p_{(\text{treat})} - p_{(\text{control})} = 0$.

Alternative hypothesis $H_a$: $p_{(\text{treat})} - p_{(\text{control})} \neq 0$

Null hypothesis $H_0$: $p_{(\text{treat})} - p_{(\text{control})} = 0$

These are independent samples, and the data for the two groups are entered as two variables. To open the dialog box for this test, select Statistics ▷ Summaries, tables, and tests ▷ Classical tests of hypotheses ▷ Proportion test. Because we have two samples, those participants in the treatment condition and those participants in the control condition, we select *Two-sample using variables* in the *Tests of proportions* section of the **Main** tab. Type `treat` in the box for the *First variable*, and type `control` in the box for the *Second variable*. That is all there is to it, and our results are

```
. prtest treat == control
Two-sample test of proportions                        treat: Number of obs =      20
                                                    control: Number of obs =      20

    Variable |      Mean    Std. Err.      z     P>|z|     [95% Conf. Interval]
    ---------+------------------------------------------------------------------
       treat |       .75    .0968246                        .5602273    .9397727
     control |       .35    .1066536                        .1409627    .5590373
    ---------+------------------------------------------------------------------
        diff |        .4    .1440486                        .1176699    .6823301
             | under Ho:    .1573213    2.54    0.011
    ---------+------------------------------------------------------------------

           diff = prop(treat) - prop(control)                       z =    2.5426
      Ho: diff = 0

      Ha: diff < 0                   Ha: diff != 0                  Ha: diff > 0
   Pr(Z < z) = 0.9945        Pr(|Z| < |z|) = 0.0110           Pr(Z > z) = 0.0055
```

These results have a layout that is similar to that of the one-sample proportion test. The difference is that we now have two groups, so we get statistical information for each group. Under **Mean** is the proportion of **1** codes in each group. We have 0.75 (75%) of the treatment group coded as cured, compared with just 0.35 (35%) of the control group. The difference between the treatment-group mean and the control-group mean is 0.40; that is, $0.75 - 0.35 = 0.40$. This appears in the table as the variable `diff` with the mean of 0.4.

Directly below the table is the null hypothesis that the difference in the proportion cured in the two groups is 0, and to the right is the computed $z$ test, `z = 2.5426`. Below this are the three hypotheses we might have selected. Using a two-tailed approach, we can say that $z = 2.54$, $p < 0.05$. If we had a one-tailed test that the treatment-group proportion was greater than the control-group proportion, our $z$ would still be 2.54, but our $p$ would be $p < 0.01$. This one-tailed test has a $p$-value, 0.0055, that is exactly one-half of the two-tailed $p$-value. If someone had hypothesized that the treatment-group success would have a lower proportion than the control group, then he or she would have the results on the far left. Here the results would not be significant because the $p = 0.9945$ is far greater than the required $p < 0.05$.

This difference-of-proportions test requires data to be entered in what is called a wide format. Each group (treatment and control) is treated as a variable with the scores on the outcome variable coded under each group, as illustrated in the listing that appeared above. Data in statistics books and related exercises often present the scores this way.

When dealing with survey data, it is common to use what is called a long format in which one variable is a grouping variable of whether someone is in the treatment group, coded 1, or the control group, coded 0. The second variable is the score on the dependent variable, which is also a binary variable coded as 1 if the person is cured and 0 if the person is not cured. This appears in the following long-format listing (`long.dta`).

```
. list
```

|      | group | cure |
|------|-------|------|
| 1.   | 1     | 1    |
| 2.   | 1     | 0    |
| 3.   | 1     | 1    |
| 4.   | 1     | 1    |
| 5.   | 1     | 1    |
| 6.   | 1     | 1    |
| 7.   | 1     | 1    |
| 8.   | 1     | 0    |
| 9.   | 1     | 1    |
| 10.  | 1     | 1    |
| 11.  | 1     | 1    |
| 12.  | 1     | 1    |
| 13.  | 1     | 1    |
| 14.  | 1     | 0    |
| 15.  | 1     | 1    |

*(output omitted)*

|      | group | cure |
|------|-------|------|
| 36.  | 0     | 0    |
| 37.  | 0     | 0    |
| 38.  | 0     | 0    |
| 39.  | 0     | 0    |
| 40.  | 0     | 0    |

When your data are entered this way, you need to use a different test for the difference of proportions. Select Statistics ▷ Summaries, tables, and tests ▷ Classical tests of hypotheses ▷ Proportion test like we did before, but this time we select *Two-group using groups* in the *Tests of proportions* section. Type `cure` under *Variable name* (the dependent variable) and `group` under *Group variable name* (the independent variable). Click on OK to obtain the following results:

```
. prtest cure, by(group)
Two-sample test of proportions                      0: Number of obs =        20
                                                    1: Number of obs =        20
```

| Variable | Mean | Std. Err. | z | P>\|z\| | [95% Conf. Interval] | |
|---|---|---|---|---|---|---|
| 0 | .35 | .1066536 | | | .1409627 | .5590373 |
| 1 | .75 | .0968246 | | | .5602273 | .9397727 |
| diff | -.4 | .1440486 | | | -.6823301 | -.1176699 |
| | under Ho: | .1573213 | -2.54 | 0.011 | | |

```
        diff = prop(0) - prop(1)                                    z =  -2.5426
  Ho: diff = 0

  Ha: diff < 0                    Ha: diff != 0                    Ha: diff > 0
Pr(Z < z) = 0.0055        Pr(|Z| < |z|) = 0.0110          Pr(Z > z) = 0.9945
```

Here we have two means: one for the group coded 0 of 0.35 and one for the group coded 1 of 0.75. These are, of course, the means for our control group and our treatment group, respectively. Below this is a row labeled `diff`, the difference in proportions, which has a value of −0.40 because the mean of the group coded as 1 is subtracted from the mean of the group coded as 0. Be careful interpreting the sign of this difference. Here a negative value means that the treatment group is higher than the control group: 0.75 versus 0.35. The $z$ test for this difference is $z = -2.54$. This is the same absolute value that we had with the test for the wide format, but the sign is reversed. Be careful interpreting this sign just like with interpreting the difference. It happens to be negative because we are subtracting the group coded as 1 from the group coded as 0.

We can interpret these results, including the negative sign on the $z$ test, as follows. The control group has a mean of 0.35 (35% of participants were cured), and the treatment group has a mean of 0.75 (75% of the participants were cured). The $z = -2.54$, $p < 0.05$ indicates that the control group had a significantly lower success rate than the treatment group. Pay close attention to the way the difference of proportions was computed so that you interpret the sign of the $z$ test correctly.

This long form is widely used in surveys. For example, in the General Social Survey 2002 data (`gss2002_chapter7.dta`), there is an item, `abany`, asking if abortion is okay anytime. The response option is binary, namely, `yes` or `no`. If we wanted to see whether more women said yes than did men, we could use a difference-of-proportions test. The grouping variable would be `sex`, and the dependent variable would be `abany`. First, we must see how they are coded. To do this, we run the command `codebook abany sex`. We see that `abany` is coded as a 1 for yes and 2 for no. We must recode `abany` and make a new variable coded 0 for no and 1 for yes. We do not need to change the coding of `sex`. The independent variable has to be binary (just two values), but does not have to be coded as 0,1. Try this and do the difference of proportions test.

# 7.7   One-sample test of means

You can do $z$ tests for one-sample tests of means, or you can do $t$ tests. We will cover only the use of $t$ tests because these are by far the most widely used. A $z$ test is appropriate when you know the population variance (which it is highly unusual to know). Unless you have a small sample, both tests yield similar results.

Several decades ago, social theorists thought that the workweek for full-time employees would get shorter and shorter. This has happened in many countries but does not seem to be happening in the United States. A full-time workweek is defined as 40 hours. Among those who say they work full time, do they work more or less than 40 hours? The General Social Survey dataset (`gss2002_chapter7.dta`) has a variable called `hrs1`, which represents the reported hours in the workweeks of the survey participants. The null hypothesis is that the mean will be 40 hours. The alternative hypothesis is that the mean will not be 40 hours. Using a two-tailed hypothesis, you can say

Alternative hypothesis $H_a$: $\mu_{(hours)} \neq 40$

Null hypothesis $H_0$: $\mu_{(hours)} = 40$

Notice that we are using the Greek letter, $\mu$, to refer to the population mean in both the alternative and the null hypotheses. To open the dialog box for this test, select Statistics ▷ Summaries, tables, and tests ▷ Classical tests of hypotheses ▷ t test (mean-comparison test). In the resulting dialog box, select *One-sample* if it is not already selected in the *t tests* section of the Main tab. Enter `hrs1` (outcome variable) under the *Variable name* and 40 (hypothesized mean for the null hypothesis) under the *Hypothesized mean*.

Because we are interested in how many hours full-time employees work, we should eliminate part-time workers. To do this, click on the by/if/in tab, and type `wrkstat == 1` as the *If: (expression)*. The variable `wrkstat` is coded as 1 if a person works full time (you can run a `codebook wrkstat` to see the frequency distribution). The resulting by/if/in tab is shown in figure 7.1.

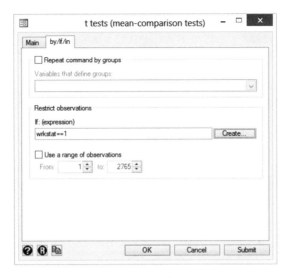

Figure 7.1. Restrict observations to those who score 1 on `wrkstat`

Clicking on OK produces the following command and results:

```
. ttest hrs1 == 40 if wrkstat == 1

One-sample t test
```

| Variable | Obs | Mean | Std. Err. | Std. Dev. | [95% Conf. Interval] |
|---|---|---|---|---|---|
| hrs1 | 1419 | 45.97111 | .3156328 | 11.88977 | 45.35195     46.59026 |

```
    mean = mean(hrs1)                                              t =   18.9179
Ho: mean = 40                                  degrees of freedom =       1418

   Ha: mean < 40                Ha: mean != 40                   Ha: mean > 40
 Pr(T < t) = 1.0000      Pr(|T| > |t|) = 0.0000          Pr(T > t) = 0.0000
```

From these results, we can see that the mean for the variable `hrs1` is $M = 45.97$ hours, and the standard deviation is SD = 11.89 hours. Because we are using a two-tailed test, we can look at the center column under the table to see that, for $t = 18.92$, $p < 0.001$. Thus full-time workers work significantly more than the traditional 40-hour workweek. In fact, they worked about 6 hours more than this standard on average in 2002. Remember, 2002 was a year of relatively full employment.

When we report a $t$ test, we need to also report degrees of freedom. Look closely at the results above. The number of observations is 1,419, but the degrees of freedom is 1,418 (found just below the $t = 18.92$). For a one-sample $t$ test, the degrees of freedom is always $N - 1$. If we report the degrees of freedom as 1,418, a reader will also know that the sample size is one more than this, or 1,419.

Different specialty areas report one-sample $t$ tests slightly differently. The format I recommend is $t(1418) = 18.918$, $p < 0.001$. This is read as saying that $t$ with 1,418 degrees of freedom is 18.918, $p$ is less than 0.001. The results show that the average full-time employee worked significantly more than the standard 40-hour week.

### Degrees of freedom

The idea of degrees of freedom was discussed in chapter 6 on cross-tabulations. For $t$ tests, the degrees of freedom is a little different. Your statistics book will cover this in detail, but here we will present just a simple illustration. Suppose that we have four cases with $M = 10$. The first three of the four cases have scores of 8, 12, and 12. Is the fourth case free? Because the mean of the four cases is 10, the sum of the four cases must be 40. Because $40 - 8 - 12 - 12 = 8$, the fourth case must be 8. It is not free. In this sense, we can say that a one-sample test of means has $N - 1$ degrees of freedom.

Those who like to use confidence intervals can see these in the results. Stata gives us a 95% confidence interval of 45.4 hours to 46.6 hours. The confidence interval is more informative than the $t$ test. We are 95% confident that the interval of 45.4 to 46.6 hours contains the mean number of hours full-time employees report working in a week. Because the value specified in the null hypothesis, $\mu = 40$, is not included in this interval, we know the results are statistically significant. The confidence interval also focuses our attention on the number of hours and helps us understand the substantive significance (importance), as well as the statistical significance.

## 7.8   Two-sample test of group means

A researcher says that men make more money than women because men work more hours a week. The argument is that a lot of women work part-time jobs, and these neither pay as well nor offer opportunities for advancement. What happens if we only consider people who say they work full time? Do men still make more than women when both the men and the women are working full time?

The General Social Survey 2002 dataset (`gss2002_chapter7.dta`) has a question asking about the respondent's income, variable `rincom98`. Like many surveys, the General Social Survey does not report the actual income but reports income in categories (for example, under $1,000, $1,000 to 2,999). Run a `tabulate` command to see the coding the surveyors used. For some reason, they have not defined a score coded as 24 as a "missing value", but this is what a code of 24 represents. Even with highly respected national datasets like the General Social Survey, you need to check for coding errors. A code of 24 was assigned to people who did not report their income. Many researchers have used `rincom98` as it is coded (I hope after defining a code of 24 as a missing value). However, this coding is problematic because the intervals are not equal. The first interval, under $1,000, is $1,000 wide, but the second interval, $1,000 to $2,999, is nearly $2,000 wide. Some intervals are as much as $10,000 wide.

Before we can compare the means for women and men, we need to recode the
rincom98 variable. We could recode it by using the dialog box as described in section 3.4, but let's type the commands instead. You could enter the following commands
in the Command window, one by one. A much better approach would be to enter them
into a do-file so that we can easily modify or reproduce the routine later if we need to.
Remember that we should add some comments at the top of the do-file that include the
name of the file and its purpose. For this example, you must use Stata/MP or Stata/SE
because the number of columns exceeds the 20-column limit of Stata/IC.

```
* recode income.do      (sample do-file)
* This is a short do-file that recodes income. It does a
* tabulation to see how income is coded (tab rincom98). People
* given a value of 24 are recoded as missing (mvdecode rincom98,
* mv(24)). We generate a new variable called inc that is equal to the
* old variable, rincom98. We recode each interval with its
* midpoint. We do a cross-tabulation of the new and old income
* variables as a check.
version 13
tabulate rincom98, missing
mvdecode rincom98, mv(24)
gen inc = rincom98
replace inc = 500    if rincom98 == 1
replace inc = 2500   if rincom98 == 2
replace inc = 3500   if rincom98 == 3
replace inc = 4500   if rincom98 == 4
replace inc = 5500   if rincom98 == 5
replace inc = 6500   if rincom98 == 6
replace inc = 7500   if rincom98 == 7
replace inc = 9000   if rincom98 == 8
replace inc = 11250 if rincom98 == 9
replace inc = 13250 if rincom98 == 10
replace inc = 16250 if rincom98 == 11
replace inc = 18750 if rincom98 == 12
replace inc = 21250 if rincom98 == 13
replace inc = 23750 if rincom98 == 14
replace inc = 27500 if rincom98 == 15
replace inc = 32500 if rincom98 == 16
replace inc = 37500 if rincom98 == 17
replace inc = 45000 if rincom98 == 18
replace inc = 55000 if rincom98 == 19
replace inc = 67500 if rincom98 == 20
replace inc = 82500 if rincom98 == 21
replace inc = 100000 if rincom98 == 22
replace inc = 110000 if rincom98 == 23
tabulate inc rincom98, missing
```

There are several lines commenting on what the do-file will do. The first line after
the version command runs a tabulation (tabulate), including missing values. The
results help us understand how the variable was coded. The next line makes the code of
24 into a missing value so that anybody who has a score on rincom98 of 24 is defined as
having missing values (mvdecode). The next line generates (gen) a new variable called
inc that is equal to the old variable rincom98. Following this command is a series of
commands to replace (replace) each interval with the value of its midpoint. Thus a
code of 8 for income98 is given a value of 9000 on inc. The final command does a cross-
tabulation (tabulate) of the two variables to check for coding errors. Economists and

demographers may not be happy with this coding system. Those who have a `rincom98` code of 1 may include people who lost a fortune, so substituting a value of 500 may not be ideal. The commands make no adjustment for these possible negative incomes. Those who have a code of 23 include people who make $110,000 but also may include people who make $1,000,000 or more. We hope that there are relatively few such cases at either end of the distribution, so the values we use here are reasonable.

Now that we have income measured in dollars, we are ready to compare the income of women and men who work full time by using a two-sample *t* test. Open the dialog box by selecting Statistics ▷ Summaries, tables, and tests ▷ Classical tests of hypotheses ▷ t test (mean-comparison test). In this dialog box, select *Two-sample using groups* in the *t tests* section of the **Main** tab because the data are arranged in the common long format. Type `inc` (outcome variable) as the *Variable name* and `sex` as the *Group variable name*. This dialog box is shown in figure 7.2.

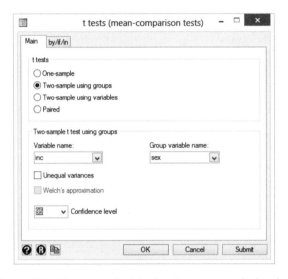

Figure 7.2. Two-sample *t* test using groups dialog box

Statistics books discuss assumptions for doing this *t* test, one of which is that the variance of the outcome variable, `inc`, is equal for both categories of the grouping variable, `sex`. That is, the variance in income is the same for women as it is for men. If we believed that the variances were unequal, we could click on the *Unequal variances* box, and Stata would automatically adjust everything accordingly.

We can click on **Submit** at the bottom of the dialog box at this point, and we will find a huge difference between women and men, with men making much more on average than women. However, remember that we only wanted to include people who work full time. To implement this restriction, click on the by/if/in tab in the dialog box. In the *Restrict observations* section, type `wrkstat == 1` (remember to use the double equal signs) in the *If: (expression)* box. Here are the results:

```
. ttest inc if wrkstat == 1, by(sex)

Two-sample t test with equal variances
```

| Group | Obs | Mean | Std. Err. | Std. Dev. | [95% Conf. Interval] | |
|---|---|---|---|---|---|---|
| male | 671 | 44567.81 | 1054.665 | 27319.7 | 42496.96 | 46638.66 |
| female | 589 | 33081.07 | 895.9353 | 21743.74 | 31321.45 | 34840.69 |
| combined | 1260 | 39198.21 | 718.7267 | 25512.27 | 37788.18 | 40608.25 |
| diff | | 11486.74 | 1404.217 | | 8731.874 | 14241.61 |

```
     diff = mean(male) - mean(female)                              t =    8.1802
Ho: diff = 0                                    degrees of freedom =     1258

    Ha: diff < 0                 Ha: diff != 0                 Ha: diff > 0
 Pr(T < t) = 1.0000       Pr(|T| > |t|) = 0.0000       Pr(T > t) = 0.0000
```

The layout of these results is similar to what we had for the one-sample $t$ test. Using a two-tailed hypothesis (center column, below the main table), we see that the $t = 8.1802$ has a $p < 0.001$ ($p = 0.0000$). The $N = 671$ men who work full time have an average income of just under \$44,568, compared with just over \$33,081 for the $N = 589$ women who are employed full time. Notice that the degrees of freedom, 1,258, is two fewer than the total number of observations, 1,260, because we used two means and lost one degree of freedom for each of them. For this two-sample $t$ test, we have $N - 2$ degrees of freedom. A good layout for reporting a two-sample $t$ test is $t(1258) = 8.18$, $p < 0.001$.

Is this result substantively significant? This question is pretty easy to answer because we all understand income. Men make about \$11,487 more on average than do women, and this is true even though both our men and our women are working full time. It is sometimes helpful to compare the difference of means with the standard deviations. The \$11,487 difference of means is roughly one-half of a standard deviation if we use the average of the two standard deviations as a guide. A difference of less than 0.2 standard deviations is considered a weak effect, a difference of 0.2 to 0.49 is considered a moderate effect, and a difference of 0.5 or more is considered a strong effect. If your statistics book covers the delta statistic, $\delta$, you can get a precise number, but here we have just eyeballed the difference.

So far, we have been using the group-comparison $t$ test, which assumes that the data are in the long format. The dependent variable `income` is coded with the income of each participant. `income` is compared across groups by `sex`, which is coded 1 for each participant who is a man and 2 for each participant who is a woman. If the data were arranged in the wide format, we would still have two variables, but they would be different. One variable would be the income for each man and would have 671 entries; the other variable would be the income for each woman and would have 589 entries. This would look like the wide format shown previously for comparing proportions, except that the variables would be called `maleinc` and `femaleinc`.

When our data are in a wide format, we again select Statistics ▷ Summaries, tables, and tests ▷ Classical tests of hypotheses ▷ t test (mean-comparison test) to open the dialog box; however, we now select *Two-sample using variables* in the *t tests* section of the Main tab. We simply enter the names of the two variables, `maleinc` and `femaleinc`. The resulting command would be `ttest maleinc == femaleinc, unpaired`. I cannot illustrate this process here because the data are in the long format. If you are interested, you can use Stata's `reshape` command to convert between formats; see `help reshape`.

**Effect size**

There are two measures of effect size that are sometimes used to measure the strength of the difference between means. These are $R^2$ and Cohen's $d$ ($\delta$). At the time this book was written, $R^2$ is not directly computed by Stata, but it can be computed using Stata's built-in calculator. The formula is $R^2 = t^2/(t^2 + \mathrm{df})$. Using the results of the two-sample $t$ test comparing income of women and men, $R^2 = 8.1802^2/(8.1802^2 + 1258)$. We can compute this with a hand calculator or with the Stata command

```
. display "r-squared = " 8.1802^2/(8.1802^2 + 1258)
r-squared = .05050561
```

The square root of this value is sometimes called the point biserial correlation. A value of 0.01 to 0.09 is a small effect, a value of 0.10 to 0.25 is a medium effect, and a value of over 0.25 is a large effect. If you use the Stata calculator with a negative $t$, it is important to insert parentheses correctly so that Stata does not see the negative sign as making the $t^2$ a negative value. If we had a $t = -4.0$ with 100 degrees of freedom, the Stata command would be

```
. display "r-squared = " (-4.0)^2/((-4.0)^2 + 100)
```

or you could simply use the absolute value of $t$ when doing the calculations.

Cohen's $d$ measures how much of a standard deviation separates the two groups.

$$\text{Cohen's } d = \frac{\text{mean difference}}{\text{pooled standard deviation}}$$

$$s_p = \sqrt{\frac{(N_1 - 1)s_1^2 + (N_2 - 1)s_2^2}{\mathrm{df}}}$$

Stata does have an effect-size command to compute Cohen's $d$, `esize`. This command has a similar structure to the two-sample `ttest` command: `esize twosample inc if wrkstat==1, by(sex)`. This command results in Cohen's $d = 0.461$ and also reports a 95% confidence interval. When you read a study that reports the means and standard deviations for each of two groups but does not report Cohen's $d$, you can compute the $d$ using an immediate form of `esize`, that is, with the `esizei` command. For example, you can type `esizei` #obs1 #mean1 #sd1 #obs2 #mean2 #sd2. Suppose we read that in a study with 100 participants in each group the means for the two groups were 60.2 and 65.3 and their respective standard deviations were 9.0 and 10.1. For our example, we can type `esizei 100 60.2 9.0 100 65.3 10.1`. Stata will report that the effect size is Cohen's $d = -0.533$. You need to be careful to interpret the sign. In our example, group two has a larger mean, 65.3, than group one, 60.2. Hence, group one's mean is less than group two's mean. You can type `help esize` to find other applications of the effect-size command.

## 7.8.1   Testing for unequal variances

In the dialog box for the group-comparison $t$ test, we did not click on the option that would adjust for unequal variances. If the variances or standard deviations are similar in the two groups, there is no reason to make this adjustment. However, when the variances or standard deviations differ sharply between the two groups, we may want to test whether the difference is significant. Before doing the $t$ test, some researchers will do a test of significance on whether variances are different. This test is problematic because the test of equal variances will show small differences in the variances to be statistically significant in large samples (where the lack of equal variance is less of a problem), but the test will not show large differences to be significant in small samples (where unequal variance is a bigger problem). To open the dialog box, select **Statistics** ▷ **Summaries, tables, and tests** ▷ **Classical tests of hypotheses** ▷ **Variance-comparison test**. Here we select *Two-sample using groups* in the *Variance-comparison tests* section. We enter our dependent variable, `inc`, and the grouping variable, `sex`, just like we did for the $t$ test. We also should use the **by/if/in** tab to restrict the sample to those who work full time, `wrkstat == 1`. The command and results are

```
. sdtest inc if wrkstat == 1, by(sex)
Variance ratio test
```

| Group | Obs | Mean | Std. Err. | Std. Dev. | [95% Conf. Interval] | |
|---|---|---|---|---|---|---|
| male | 671 | 44567.81 | 1054.665 | 27319.7 | 42496.96 | 46638.66 |
| female | 589 | 33081.07 | 895.9353 | 21743.74 | 31321.45 | 34840.69 |
| combined | 1260 | 39198.21 | 718.7267 | 25512.27 | 37788.18 | 40608.25 |

```
     ratio = sd(male) / sd(female)                                    f =    1.5786
Ho: ratio = 1                                        degrees of freedom = 670, 588

    Ha: ratio < 1                  Ha: ratio != 1                    Ha: ratio > 1
  Pr(F < f) = 1.0000        2*Pr(F > f) = 0.0000           Pr(F > f) = 0.0000
```

This test produces an $F$ test that tests for a significant difference in the variances (although the output reports only standard deviations). The $F$ test is simply the ratio of the variances of the two groups. Just below the table, we can see on the left side that the null hypothesis is `Ho: ratio = 1`. Below the table on the right side, we see that the `f = 1.5786` (note that this is the ratio of the variances, i.e., $27319.7^2/21743.74^2$) and we see two degrees of freedom, `degrees of freedom = 670, 588`. The first number, 670, is the degrees of freedom for males ($N_1 - 1$), and the second number is the degrees of freedom for females ($N_2 - 1$). The variance for males is 1.58 times as great as the variance for females. In the middle bottom of the output, we can see the two-tailed alternative hypothesis, `Ha: ratio != 1`, and a reported probability of 0.0000. Because the $F$ test has two numbers for degrees of freedom, we would report this as $F(670, 588) = 1.58, p < 0.001$.

Although the variances are significantly different, remember that this test is sensitive to large sample sizes, and in such cases, the assumption is less serious. If the groups have roughly similar sample sizes and if the standard deviations are not dramatically different, most researchers choose to ignore this test. If you want to adjust for unequal variances, however, look back at the dialog box for the two-sample $t$ test. All you need to do is check the box for *Unequal variances*. When you get the results, the degrees of freedom are no longer $N - 2$ but are based on a complex formula that Stata will compute for you.

## 7.9   Repeated-measures t test

The repeated-measures $t$ test goes by many names. Some statistics books call it a repeated-measures $t$ test, some call it a dependent-sample $t$ test, and some call it a paired-sample $t$ test. A repeated-measures $t$ test is used when one group of people is measured at two points. An example would be an experiment in which you measured everybody's weight at the start of the experiment and then measured their weight a second time at the end of the experiment. Did they have a significant loss of weight? As a second example, we may want to know if it helps to retake the GRE. How much better do students perform when they repeat the GRE? We have one group of students, but we measure them twice—once the first time they take the GRE and again the second time they take it. The idea is that we measure the group on a variable, something happens, and then we measure them on that variable a second time. A control group is not needed in this approach because the participants serve as their own control.

An alternative use of the repeated-measures $t$ test is to think of the group as being related people, such as parents. Husbands and wives would be paired. An example would be to compare the time a wife spends on household chores with the time her husband spends on household chores. Here each husband and each wife would have a score on time spent for the wife and a score on time spent for the husband.

Who spends more time on chores—wives or husbands? Here we do not measure time spent on chores twice for the same person, but we measure it twice for each related pair of people. The way the data are organized involves having two scores for each case. With paired data, the case consists of two related people (see `chores.dta`). Here is what five cases (couples) might look like:

```
. list

     | husband    wife |
     |-----------------|
  1. |      11      31 |
  2. |      10      40 |
  3. |      21      44 |
  4. |      15      36 |
  5. |      12      29 |
```

We will illustrate the related-sample $t$ test with data from the General Social Survey 2002 dataset `gss2002_chapter7.dta`. Each participant was asked how much education his or her mother and his or her father had completed. This survey is not like the example of mothers and fathers because we have two measures for each participant. We know how much education each person reports for his or her mother and for his or her father. Do these respondents report more education for their mothers or for their fathers? Here we will use a two-tailed test with a null hypothesis that there is no difference on average and an alternative hypothesis that there is a difference:

Alternative hypothesis: $H_a$: $\mu_{\text{diff}} \neq 0$

Null hypothesis: $H_0$: $\mu_{\text{diff}} = 0$

Remembering that we use $M$ for a sample mean and $\mu$ for a population mean, we express these hypotheses as mean differences in the population ($\mu_{\text{diff}}$). To open the dialog box, select Statistics ▷ Summaries, tables, and tests ▷ Classical tests of hypotheses ▷ t test (mean-comparison test). This is the same dialog box that we have been using, but this time we select *Paired* in the *t tests* section. We enter the two variables. It does not really matter which we enter as the *First variable* and which as the *Second variable*, but we need to remember this order because changing the order will reverse the signs on all the differences. Type `paeduc` as the *First variable* and `maeduc` as the *Second variable*. Remember that different statistics books will call this test the repeated-measures $t$ test, paired-sample $t$ test, or dependent-sample $t$ test. All of these tests are the same.

Here are the command and results:

```
. ttest paeduc == maeduc

Paired t test
```

| Variable | Obs | Mean | Std. Err. | Std. Dev. | [95% Conf. Interval] | |
|---|---|---|---|---|---|---|
| paeduc | 1903 | 11.42091 | .0917118 | 4.000779 | 11.24105 | 11.60078 |
| maeduc | 1903 | 11.58276 | .0781763 | 3.410314 | 11.42944 | 11.73608 |
| diff | 1903 | -.1618497 | .0729315 | 3.181518 | -.3048838 | -.0188156 |

```
       mean(diff) = mean(paeduc - maeduc)                          t =  -2.2192
 Ho: mean(diff) = 0                               degrees of freedom =      1902

 Ha: mean(diff) < 0            Ha: mean(diff) != 0            Ha: mean(diff) > 0
 Pr(T < t) = 0.0133        Pr(|T| > |t|) = 0.0266          Pr(T > t) = 0.9867
```

Mothers have more education on average than fathers. The mean for mothers is $M = 11.58$ years, compared with $M = 11.42$ years for fathers. Before going further, we need to ask if this difference (father's education − mother's education = −0.16) is important. Because we entered father's education as the first variable and mother's education as the second variable, the negative difference indicates that the mothers had slightly more education than did the fathers. However, 0.16 years of education would amount to $0.16 \times 12 = 1.92$ months of education, which does not sound like an important difference, even though it is statistically significant.

Just below the table, on the left, is the null hypothesis that there is no difference, Ho: mean(diff) = 0, and on the right is the *t* test, t = -2.2192, and its degrees of freedom, 1,902. The middle column under the main table has the alternative hypothesis, namely, that the mean difference is not 0. The $t = -2.22$ has a $p < 0.05$; thus, the difference is statistically significant.

This example illustrates a problem with focusing too much on the tests of significance. With a large sample, almost any difference will be statistically significant; it still may not be important. We have seen that this difference represents fewer than 2 months of education ($M_{diff} = 1.92$ months), so it might be fair to say that we have a statistically significant difference that is not substantively significant.

## 7.10   Power analysis

You can think of the power of a test as its ability to do what you want to do. Normally, using your sample, you want to reject the null hypothesis in favor of the alternative hypothesis when the alternative hypothesis is correct in the larger population. If the power of a test is 0.20, this means that you have only a 20% chance of rejecting the null hypothesis when the alternative hypothesis is correct. Before a researcher begins collecting data to do a study, he or she should have a good idea of the power of the tests that will be run. A pharmaceutical company would not want to fund a $1,000,000 study that had only a 20% chance of finding a statistically significant result for a new drug that actually was effective. Nor would a thesis committee want to support such a study. Certainly, you would not want to invest your own time in such a project because it is virtually doomed to failure.

Power analysis can be done a priori, that is, before the study is done, or a posteriori, that is, after the study is done. It makes the most sense to do power analysis before you do the study so that you are confident that your sample size will be sufficient to have adequate power. However, the procedures for doing power analysis can be reversed so that they are done after a study is completed. Doing power analysis a posteriori is controversial but may be appropriate in some cases. Suppose that you read a study that had a very small sample and, in spite of having means that were very different, did not obtain a statistically significant alpha. Was this because the means really were not different in the population as specified by the null hypothesis? Or was this because the means in the population really were different as specified by the alternative hypothesis, but the sample size or alpha level was too small to have a powerful test? A posteriori power analysis could tell you that given the study design, say, a two-sample *t* test a sample size, say, $n_1 = n_2 = 10$; and an alpha value, say, 0.01, the power was extremely low. This information would help you evaluate the importance of the statistically insignificant finding. With Stata, we can do either an a priori or an a posteriori power analysis.

There are seven things we need to know or estimate to do a power analysis for a $t$ test:

1. Is it a repeated or paired $t$ test, or are there two independent samples?

2. Is it a one-tailed or two-tailed test?

3. What alpha value is used? Usually, $\alpha = 0.05$.

4. How big of a difference do we need for it to be considered a substantively significant difference? This is called the critical effect size (Cohen 1988).

5. How much power do we need? Usually, a power of 0.80 is minimally adequate for exploratory research, but a power of 0.90 or 0.95 may be required in highly critical areas.

6. If there are two independent samples, what is the ratio of the $N$s? (Do we have the same number of participants in each group? Do we have twice as many participants in one group as in the other?)

7. We need an estimate of the standard deviation of the two groups.

We have already covered most of these requirements, but the idea of an effect size is new. The effect size is simply how much difference we need to be able to consider it substantively significant. There are many measures of effect size, depending on what test you are using. We will illustrate effect size with a measure called Cohen's $d$ that is used when there are two independent samples. Cohen's $d$ is calculated using

$$d = \frac{M_t - M_c}{\text{SD}_{\text{pooled}}}$$

where

$$\text{SD}_{\text{pooled}} = \sqrt{\frac{\text{SD}_t^2(N_t - 1) + \text{SD}_c^2(N_c - 1)}{N_c + N_t - 2}}$$

and where $M_t$ and $M_c$ are the means for the treatment group and the control group, respectively; $\text{SD}_t^2$ and $\text{SD}_c^2$ are the square of the respective standard deviations; and $N_t$ and $N_c$ are the respective sample sizes.

Cohen's $d$ tells us how far the two means need to be separated for the results to be substantively significant. Sometimes we do not need the pooled standard deviation that is in the denominator to say a difference is substantively significant. If, in comparing women and men on income, we said the mean for men was \$90,000 and the mean for women was \$50,000, we would all be satisfied that this was a big difference. Other times, our variables are measured in ways where it is hard to compare the two means directly.

Suppose we said that one group had a mean score on depression of 49 and the other group had a mean score of 40. Is this an important difference? One way to judge the difference is in terms of how many standard deviations separate the two groups. This is precisely what Cohen's *d* does. If we have a mean of 100 in one group and a mean of 115 in another group and the pooled standard deviation is 15, then we would say that the groups are one standard deviation apart. This would be a very big difference. Imagine two people, one with an IQ of 100 and the other with an IQ of 115, where the standard deviation for IQ scores is about 15. These two people are about one standard deviation apart. With an IQ of 100, a person should be able to complete high school without much trouble but might have trouble completing a college degree. The person with an IQ of 115 would be able to complete college without much trouble. A difference of one standard deviation is definitely a big difference.

When Cohen defined his Cohen's *d*, he said that a $d = 0.2$ would be considered a small effect, a $d = 0.5$ would be a moderate effect, and a $d = 0.8$ would be a large effect. The following three figures give you a visual of these three *d* values. Figure 7.3a has a Cohen's *d* of 0.2 and shows the distribution of two groups. Clearly, there is a lot of overlap and this is a small effect. Those in the treatment group are the distribution under the solid line and those in the control group are the distribution under the dashed line.

Figure 7.3b shows a medium effect, where there is still a lot of overlap. Even in figure 7.3c, where the effect is described as large, there is a lot of overlap. Indeed, the majority of people under the control-group distribution would also fall under the treatment-group distribution, and vice versa.

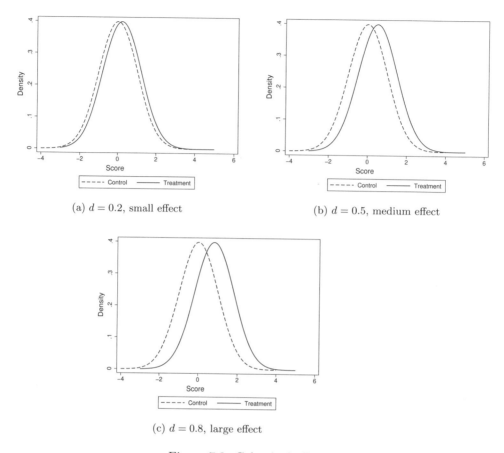

(a) $d = 0.2$, small effect

(b) $d = 0.5$, medium effect

(c) $d = 0.8$, large effect

Figure 7.3. Cohen's $d$ effects

You can see how Cohen's $d$ is important to power. It would take a very large sample to show a statistically significant difference between the two groups if you insist on being able to detect a small effect. It would take a much smaller sample to be able to detect a statistically significant effect for the large effect.

To estimate power, we do not need to first estimate the effect size, but we need to decide whether we want to detect a small, medium, or large effect. If a cost-benefit analysis shows that a small difference would be important, then we need a big enough sample to detect a small effect. If any improvement or any difference between the null and the alternative hypothesis would be important, then a small effect size is what you have. Otherwise, you may only be interested in demonstrating a difference if the effect size is moderate or large. This is a decision you need to make.

Looking at the equation defining Cohen's $d$, we need to estimate the pooled standard deviation, which is often challenging to do. Suppose that you have an intervention designed to reduce depression. You know that the depression score runs from 0 to 50.

Reading literature where this scale has been used, you may find that several authors reported standard deviations for their samples. You could use one of those values as your standard deviation. If you are unable to find a published standard deviation, you might run a pretest on 30 people and compute the standard deviation for this sample. If you cannot do that, you can make an intelligent guess. A good guess would be the interquartile range divided by 1.35, but you are not likely to know the interquartile range. So here is a very crude way to guess what the standard deviation is. We know that 4 standard deviations covers about 95% of a normal distribution. We might think that very few people would score below 5 or above 45 on our 0 to 50 scale of depression. If that is the case, then perhaps about 95% of the people would have a score between 5 and 45. Our guess of the standard deviation would be $(45 - 5)/4 = 10$. Then we can use this value of 10 in our estimation of power.

We also need some way to estimate the means for the two groups. Again, if we have any past research or a pilot study, we might get good estimates of the mean. But let's suppose that we have no available information to help us estimate the means. We can use Cohen's $d$ to give us a difference of means. The difference of means, $(M_c - M_t)/10$, is 0.2 for a small effect size, 0.5 for a medium effect size, and 0.8 for a large effect size. Thus $\text{SD} \times 0.2 = 10 \times 0.2 = 2$ for a small difference of means, $10 \times 0.5 = 5$ for a medium difference of means, and $10 \times 0.8 = 8$ for a large difference of means.

If we want to be able to detect a small difference of means on a depression scale that ranges from 0 to 50, we could pick as examples reasonable values that are 2 points apart, for example, 35 and 37, 30 and 32, or 38 and 40. If we have a large effect size, then we would pick means that are within our scale range and are 8 points apart, such as 30 and 38, 25 and 33, or 35 and 43.

With this background on picking a mean and a standard deviation, we are ready for our power analysis. Here is what we know or assume:

1. Two-sample comparison of means (two independent samples)

2. Two-tailed test

3. $\alpha = 0.05$

4. Small effect size, $M_c = 35$, $M_t = 37$; $\text{SD} = 10$

5. Desired power $= 0.90$

6. $N_c = N_t$, so the ratio is 1

7. SD for both groups approximately 10

Beginning with Stata 13, the new **power** command can be used to estimate our two-sample means test. Let's open the dialog box for power analysis; select Statistics ▷ Power and sample size. A control panel with many options opens instead of a dialog box like we have previously seen; see figure 7.4. From this control panel, we can easily open any **power** dialog box.

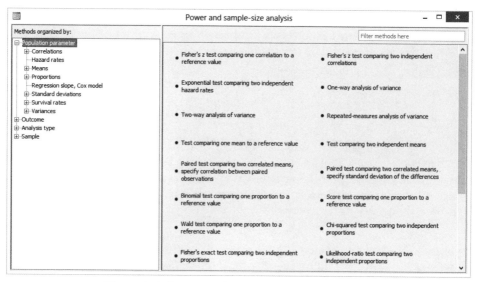

Figure 7.4. Power and sample-size control panel

The left pane organizes the methods, and the right pane displays the methods corresponding to the selection in the left pane. Because we are comparing two means from independent samples, we will use the **power twomeans** command. To open the related dialog box, select Population parameter ▷ Means from the left pane, and then select Test comparing two independent means from the right pane. After you read the rest of this section and perhaps the *Stata Power and Sample-Size Reference Manual* (StataCorp 2013b), you will be able to apply the following ideas to other types of power analysis. For now, we will focus on the commands rather than using the menu system.

How many people do we need, combined across the two groups, to have a power of 0.80, assuming our means are $M_1 = 35$ and $M_2 = 37$, $\text{SD}_1 = \text{SD}_2 = 10$? Because both standard deviations are assumed to be equal, we will enter only one of them. (If we had reason to believe the standard deviations were different, say, $\text{SD}_1 = 10$ and $\text{SD}_2 = 15$, our command would be **power twomeans 35 37, sd1(10) sd2(10) power(0.80)** and the Satterthwaite's $t$ test assuming unequal variances adjustment would be made). Here are the command and results:

```
. power twomeans 35 37, sd(10) power(0.80)

Performing iteration ...

Estimated sample sizes for a two-sample means test
t test assuming sd1 = sd2 = sd
Ho: m2 = m1   versus   Ha: m2 != m1

Study parameters:
            alpha =     0.0500
            power =     0.8000
            delta =     2.0000
               m1 =    35.0000
               m2 =    37.0000
               sd =    10.0000

Estimated sample sizes:
                N =        788
    N per group =        394
```

In the above command, 35 37 are our guesses on what the means might be, sd(10) is our guess on what the standard deviation might be in both groups, and power(0.08) is provided as is commonly done in power analysis because a power of 0.08 is considered adequate in most applications. How do you pick the mean values for the command? These should be values that represent the minimum difference that would be clinically important. As explained above, this combinations of means and standard deviations would represent a small effect size of $d = 0.20$, that is, $(35 - 37)/10$. We could use any values for the means that are about a fifth of a standard deviation apart if a small effect size would have clinical or substantive importance.

The results indicate that we are using a $t$ test and the $SD_1 = SD_2$. We are using a two-tailed test (the default) because our alternative hypothesis is that the two means are not equal (that is, != is read as not equal). The results show our assumed alpha is 0.05 (the default). If you wanted a different alpha, say, we wanted alpha $< 0.001$, we would add the option alpha(0.001). We have specified a power of 0.80 with $M_1 = 35$ and $M_2 = 37$. Both samples are assumed to have a standard deviation of 10. The results also show the estimated sample sizes of $N = 788$ with 394 people in each of our groups.

If we had typed power twomeans 35 37, sd(10) power(0.80) onesided, Stata would have estimated the sample sizes based on an alternative hypothesis of $M_2 > M_1$ for a one-sided test. Try this, and you should get a total $N = 620$ with 310 people needed in each group.

Sometimes, it is much easier to obtain participants for the control group than it is for the intervention group or vice versa. Suppose you wanted a power of 0.90 and you believe you can get three people in the control group for every person in the intervention group. Unequal sample size can lead to misleading results, especially when the variances are also unequal, but let's estimate the power where the ratio of the control group to the intervention group is three to one. We use the nratio(3) option. This will give us the total sample size we need and the size we need in each group with the restriction that we have three times as many participants in the control group as we have in the intervention group:

```
. power twomeans 35 37, sd(10) power(0.80) nratio(3)

Performing iteration ...

Estimated sample sizes for a two-sample means test
t test assuming sd1 = sd2 = sd
Ho: m2 = m1   versus   Ha: m2 != m1

Study parameters:

        alpha =     0.0500
        power =     0.8000
        delta =     2.0000
           m1 =    35.0000
           m2 =    37.0000
           sd =    10.0000
        N2/N1 =     3.0000

Estimated sample sizes:
          N =        1052
         N1 =         263
         N2 =         789
```

Suppose you wanted to know how much power you would have if you could afford a sample of 1,000 people with 500 in each group.

```
. power twomeans 35 37, sd(10) n(1000)

Estimated power for a two-sample means test
t test assuming sd1 = sd2 = sd
Ho: m2 = m1   versus   Ha: m2 != m1

Study parameters:

        alpha =     0.0500
            N =        1000
  N per group =         500
        delta =     2.0000
           m1 =    35.0000
           m2 =    37.0000
           sd =    10.0000

Estimated power:
        power =     0.8848
```

The results show that if you had 1,000 people with 500 in each group you would have a power of 0.88.

In some applications, you have little control over your group size. You might want to know how great the difference would need to be to give you a power of 0.80 when you know you will have 100 people in each of your groups. We would only specify one of the group means, and Stata will give us the value the other group mean would need to be to have the power of 0.80 with 200 people divided so that 100 of them were in each group.

```
. power twomeans 35, sd(10) power(0.80) n(200)

Performing iteration ...

Estimated experimental-group mean for a two-sample means test
t test assuming sd1 = sd2 = sd
Ho: m2 = m1  versus  Ha: m2 != m1; m2 > m1

Study parameters:
            alpha =     0.0500
            power =     0.8000
                N =        200
    N per group =        100
               m1 =    35.0000
               sd =    10.0000

Estimated effect size and experimental-group mean:
            delta =     3.9814
               m2 =    38.9814
```

The results show that the group mean for the treatment group would need to be 38.98 or more to have a power of 0.80 with 100 participants in each group and both groups having a standard deviation of 10.00. You could convert this to a Cohen's $d$ measure of effect size, $d = (38.981 - 35.000)/10 = 0.398$. You would have a power of 0.80 to detect a difference with this effect size. If only a moderate ($d = 0.50$) or large ($d = 0.80$) effect would be clinically important, then this study should be alright.

It is generally not recommended to do a power analysis after you have run an experiment. However, this may sometimes be useful. Some researchers conduct poorly designed studies that are underpowered. That is, there could be a clinically important difference between the means, but because the sample size was too small, the difference is not statistically significant. So far, we have been using a fairly small difference $37 - 35 = 2$. This is a difference of just a fifth of a standard deviation, $d = 2/10 = 0.20$. Suppose there was a substantial difference of half a standard deviation. The treatment group would need to have a mean of 40 in this case, and the difference would be $d = 5/10 = 0.50$ or half a standard deviation.

Unfortunately, the researchers did not do a power analysis prior to conducting their study and settled on having just 10 participants in the control group and 10 in the treatment group for a total $N = 20$. Did they have a reasonable power to detect a substantial difference of half a standard deviation? Let's see.

```
. power twomeans 35 40, n1(10) n2(10) sd(10)

Estimated power for a two-sample means test
t test assuming sd1 = sd2 = sd
Ho: m2 = m1   versus   Ha: m2 != m1

Study parameters:
        alpha =     0.0500
            N =         20
           N1 =         10
           N2 =         10
        delta =     5.0000
           m1 =    35.0000
           m2 =    40.0000
           sd =    10.0000

Estimated power:
        power =     0.1851
```

Here we specify the means, 35 and 40, and then for the options we specify the *N*'s for each group and the standard deviation. The result is `power = 0.185`. Remember what power means: the researchers had only an 18.5% chance of finding a significant difference given that there was a substantial difference of half a standard deviation. I would not want to reject the efficacy of the intervention the researchers found to be insignificance because their design was grossly underpowered.

Power analysis is highly dependent on the assumptions you make:

- What is a clinically important difference?

- What is the standard deviation of each group?

- How much power do I want to have?

- What alpha level am I using?

If you make small changes in how you answer any of these questions, you will get very different results. We used an effect size of $d = 0.20$ with an $SD_1 = SD_2 = 10.0$ to select our means of 35 and 37. Try slightly different means, say, 35 and 36 or 35 and 38. These small changes in the minimum difference you consider clinically important make a huge difference in the sample sizes you would need. Increasing your desired power has a similar effect. You can obtain the results by typing

```
. power twomeans 35 (36 37 38), sd(10) power(0.80)
```

The positive value of power analysis is it helps us to plan our studies and to justify what we are doing when we are requesting funding for our project. If a funding agency considers a small effect to be substantively important, they will not want to fund us if we have too small of a sample to detect such an effect. On the other hand, if we request funding for a very large sample when we have adequate power for a much smaller sample, the funding agency will not want to waste its money on the excessively large sample.

# 7.11 Nonparametric alternatives

The examples in this chapter for *t* tests have assumed that the outcome or dependent variable is measured on at least the interval level. The examples also assumed that both groups have equal variances in their populations and are normally distributed. The tests we have covered are remarkably robust against violations of these assumptions, but sometimes we want to use tests that make less-challenging assumptions. Nonparametric alternatives to conventional *t* tests are such tests. These tests have some limitations of their own, including being somewhat less powerful than the *t* test. Here are examples of nonparametric alternatives available in Stata.

## 7.11.1 Mann–Whitney two-sample rank-sum test

If we assume that we have ordinal measurement rather than interval measurement, we can do the rank-sum test. We can test the hypothesis that two independent samples (that is, unmatched data) are from a population distribution by using the Mann–Whitney two-sample rank-sum test. The rank-sum test works for both ordinal and interval data.

This test involves a simple idea. It combines the two groups into one group and ranks the participants from top to bottom. The highest score gets a rank of $N$ (8,871 in this example), and the lowest score gets a rank of 1. If one group has higher scores than the other group, that group should have a predominance of higher ranks. If girls report that more of their friends smoke, they should have more of the top ranks, and boys should have more of the lower ranks. If girls and boys have the same distribution, then both groups should have about the same number of high ranks and low ranks. In other words, the sum of the girls' ranks should be about the same as the sum of the boys' ranks.

The test computes the sum of ranks for each group and compares this with what we would expect by chance. Using data from the National Longitudinal Survey of Youth 1997 (`nlsy97_chapter7.dta`), we can compare the answers girls and boys give on how many of their friends smoke. They were asked what percentage of their friends smoke, and they answered in broad categories that were coded 1–5. Treating these categories as ordinal, we can do a rank-sum test. I will not show the dialog box for this, but the command and results are

```
. ranksum psmoke97, by(gender97)

Two-sample Wilcoxon rank-sum (Mann-Whitney) test
     gender97 |      obs    rank sum    expected
 -------------+---------------------------------
            1 |     4540    19130393    20139440
            2 |     4331    20221363    19212316
 -------------+---------------------------------
     combined |     8871    39351756    39351756

unadjusted variance     1.454e+10
adjustment for ties    -7.360e+08
                      -----------
adjusted variance       1.380e+10

Ho: psmoke97(gender97==1) = psmoke97(gender97==2)
             z =  -8.589
    Prob > |z| =   0.0000
```

This output shows that there is a significant difference of $z = -8.589$. We should also report the medians and means so that the direction of this difference is clear. The rank test compares the entire distribution rather than a particular parameter, such as the mean or median.

## 7.11.2   Nonparametric alternative: Median test

Sometimes you will want to compare the medians of two groups rather than the means. Whereas the Mann–Whitney rank-sum test compares entire distributions, the median test performs a nonparametric $K$-sample test on the equality of medians. It tests the null hypothesis that the $K$ samples were drawn from populations with the same medians. In the case of two samples ($K = 2$), a chi-squared statistic is calculated with and without a continuity correction.

Ideally, there would be no ties, and a median could be identified that had exactly 50% of the observations above it and 50% below it. With this example, the dependent variable is on a 1-to-5 scale, and there are many ties; for example, hundreds of students have a score of 2. This means that there is not an equal number of observations that are above and below the median for either group. I will not illustrate the dialog box for this command, but the command and results are

```
. median psmoke97, by(gender97) exact medianties(split)
Median test

    Greater
   than the    youth gender 1997
    median         1           2    |    Total
  ------------+--------------------+----------
        no       2,904       2,470       5,374
        yes      1,636       1,861       3,497
  ------------+--------------------+----------
     Total       4,540       4,331       8,871

              Pearson chi2(1) =  44.6269    Pr = 0.000
              Fisher's exact =                   0.000
      1-sided Fisher's exact =                   0.000

     Continuity corrected:
              Pearson chi2(1) =  44.3371    Pr = 0.000
```

The median test makes sense for comparing skewed variables, such as income. We would report the example as $\chi^2(1) = 44.63$, $p < 0.001$. If we set up a hypothesis, we would reject the null hypothesis, $H_0$: $\text{Med}_1 = \text{Med}_2$, in favor of the alternative hypothesis, $H_a$: $\text{Med}_1 \neq \text{Med}_2$.

## 7.12  Summary

The tests presented in this chapter are among the most frequently used statistical tests. As you read this chapter, you may have thought of applications in which you can use a $t$ test or a $z$ test for testing individual means and proportions or for testing differences. These are extremely useful tests for people who work in an applied setting and need to make decisions based on statistics. It is becoming a requirement to demonstrate that any new program has advantages over older programs; it is not enough to say that you believe that the new program is better or that you can "see" the difference. You need to find appropriate outcomes, such as reading readiness, retention rate, loss of weight, increased skill, or participant satisfaction. Then you need to show that the new program has a statistically significant influence on the outcomes you select. If you are designing an exercise program for older adults, there is not much reason to implement your program unless you can demonstrate that it has a good retention rate, that participants are satisfied with the program, and that behavioral outcome goals are met.

In this chapter, we discussed statistical significance and substantive significance, both of which are extremely important. If something is not statistically significant, we do not have confidence that there is a real effect or difference; what we observed in our data may be something that could have happened just by chance. There are two major reasons why a result may not be statistically significant. The first is that the result represents a small effect where there is little substantive difference between the groups. The second is that we have designed our study with too few observations to show significance, even when the actual difference is important. Therefore, I introduced the basic Stata commands to estimate the sample size needed to do a study with sufficient power to show that an important result is statistically significant. We only touched the surface of what Stata can do with power analysis, and I encourage you to check the Stata reference manuals for more ways to estimate power and sample-size requirements.

When we have statistical significance, we are confident that what we observed in our data represents a real effect that should not be attributed to chance. It is still essential to evaluate how substantively significant the result is. With a large sample, we may find a statistically significant difference of means when the actual difference is small and substantively insignificant. Finding a significant result begs the question of how important the result is. A statistically significant result may or may not be large enough to be important.

Finally, we learned briefly about nonparametric alternatives to $z$ tests and $t$ tests. These alternatives usually have less power but may be more easily justified in the assumptions they make.

Sometimes we have more than two groups, and the $t$ test is no longer adequate. Chapter 9 discusses analysis of variance (ANOVA), which is an extension of the $t$ test that allows us to work with more than two groups. Before we do analysis of variance, however, we will cover bivariate correlation and regression in chapter 8.

# 7.13   Exercises

When doing these exercises, you should create a do-file for each assignment; you might name the do-file for the first exercise c7_1.do. Put these do-files in the directory where you are keeping the Stata programs related to this book. Having these do-files will be useful in the future if you need to redo one of these examples or want to do a similar task. For example, you could use the do-file for the second exercise anytime you need to do randomization of participants in a study.

1. According to the registrar's office at your university, 52% of the students are women. You do a web-based survey of attitudes toward student fees supporting the sports program. You have 20 respondents, 14 of whom are men and 6 of whom are women. Is there a gender bias in your sample? To answer this, create a new dataset that has one variable, namely, sex. Enter a value of 1 for your first 14 observations (for your 14 males) and a value of 0 for your last 6 observations (for your 6 females). Your data will have one column and 20 rows. Then do a one-sample $z$ test against the null hypothesis that $p = 0.52$. Explain your null hypothesis. Interpret your results.

2. You have 30 volunteers to participate in an exercise program. You want to randomly assign 15 of them to the control group and 15 to the treatment group. You list them by name—the order being arbitrary—and assign numbers from 1 to 30. What are the Stata commands you would use to do the random assignment (randomization without replacement)? Show how you would do this.

3. You learned how to do a random sample without replacement. To do a random sample with replacement, you use the `bsample` command. Repeat exercise 2, but use the command `bsample 15`. Then do a tabulation to see if any observations were selected more than once.

4. Using the approach you used in exercise 1, draw a random sample of 10 students from a large class of 200 students. You use the class roster in which each student has a number from 1 to 200. Show the commands you use, and set the seed at 953. List the numbers for the 10 students you select.

5. Open `nlsy97_chapter7.dta`. A friend says that Hispanic families have more children than do other ethnic groups. Use the variables `hh18_97` (number of children under age 18) and `ethnic97` (0 being non-Hispanic and 1 being Hispanic) to test whether this is true. Are the means different? If the result is statistically significant, how substantively significant is the difference?

6. Use the same variables as exercise 5. Use `summarize`, `detail`, and `tabulate` to check the distribution of `hh18_97`. Do this separately for Hispanics and for non-Hispanics. What are the medians of each group? Run a median test. Is the difference significant? How can you reconcile this with the medians you computed? (Think about ties and the distributions.)

7. You want to compare Democrats and Republicans on their mean scores on abortion rights. From earlier uses of the scale, you know that the mean is somewhere around 50 and the standard deviation is about 15. Select an alpha level and the minimum difference that you would find important. Justify your minimum difference (this is pretty subjective, but you might think in terms of a proportion of a standard deviation difference). How many cases do you need to have 80% power? How many cases do you need to have 90% power? How many cases do you need to have 99% power?

8. A friend believes that women are more likely to feel that abortion is okay under any circumstances than are men because women have more at stake in a decision about whether to have an abortion. Use the General Social Survey 2002 dataset (`gss2002_chapter7.dta`), and test whether there is a significant difference between women and men (`sex`) on whether abortion is acceptable in any case (`abany`).

9. You are planning to evaluate the effectiveness of Overeaters Anonymous to reduce the body mass index (BMI) of participants. You will weigh each person at the start of the intervention and then again after they have participated for 5 weeks. From past research, you expect the average BMI to be 30 prior to the intervention. The standard deviation is about 4. You want to be able to detect a difference if the effect size is medium. How many people do you need to have in your intervention?

10. You are planning to compare how satisfied women and men are with a particular weight loss program. You have 20 items measuring satisfaction and each item is

scored from 1 for very dissatisfied to 5 for very satisfied. You are interested in being able to detect a small effect size with alpha = 0.05 and power = 0.90. How big of a sample will you need?

11. A researcher is planning a study of the effectiveness of a new method of teaching a course on statistics. Students in one class get a traditional course, where all the work is done without computers. Students in a different course (new method) do half the work by hand and half using Stata. If you want a moderate effect size, alpha of 0.05, and a power of 0.80, how many students are needed in each class? The final examination is used to evaluate performance, and the mean score has been around 80 with a standard deviation of 10.

12. You are planning an intervention to reduce test-taking anxiety. You have a 100-point scale to measure anxiety and a pretest you did without any intervention had a mean of 70 and a standard deviation of 15. You want to be able to show that a true difference of means between your intervention group and your control group should lower the anxiety by one third of a standard deviation or 5 points. You want a power of 0.80.

    a. How many students do you need in each group (control group and intervention group) using a two-tail test?

    b. How many using a one-tail test?

13. Suppose you are doing question 12, it is much more expensive to get measures for the intervention group. Because of the cost of running the intervention, let's say it is two times as expensive to obtain students' scores for the intervention group. How many students do you need in the intervention group? The control group?

    a. For a two-tail test?

    b. For a one-tail test?

# 8  Bivariate correlation and regression

## 8.1   Introduction to bivariate correlation and regression

Bivariate correlation and regression are used to examine the relationship between two variables. They are usually used with quantitative variables that we assume are measured on the interval level. Some social-science statistics books present correlation first and then regression, whereas others do just the opposite. You need to understand both of these, but which comes first is sort of like asking whether the chicken or the egg came first. In this chapter, I present topics in the following order:

1. Construct a scattergram. This is a graphic representation of the relationship between two variables.

2. Superimpose a regression line. This is a straight line that best describes the linear form of the relationship between two variables.

3. Estimate and interpret a correlation. This tells us the strength of the relationship.

4. Estimate and interpret the regression coefficients. This tells us the functional form of the relationship.

5. Estimate and interpret Spearman's rho as a rank-order correlation for ordinal data.

6. Estimate and interpret the reliability coefficient called alpha. This uses correlation to assess the reliability of a measure.

7. Estimate and interpret kappa as a measure of agreement for categorical data.

In chapter 10, we will expand this process to include multiple correlation and multiple regression. If you have a good understanding of bivariate correlation and regression, you will be ready for chapter 10.

## 8.2   Scattergrams

Suppose that we are interested in the relationship between an adult man's education and the education his father completed. We believe that the advantages of education are transmitted from one generation to the next. Fathers who have limited education will have sons who have restricted opportunities, and hence they too will have limited education. On the other hand, fathers who have high education will offer their sons opportunities to get more education. When examining a relationship like this, we know that it is not going to be "perfect". We all know adult men who have far more education than their own fathers, and we may know some who have far less.

To understand the relationship between these two variables, let's use the dataset `gss2006_chapter8.dta` and create a graph. Because this is a large dataset, a scattergram will have so many dots that it will not make much sense. Suppose that we have 20 father–son dyads that have a father with a score of 15 and a son with a score of 15, but just 1 father–son dyad that has a father with a score of 20 and a son with a score of 10. The 20 dyads and the 1 dyad would each appear as one dot on the graph, creating a visual distortion of the relationship. (We will discuss ways to work around this later.) To manage this problem, limit the scattergram to a random sample of 100 observations by using the command `sample 100, count`.

### How can you get the same result each time?

When you take a random sample of your data for a scattergram or for any other purpose, you may want to make sure you get the same sample if you repeat the process later. Random sampling is done by first generating a *seed* that instructs Stata where to begin the random process. If you have a table of random numbers, generating a seed is equivalent to picking where you will start in the table. By setting the seed at a specific value, you can replicate your results. The easiest way to select your subsample and make sure that you get the same subsample if you repeat the process is with this pair of commands:

```
. set seed 123
. sample 100, count
```

Here I used a seed of 123. You can pick any number, and when you repeat the command using the same number, you will get the same random sample.

Because a father's education comes before his son's, the father's education will be the predictor (the $X$ or independent variable). The son's education will be the outcome (the $Y$ or dependent variable). By convention, a scattergram places the predictor on the horizontal axis and the outcome on the vertical axis.

The interface is helpful when working with scattergrams because scattergrams have so many options that it would be hard to remember them all. Open the dialog box by selecting Graphics ▷ Twoway graph (scatter, line, etc.). The resulting dialog box is shown in figure 8.1.

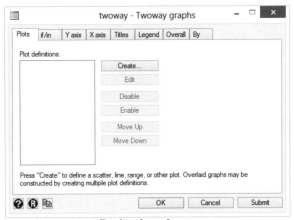

Figure 8.1. Dialog box for a scattergram

Under the Plots tab, click on *Create....* We will use the *Basic plots* and make sure *Scatter* is highlighted. Type educ ($Y$ variable) and paeduc ($X$ variable), and click on Accept. This returns you to the main Plots tab. Here you need to make sure that *Plot 1* is highlighted, and you can add features to the graph by using the other tabs. Click on the if/in tab and type sex==1 in the *If: (expression)* box, because we want to limit this graph to sons and sex==1 is the code for males. Click on the Y axis tab and, under *Title*, type Son's education. Click on the X axis tab and, under *Title*, type Father's education. Using the Titles tab, you can give the graph a title of Scattergram relating father's education to his son's education and add a note at the bottom of the graph, Data source: GSS 2006; random sample of N = 100. If you click on Submit, you will probably decide the title is too wide for the graph. To make it fit better, you can click on *Properties* (to the right of the title) in the dialog box and change the *Size* to *Medium*.

Click on Submit. This gives a nice graph, but the $y$-axis title might be too close to the numbers on the $y$ axis. If you want to move that title away from the numbers, then go back to the Y axis tab and click on *Properties*, which is to the right of the title. This gives you several options. Go to the Advanced tab to change the *Inner gap*. Try putting a 5 here. Click on Submit and see how you like the graph.

These dialog boxes are complex, but remember that you are not going to hurt Stata or your data by trying different options. If you do not understand what an option does, just try it and see what happens. You can always return to the default value for the option in the dialog box by clicking on the Ⓡ icon in the bottom left of the dialog box.

Here is the resulting Stata command. (Notice that in addition to what we have done with the dialog box, I turned the legend off in the **Legend** tab. Also the command is listed as you would see it in a do-file with the three forward slashes, ///, used as a line break.) The resulting graph is shown as figure 8.2.

```
twoway (scatter educ paeduc) if sex==1, ytitle(Son's education) ///
  yscale(titlegap(5)) xtitle(Father's education) ///
  title(Scattergram relating father's education to his son's education, ///
  size(medium)) note(Data source: GSS 2006; random sample of N = 100) legend(off)
```

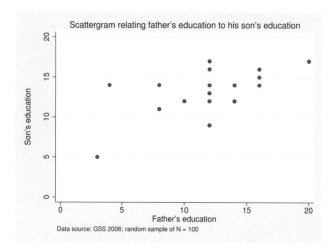

Figure 8.2. Scattergram of son's education on father's education

How do we interpret the resulting scattergram? There is a general pattern that the higher the father's education, the higher his son's education, but there are many exceptions. The data are for the United States. Some countries have a virtual caste system, which permits little opportunity for upward mobility and little risk of downward mobility. In those countries, the scattergram would fall close to a straight line, meaning the father's education largely determined his son's education. In liberal democracies, such as the United States, the relationship is only moderately strong, and there is a lot of room for intergenerational mobility.

Can you think of relationships for which a scattergram would be useful? A gerontologist might want to examine the relationship between a score on the frequency of exercise and the number of accidental falls an older person has had in the last year. Unlike the results in figure 8.2, the older people who exercise more often might have fewer falls. The scatter would start in the upper left and go down if those who exercise more regularly are less prone to accidental falls. An educational psychologist might examine the relationship between parental support and academic performance. All you need is a pair of variables that are interval level or that you are willing to treat as interval level.

What happens to the relationship between self-confidence among college students and how long they have been in college? Is the initial level of self-confidence high? Does it drop to a low point during the freshman year, only to gradually increase during the sophomore, junior, and senior years? This is an empirical question whose answer could be illustrated with a scattergram.

Rarely does a scattergram help when you have a very large sample, say, 500 or more observations. However, there is a clever approach in Stata that tries to get around this problem and that works for reasonably sized datasets: the `sunflower` command. Sunflowers are used to represent clusters of observations. Sunflowers vary in the number of "petals" they have. The more petals there are, the more observations are at that particular coordinate point. Even the `sunflower` command is of limited value with really large samples. If you want to try the `sunflower` command, you can enter `help sunflower` to see the various options. When you are in the Viewer window with the help file for the `sunflower` command open, you can select Dialog ▷ sunflower from the menu at the top right corner to open the `sunflower` dialog box. If you have trouble finding a dialog box but you know the name of the command, you can always access the dialog box through the help file in this way.

A more common approach is to add "jitter", which is spherical random noise, to the markers (dots) where there are several observations with the same score on both the $X$ variable and the $Y$ variable. To do this, return to the `twoway` dialog box and click on the Plots tab. Make sure that *Plot 1* is highlighted, and click on *Edit*. Next you will click on *Marker properties* and click on the Advanced tab. Check the box for *Add jitter to markers* and type 6 for the *Noise factor*. You might want to set the seed at some number, here 222. The resulting graph (figure 8.3) is a more realistic representation of our 100 observations.

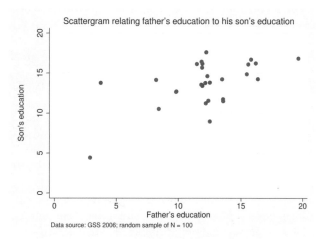

Figure 8.3. Scattergram of son's education on father's education with "jitter"

The command that results from the **twoway** dialog box and produces the above graph is

```
twoway (scatter educ paeduc, jitter(6) jitterseed(222)) if sex==1,
  ytitle(Son's education) yscale(titlegap(5)) xtitle(Father's education)
  title(Scattergram relating father's education to his son's education,
  size(medium)) note(Data source: GSS 2006; random sample of N = 100) legend(off)
```

## Predictors and outcomes

Different statistics books and different substantive areas vary in the terms they use for predictors and outcomes. We need to be comfortable with all these different terms. A scattergram is a great place to really understand the terminology because we must have one variable on the horizontal $x$ axis and another variable on the vertical $y$ axis. Sometimes it helps to think of the $X$ variable as a cause and the $Y$ variable as an effect. This works if we can make the statement "$X$ is a cause of $Y$". It makes sense to say that the father's education is a cause of his son's education rather than the other way around. We say "a" cause rather than "the" cause because many variables will contribute to how much education a son will achieve.

People who are uncomfortable with cause–effect terminology often call the $X$ variable the independent variable and the $Y$ variable the dependent variable. Which variable is dependent? In the scattergram, we would say that the son's education is dependent on his father's education. We would not say that the father's education depends on his son's education. The dependent variable is the $Y$ variable, and the other variable is the $X$ variable or the independent variable—because it does not depend. That is, a father's education does not depend on his son's education.

Other people prefer to think of an outcome variable and a predictor. We do not need to know what causes something, but we can discover what predicts it. The son's education is an outcome, and the father's education is the predictor. Predictors may or may not meet a philosopher's definition of a cause. Couples who fight a lot before they get married are more likely to fight after they are married, so we would say that conflict prior to marriage predicts conflict after marriage. The premarriage conflict is the predictor, and the postmarriage conflict is the outcome.

Sometimes none of these terms makes a lot of sense. Wives who have high marital satisfaction more often have husbands who have high marital satisfaction. Both variables depend on each other and simultaneously influence each other. Sometimes we refer to these kinds of variables as having reciprocal relationships. In such a case, which variable is the $X$ variable and which is the $Y$ variable is arbitrary.

## 8.3  Plotting the regression line

Later in this chapter, we will learn how to do a regression analysis. For now, I will simply introduce the concept and show how to plot it on the scattergram. The regression line shows the form of the relationship between two variables. Ordinarily, we assume that the relationship is linear. For example, what is the relationship between income and education? To answer this question, we need to know how much income you could expect to make if you had no education (intercept or constant) and then how much more income you can expect for each additional year of education (slope). Suppose that you could expect to make $10,000 per year, even with no formal education, and for each additional year of education, you could expect to earn another $3,000. You could write this as an equation

$$\text{Estimated income} = 10000 + 3000(\texttt{educ})$$

where 10000 is the estimated income if you had no education, that is, $\texttt{educ} = 0$. This is called the intercept or constant. The 3000 is how much more you could expect for each additional year of $\texttt{educ}$, and it is called the slope. Thus a person who has 12 years of education would have an estimated income of $10000 + 3000(12) = 46000$ or $46,000.

A symbolic way of writing a regression equation is

$$\widehat{Y} = a + b(X)$$

where $a$ is the intercept or constant, and $b$ is the slope. The $\widehat{Y}$ (pronounced Y-hat) is the dependent variable ($\texttt{income}$), and the circumflex means that it is estimated (many statistics texts use a $\widehat{\phantom{a}}$ over the $Y$ for an estimated value). The $X$ is the independent variable ($\texttt{educ}$). Let's use the example relating educational attainment of fathers and sons to see how the regression line appears. We will not get the values of $a$ and $b$ until later, but we will get a graph with the line drawn for us.

Return to your Plots tab, or if you closed the dialog box by clicking on OK instead of Submit, you will need to reopen it by selecting Graphics ▷ Twoway graph (scatter, line, etc.). We already have the information entered for *Plot 1*, the scattergram. Click on *Create....* Previously, we selected *Basic plots* and *Scatter*. This time, we select *Fit plots* and make sure that *Linear prediction* is highlighted. We type educ and paeduc as the $Y$ variable and $X$ variable, respectively. Click on Submit, and the graph appears as shown in figure 8.4.

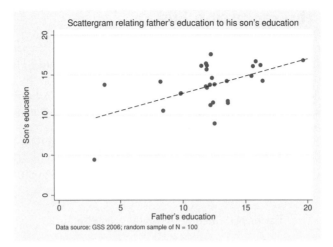

Figure 8.4. Scattergram of son's education on father's education with a regression line

The linear regression line gives you a good sense of the relationship. It shows how the higher the father's education is, the higher his son's education is on average. The observations are not extremely close to the regression line, meaning that there is substantial intergenerational mobility. Sons whose fathers have little education are expected to have more education than their fathers, but sons whose fathers have a lot of education are expected to have a little less education than their fathers. Can you see this in the scattergram? Pick a father who has 5 years of education, $X = 5$. Project a line straight up to the regression line and then straight over to the $Y$ axis, and you can see that we would expect his son to have about $Y = 10$ years of education. Pick a father who has 20 years of education, $X = 20$. Can you see that we would expect his son to have less education than he did?

## 8.4   An alternative to producing a scattergram, binscatter

Using a scattergram with a large sample is impractical, and taking a subsample of observations or adding the `jitter()` option is an imperfect solution. Michael Stepner is the author of a user-written command, `binscatter`. This command develops an earlier version by Jessica Laird and had guidance from Raj Chetty, John Friedman, and Lazlo Sandor. It produces a remarkably simple solution for conveying a relationship by plotting the mean value of the dependent variable for equal-sized bins on the continuous independent variable. Rather than having hundreds or thousands of dots on the scattergram, we have a small number of means. The set of mean outcome scores for the bins gives a clearer idea of the relationship than does a scatter when there are a large number of observations. We will illustrate this by using a few of the commands borrowed from the `binscatter` help file.

First, you need to install the command, because it is not part of official Stata. In your Command window, type `ssc install binscatter`. This will install the command

and the help file. The command can do much more than I illustrate, so you might want
to read the help file, `help binscatter`. The help file uses a dataset that is installed
with Stata, so you can type

```
. sysuse nlsw88
. keep if age > 34 & age < 45 & race < 3
```

The first command opens the 1988 National Longitudinal Survey of Women. The
second command restricts our sample to people who are between ages 35 and 44 and
who are either white or black. We want to know the relationship between tenure on a
job and a woman's wages. Now type a `scatter` command and a `binscatter` command
to see the differences:

Here is the `scatter` command, which creates the graph in figure 8.5.

```
. scatter wage tenure
```

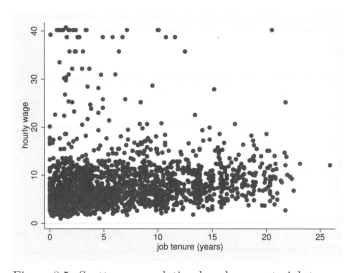

Figure 8.5. Scattergram relating hourly wage to job tenure

We have over 2,000 observations in this dataset, and you can see that it is hard to
discern any pattern in the relationship. The `binscatter` command divides the independent variable, `tenure`, into a series of bins. The default is 20 bins. (If the independent
variable has fewer than 20 distinct values, `binscatter` will compute the mean for each
value.) `binscatter` then places a dot for the mean on the outcome variable, `wage`,
for each bin. Here is the `binscatter` command, followed by the resulting graph in
figure 8.6.

```
. binscatter wage tenure
```

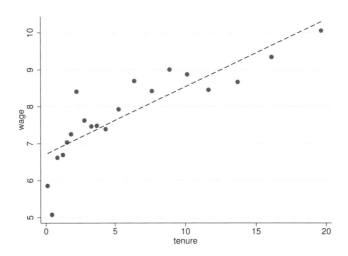

Figure 8.6. Average wage by tenure

As you can see, the **binscatter** command also places a linear regression line on the figure. Unlike the scattergram, the graph created by **binscatter** shows a consistent increase in wages over years. We would have missed this finding completely if we relied on the scattergram. Examining the graph created by **binscatter**, you might say that a curve would fit better. Fitting a curve will be covered in chapter 10, but the option for **binscatter** is **line(qfit)**. We will not show the result, but you can run the command yourself to see how the curve fits.

Can you see an alternative to using a curve? Carefully examining the **binscatter** graph, we see that there is a very rapid increase in wages the first 2 or 3 years. After that, an increase is evident, but the increase is less dramatic. We could use two linear regressions. The point of discontinuity looks to be around 3 years of tenure. Between 0 and 3 years, there is a dramatic increase, but from 3 years to 20 years, there is a much less dramatic increase in wages. Here is the **binscatter** command followed by the resulting graph in figure 8.7. Notice that the **rd(3.0)** option is the point of discontinuity where the rate of growth changes. You might try this option with slightly different values, such as **rd(2.5)**. Do you see if a different possible point of discontinuity produces similar results?

. binscatter wage tenure, rd(3.0)

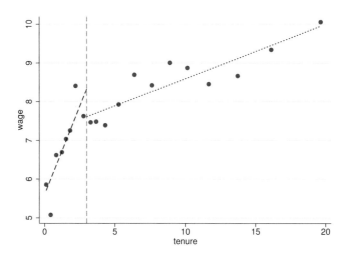

Figure 8.7. Relationship between wages and tenure with a discontinuity in the relationship at 3 years

We might want to see if this relationship varies by race. We may hypothesize that the rate of increase is less dramatic for blacks than it is for whites because of discrimination. We have already restricted our sample to blacks and whites, and we can show a separate result for each group by using

. binscatter wage tenure, rd(3.0) by(race)

Figure 8.8 shows the result. It appears that both black and white women have a positive relationship between wages and tenure and the discontinuity applies to both groups. The biggest contrast is that black women are systematically paid less than white women regardless of their tenure. The rate of increase does not appear to be greater for whites than it is for blacks, but the main take away is that black women have lower wages regardless of their tenure. In figure 8.8, I used the Graph Editor to slightly modify the graph produced by binscatter.

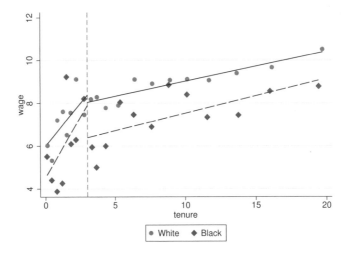

Figure 8.8. Relationship between wages and tenure with a discontinuity in the relationship at 3 years; whites shown with solid lines and blacks shown with dashed lines

## 8.5   Correlation

Your statistics textbook gives you the formulas for computing correlation, and if you have done a few of these by hand, you will love the ease of using Stata. We will not worry about the formulas. Correlation measures how close the observations are to the regression line. We need to be cautious in interpreting a correlation coefficient. Correlation does not tell us how steep the relationship is; that measure comes from the regression. You may have a steep relationship or an almost flat relationship, and both relationships could have the same correlation. Suppose that the correlation between education and income is $r = 0.3$ for women and $r = 0.5$ for men. Does this mean that education has a bigger payoff for men than it does for women? We really cannot know the answer from the correlation. The correlation tells us almost nothing about the form of the relationship. The fact that the $r$ is larger for men than it is for women is not evidence that men get a bigger payoff from an additional year of education. Only the regression line will tell us that. The $r = 0.5$ for men means that the observations for men are closer to the regression line than are the observations for women ($r = 0.3$) and that the income of men is more predictable than that of women. Correlation also tells us whether the regression line goes up ($r$ will be positive) or down ($r$ will be negative). Strictly speaking, correlation measures the strength of the relationship only for how close the dots are to the regression line.

Bivariate correlation is used by social scientists in many ways. Sometimes we will be interested simply in the correlation between two variables. We might be interested in the relationship between calorie consumption per day and weight loss. If you discover that $r = -0.5$, this would indicate a fairly strong relationship. Generally, a correlation of $|r| = 0.1$ is a weak relationship, $|r| = 0.3$ is a moderate relationship, and $|r| = 0.5$ is a strong relationship. An $r$ of $-0.3$ and an $r$ of $0.3$ are equally strong. The negative correlation means that as $X$ goes up, $Y$ goes down; the positive correlation means that as $X$ goes up, $Y$ goes up.

We might be interested in the relationship between several variables. We could compare three relationships between 1) weight loss and calorie consumption, 2) weight loss and the time spent in daily exercise, and 3) weight loss and the number of days per week a person exercises. We could use the three correlations to see which predictor is more correlated with weight loss.

Suppose that we wanted to create a scale to measure a variable, such as political conservatism. We would use several specific questions and combine them to get a score on the scale. We can compute a correlation matrix of the individual items. All of them should be at least moderately correlated with each other because they were selected to measure the same concept.

When we estimate a correlation, we also need to report its statistical significance level. The test of statistical significance of a correlation depends on the size or substantive significance of a correlation in the sample and depends on the size of the sample. An $r = 0.5$ might be observed in a very small sample, just by chance, even though there were no correlations in the population. On the other hand, an $r = 0.1$, although a weak substantive relationship, might be statistically significant if we had a huge sample.

## Statistical and substantive significance

It is easy to confuse statistical significance and substantive significance. Usually, we want to find a correlation that is substantively significant ($r$ is moderate or strong in our sample) and statistically significant (the population correlation is almost certainly not zero). With a very large sample, we can find statistical significance even when $r = 0.1$ or less. What is important about this is that we are confident that the population correlation is not zero and that it is very small. Some researchers mistakenly assume that a statistically significant correlation automatically means that it is important when it may mean just the opposite—we are confident it is not very important.

With a very small sample, we can find a substantively significant $r = 0.5$ or more that is not statistically significant. Even though we observe a strong relationship in our small sample, we are not justified in generalizing this finding to the population. In fact, we must acknowledge that the correlation in the population might even be zero.

- Substantive significance is based on the size of the correlation.

- Statistical significance is based on the probability that you could get the observed correlation by chance if the population correlation is zero.

Now let's look at an example that we downloaded from the UCLA Stata Portal. As mentioned at the beginning of the book, this portal is an exceptional source of tutorials, including movies on how to use Stata. The data used are from a study called High School and Beyond. Here you will download a part of this dataset used for illustrating how to use Stata to estimate correlations. Go to your command line, and enter the command

```
. use http://www.ats.ucla.edu/stat/stata/notes/hsb2, replace
```

You will get a message back that you have downloaded 200 cases, and your listing of variables will show the subset of the High School and Beyond dataset. If your computer is not connected to the Internet, you should use one that is connected, download this file, save it to a flash disk, and then transfer it to your computer. This dataset is also available from this book's webpage.

Say that we are interested in the bivariate correlations between `read`, `write`, `math`, and `science` skills for these 200 students. We are also interested in the bivariate relationships between each of these skills and each students' socioeconomic status and between each of these skills and each students' gender. We believe that socioeconomic status is more related to these skills than gender is.

It is reasonable to treat the skills as continuous variables measured at close to the interval level, and some statistics books say that interval-level measurement is a critical assumption. However, it is problematic to treat socioeconomic status and gender as continuous variables. If we run a tabulation of socioeconomic status and gender, `tab1 ses female`, we will see the problem. Socioeconomic status has just three levels (low, middle, and high) and gender has just two levels (male and female). This dataset has all values labeled, so the default tabulation does not show the numbers assigned to these codes. We can run `codebook ses female` and see that `female` is coded 1 for girls and 0 for boys. Similarly, `ses` is coded 1 for low, 2 for middle, and 3 for high. If you have installed the `fre` command using `ssc install fre`, you can use the command `fre female ses` to show both the value labels and the codes, as we discussed in section 5.4. We will compute the correlations anyway and see if they make sense.

Stata has two commands for doing a correlation: `correlate` and `pwcorr`. The `correlate` command runs the correlation using a casewise deletion (some books call this listwise deletion) option. Casewise deletion means that if any observation is missing for any of the variables, even just one variable, the observation will be dropped from the analysis. Many datasets, especially those based on surveys, have many missing values. For example, it is common for about 30% of people to refuse to report their income. Some survey participants will skip a page of questions by mistake. Casewise deletion can introduce serious bias and greatly reduce the working sample size. Casewise deletion is a problem for external validity or the ability to generalize when there are a lot of missing data. Many studies using casewise deletion will end up dropping 30% or more of the observations, and this makes generalizing a problem even though the total sample may have been representative.

The `pwcorr` command uses a pairwise deletion to estimate each correlation based on all the people who answered each pair of items. For example, if Julia has a score on `write` and `read` but nothing else, she will be included in estimating the correlation between `write` and `read`. Pairwise deletion introduces its own problems. Each correlation may be based on a different subsample of observations, namely, those observations who answered both variables in the pair. We might have 500 people who answered both `var1` and `var2`, 400 people who answered both `var1` and `var3`, and 450 people who answered both `var2` and `var3`. Because each correlation is based on a different subsample, under extreme circumstances it is possible to get a set of correlations that would be impossible for a population.

To open the `correlate` dialog box, select Statistics ▷ Summaries, tables, and tests ▷  Summary and descriptive statistics ▷ Correlations and covariances. To open the `pwcorr` dialog box, select Statistics ▷ Summaries, tables, and tests ▷ Summary and descriptive statistics ▷ Pairwise correlations. Because the command is so simple, we can just enter the command directly.

```
. correlate read write math science ses female
(obs=200)
                   read    write    math  science      ses   female

       read    1.0000
      write    0.5968   1.0000
       math    0.6623   0.6174   1.0000
    science    0.6302   0.5704   0.6307   1.0000
        ses    0.2933   0.2075   0.2725   0.2829   1.0000
     female   -0.0531   0.2565  -0.0293  -0.1277  -0.1250   1.0000
```

We can read the correlation table going either across the rows or down the columns. The $r = 0.63$ between `science` and `read` indicates that these two skills are strongly related. Having good reading skills is probably helpful to having good science skills. All the skills are weakly to moderately related to socioeconomic status, `ses` ($r = 0.21$ to $r = 0.29$). Having a higher socioeconomic status does result in higher expected scores on all the skills for the 200 adolescents in the sample.

A dichotomous variable, such as gender, that is coded with a 0 for one category (man) and 1 for the other category (woman) is called a dummy variable or indicator variable. Thus `female` is a dummy variable (a useful standard is to name the variable to match the category coded as 1). When you are using a dummy variable, the stronger the correlation is, the greater impact the dummy variable has on the outcome variable. The last row of the correlation matrix shows the correlation between `female` and each skill. The $r = 0.26$ between being a girl and writing skills means that girls (they were coded 1 on `female`) have higher writing skills than boys (they were coded 0 on `female`), and this is almost a moderate relationship. You have probably read that girls are not as skilled in math as are boys. The $r = -0.03$ between `female` and `math` means that in this sample, the girls had just slightly lower scores than boys (remember an $|r| = 0.1$ is weak, so anything close to zero is very weak). If, instead of having 200 observations, we had 20,000, this small of a correlation would be statistically significant. Still, it is best described as very weak, whether it is statistically significant or not. The math advantage that is widely attributed to boys is very small compared with the writing advantage attributed to girls.

Stata's `correlate` command does not give us the significance of the correlations when using casewise deletion. The `pwcorr` command is a much more general command to estimate correlations because it has several important options that are not available using the `correlate` command. Indeed, the `pwcorr` command can do casewise/listwise deletion as well as pairwise deletion. When you are generating a set of correlations, you usually want to know the significance level, and it would be nice to have an asterisk attached to each correlation that is significant at the 0.05 level. You can use the dialog box or simply enter the command directly. We use the same command as we did for `correlate`, substituting `pwcorr` for `correlate` and adding `listwise`, `sig`, and `star(5)` as options:

```
. pwcorr read write math science socst ses female, listwise sig star(5)
                   read     write     math  science    socst      ses   female

      read |   1.0000


     write |   0.5968*   1.0000
           |   0.0000


      math |   0.6623*   0.6174*   1.0000
           |   0.0000    0.0000


   science |   0.6302*   0.5704*   0.6307*   1.0000
           |   0.0000    0.0000    0.0000


     socst |   0.6215*   0.6048*   0.5445*   0.4651*   1.0000
           |   0.0000    0.0000    0.0000    0.0000


       ses |   0.2933*   0.2075*   0.2725*   0.2829*   0.3319*   1.0000
           |   0.0000    0.0032    0.0001    0.0000    0.0000


    female |  -0.0531    0.2565*  -0.0293   -0.1277    0.0524   -0.1250   1.0000
           |   0.4553    0.0002    0.6801    0.0714    0.4614    0.0778
```

*p value* (handwritten annotation pointing to the second row under science)

In this table, the listwise correlation between science and reading is $r = 0.63$. The asterisk indicates this is significant at the 0.05 level. Below the correlation is the probability and we can say that the correlation is significant at the $p < 0.001$ level. The reported probability is for a two-tailed test. If you had a one-tailed hypothesis, you could divide the probability in half.

If you want the correlations using pairwise deletion, you would also want to know how many observations were used for estimating each correlation. The command for pairwise deletion that gives you the number of observations, the significance, and an asterisk for correlations significant at the 0.05 level is

```
. pwcorr read write math science ses female, obs sig star(5)
```

Notice that the only change was to replace the **listwise** option with the **obs** option.

### Multiple-comparison procedures with correlations

When you are estimating several correlations, the reported significance level given by the `sig` option can be misleading. If you made 100 independent estimates of a correlation that was zero in the population, you would expect to get five significant results by chance (using the 5% level). In this example, we had 21 correlations, and because we are considering all of them, we might want to adjust the probability estimate. One of the ways to adjust the probability estimate in the `pwcorr` command is with the option `bon`, short for the Bonferroni multiple-comparison procedure. You can get this procedure simply by adding the `bon` option at any point after the comma in the `pwcorr` command. The complete command would be (add the `listwise` option if you want to use casewise/listwise deletion)

```
. pwcorr read write math science socst ses female, bon obs sig star(5)
```

Without this correction, the correlation between `write` and `ses`, $r = 0.21$, had a $p = 0.0032$ and was significant at the 0.01 level. With the Bonferroni adjustment, the $r = 0.21$ does not change, but the correlation now has a $p = 0.067$ and is no longer statistically significant. (An alternative multiple-comparison procedure uses the `sidak` option, which produces the Šidák-adjusted significance level.) It is difficult to give simple advice on when you should or should not use a multiple-comparison adjustment. If your hypothesis is that a certain pattern of correlations will be significant and this involves the set of all the correlations (here, 21), the multiple-comparison adjustment is appropriate. If your focus is on individual correlations, as it probably is here, then the adjustment is not necessary.

## 8.6   Regression

Earlier you learned how to plot a regression line on a scattergram. Now we will focus on how to estimate the regression line itself. Suppose that you are interested in the relationship between how many hours per week a person works and how much occupational prestige he or she has. You expect that careers with high occupational prestige require more work rather than less. Therefore, you expect that the more hours respondents work, the more occupational prestige they will have. This is certainly not a perfect relationship, and we have all known people who work many hours, even doing two jobs, who do not have high occupational prestige. We will use the General Social Survey 2006 dataset (`gss2006_chapter8_selected.dta`) for this section. It has variables called `prestg80`, which is a scale of occupational prestige, and `hrs1`, which is the number of hours respondents worked last week in their primary jobs.

Before doing the regression procedure, we should summarize the variables:

```
. summarize prestg80 hrs1
    Variable │       Obs        Mean    Std. Dev.        Min        Max
─────────────┼─────────────────────────────────────────────────────────
     prestg80 │      4270    44.16745    13.99946         17         86
         hrs1 │      2739    42.07631    14.23166          1         89
```

This summary gives us a sense of the scale of the variables. The independent variable, hrs1, is measured in hours with a mean of $M = 42.08$ hours and a standard deviation of SD $= 14.23$ hours. The outcome variable, prestg80, has corresponding values of $M = 44.17$ and SD $= 14.00$. A scattergram does not help because there are so many cases that no pattern is clear. A correlation—pwcorr prestg80 hrs1, obs sig—tells us that $r = 0.16$, $p < 0.001$. We can interpret this as a fairly weak relationship that is statistically significant.

To estimate the regression, open the <u>regress dialog box</u> by selecting Statistics ▷ Linear models and related ▷ Linear regression. This dialog box is shown in figure 8.9.

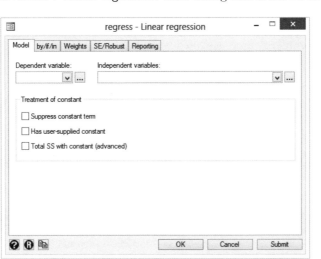

Figure 8.9. The Model tab of the regress dialog box

Enter the *Dependent variable*, prestg80, and the *Independent variable*, hrs1. Click on the Reporting tab and check *Standardized beta coefficients*. Click on Submit to obtain the following command and results:

```
. regress prestg80 hrs1, beta

     Source |       SS       df       MS              Number of obs =    2715
------------+------------------------------           F(  1,  2713) =   67.32
      Model | 13125.035        1   13125.035          Prob > F      =  0.0000      Std. Sign
   Residual | 528901.224    2713  194.950691          R-squared     =  0.0242
------------+------------------------------           Adj R-squared =  0.0239
      Total | 542026.259    2714  199.714908          Root MSE      =   13.962

------------+------------------------------------------------------------------
    prestg80 |     Coef.   Std. Err.      t    P>|t|                       Beta
------------+------------------------------------------------------------------
       hrs1 | .1546416   .0188468     8.21   0.000                    .1556109
      _cons | 38.14733   .8362129    45.62   0.000                           .
------------------------------------------------------------------------------
```

Understanding the format of this regression command is important because more-advanced procedures that generalize from this command follow the same command structure. The first variable after the name of the command, that is, `prestg80`, is always the dependent variable. The second variable, `hrs1`, is the independent variable. When we do multiple regression in chapter 10, we will simply add more independent variables. When we do logistic regression in chapter 11, we will simply change the name of the command. After the comma, we have the option `beta`. This will give us beta weights, which are represented as $\beta$ and will be interpreted below.

The table in the upper left of the output shows the `Source`, `SS`, `df`, and `MS`. This is an analysis of variance (ANOVA) table summarizing the results from an ANOVA perspective (ANOVA is covered in chapter 9), and we will ignore this table for now. In the upper right corner is the number of observations, $N = 2715$, representing the number of people who have been measured on both variables. We also have an $F$ test, which we will cover more fully in the ANOVA chapter. The larger the $F$ ratio is, the greater the significance. Like the $t$ test and chi-squared, $F$ also involves the idea of degrees of freedom. There are two values for the degrees of freedom associated with an $F$ test: the number of predictors (here, 1) and $N - 2$ (here, 2,713). Just below the number of observations are $F(1, 2713) = 67.32$ and the probability level (Prob $> F = 0.0000$). Any probability less than 0.0001 is reported as 0.0000 by Stata. We could write this as $F(1, 2713) = 67.32$, $p < 0.001$. Thus there is a statistically significant relationship between hours worked and the prestige of the job.

Is this relationship strong? We have two values, namely, $R^2$ and the adjusted $R^2$, that serve to measure the strength of the relationship. When we are doing bivariate regression, $R^2$ is simply $r \times r$. Similarly, $r = \sqrt{R^2}$, but we need to decide on the sign of the $r$-value—whether it is positive or negative. For our model, $R^2 = 0.02$, meaning that the hours worked explain 2% of the variation in the prestige rating. Because of the large sample, this $R^2$, although obviously weak because it does not explain 98% of the variation in prestige, is still statistically significant. When there are many predictors and a small sample, neither of which apply here, some report the adjusted $R^2$. The adjusted $R^2$ removes the part of $R^2$ that would be expected just by chance. Whenever the adjusted $R^2$ is substantially smaller than the $R^2$ because there is a small sample relative to the number of predictors, it is best to report both values.

The root mean squared error, `Root MSE = 13.96`, has a strange name, but it is a useful piece of information. The `summarize` command showed SD = 14.00 for our dependent variable, `prestg80`. The root mean squared error is the standard deviation around the regression line. Recall when we did a plot of a regression line. If the observations are close to this line, the standard deviation around it should be much smaller than the standard deviation around the overall mean. It is not surprising that our $R^2$ is so small, given that the standard deviation around the regression line, 13.96, is nearly as big as the standard deviation around the mean, 14.00. In other words, the regression line does little to improve our prediction.

The bottom table gives us the regression results. The first column lists the outcome variable, `prestg80`, followed by the predictor, `hrs1`, and the constant, called `_cons`. This last variable is the constant, or intercept. The equation would be written as

$$\text{Estimated prestige} = 38.15 + 0.15(\text{hours})$$

Remember that $M = 44.17$ and SD $= 14.00$ for prestige. If we did not know how many hours a person worked per week, we might guess that they had a prestige score of 44.17 given that 44.17 is the mean prestige score. The regression equation lets us estimate prestige differently depending on how many hours a person works per week. This equation tell us that a person who worked 0 hours would have a prestige score of 38.15 (the intercept or constant), and for each additional hour he or she worked, the prestige score would go up 0.15. For example, a person who worked 40 hours a week would have a predicted prestige score of (approximately, because we have rounded) $38.15 + 0.15(40) = 44.15$, but a person who worked 60 hours a week would have a predicted prestige score of $38.15 + 0.15(60) = 47.15$. This shows a small payoff in prestige for working longer hours. The payoff is statistically significant but not very big.

The column labeled `Std. Err.` is the standard error. The `t` is computed by dividing the regression coefficient by its standard error; for example, $0.1546416/0.0188468 = 8.21$. The `t` is evaluated using $N-2$ degrees of freedom. We do not need to look up the $t$-value because Stata reports the probability as 0.000. We would report this as $t(2713) = 8.21$, $p < 0.001$.

The final column gives us a beta weight, $\beta = 0.16$. When we have just one predictor, the $\beta$ weight is always the same as the correlation (this is not the case when there are multiple predictors). The $\beta$ is a measure of the effect size and is interpreted much like a correlation with $\beta = 0.10$ being weak, 0.30 being moderate, and 0.50 being strong.

Your statistics book may also show you how to do confidence intervals. There are two types of these: a confidence interval around the regression coefficient and a confidence interval around the regression line. Stata gives you the former as an option for regression and the latter as an option for the scattergram. First, let's do the confidence interval around the regression coefficient. Reopen the `regress` dialog box and click on the Reporting tab. This time, make sure that the *Standardized beta coefficients* box is not checked. By unchecking this option, you automatically get the confidence interval in place of $\beta$. The command is the same except there is no option.

```
. regress prestg80 hrs1      no Beta

      Source |       SS       df       MS              Number of obs =    2715
-------------+------------------------------           F(  1,  2713) =   67.32
       Model |  13125.035     1   13125.035            Prob > F      =  0.0000
    Residual |  528901.224  2713  194.950691           R-squared     =  0.0242
-------------+------------------------------           Adj R-squared =  0.0239
       Total |  542026.259  2714  199.714908           Root MSE      =  13.962

-----------------------------------------------------------------------------
    prestg80 |     Coef.   Std. Err.      t    P>|t|    [95% Conf. Interval]
-------------+---------------------------------------------------------------
        hrs1 |  .1546416   .0188468     8.21   0.000    .117686    .1915972
       _cons |  38.14733   .8362129    45.62   0.000    36.50765   39.78701
-----------------------------------------------------------------------------
```

All of this is identical, except that where we had $\beta$, we now have a 95% confidence interval for each regression coefficient. The effect of an additional hour of work on prestige is $b = 0.15$. (Do not confuse $b = 0.15$ with $\beta = 0.16$ from the previous example.) The $b$s and $\beta$s are usually different values. The 95% confidence interval for $b$ has a range from 0.12 to 0.19. Because a value of zero is not included in the confidence interval (a zero value signifies no relationship), we know that the slope is statistically significant. It could have been as little as 0.12 or as much as 0.19. We would report this by stating that we are 95% confident that the interval of 0.12 to 0.19 contains the true slope.

The second type of confidence interval is on the overall regression line. We will run into trouble using the regression line to predict cases that are very high or very low on the predictor. Usually, there are just a few people at the tails of the distribution, so we do not have as much information there as we do in the middle. We also get to areas that make no sense or where there is likely a misunderstanding. For example, predicting the occupational prestige of a person who works 0 hours a week makes no sense, but it does not make much more sense to predict it for somebody who works just 1 or 2 hours a week. Similarly, if a person said they worked 140 hours a week, they probably misunderstood the question because hardly anybody actually works 140 hours a week on a regular basis given that there are only 168 hours in a week. That would be 20 hours a day, 7 days a week! On the other hand, there are lots of people who work 30–50 hours a week, and here we have much more information for making a prediction. Thus, if we put a band around the regression line to represent our confidence, it would be narrowest near the middle of the distribution on the independent variable and widest at the ends.

This relationship is easy to represent with a scattergram. If we use the menu system, selecting Graphics ▷ Twoway graph (scatter, line, etc.) opens the dialog box we need. Next click on *Create....* Then click on *Fit plots* and highlight *Linear prediction w/CI*. Finally, we enter our two variables. The results are shown in figure 8.10.

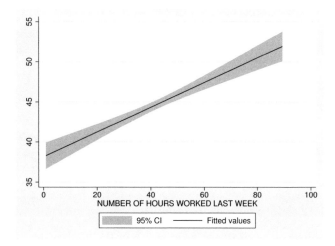

Figure 8.10. Confidence band around regression prediction

## 8.7   Spearman's rho: Rank-order correlation for ordinal data

Spearman's rho, $\rho_s$, is the correlation between two variables that are treated as ranks. This procedure converts the variables to ranks (1st, 2nd, ..., $n$th) before estimating the correlation. For example, if the ages of five observations are 18, 29, 35, 61, and 20, these five participants would be assigned ranks of 1 for 18, 3 for 29, 4 for 35, 5 for 61, and 2 for 20. If we had a measure of liberalism for these five cases, we could convert that to a rank as well. Here are the data (`spearman.dta`) for this simple example:

```
. list
```

|     | age | liberal | rankage | ranklib |
|-----|-----|---------|---------|---------|
| 1.  | 18  | 90      | 1       | 4.5     |
| 2.  | 29  | 90      | 3       | 4.5     |
| 3.  | 35  | 80      | 4       | 2       |
| 4.  | 61  | 50      | 5       | 1       |
| 5.  | 20  | 89      | 2       | 3       |

Ideally, all the scores (and hence, the ranks) would be unique. Stata will assign an average value when there are ties. The two observations who have a score of 90 on `liberal` occupy the fourth and fifth (highest) rank. Because they are tied, they are both assigned a rank score of 4.5. Computing the correlation between `age` and `liberal` by using the command `corr age liberal`, we obtain an $r$ of $-0.97$. Computing the correlation between the ranked data by using the command `corr rankage ranklib`, we obtain an $r$ of $-0.82$. This is lower because the one extreme case (`age` = 61, `liberal` = 50) inflates the Pearson correlation, but this case is not extreme when ranked data are used.

Spearman's rho is a correlation of ranked data. To save the time of converting the variables to ranks and then doing a Pearson's correlation, Stata has a special command: `spearman`. Here we could run the command `spearman age liberal` to yield $\rho_s = -0.82$.

## 8.8   Summary

Scattergrams, correlations, and regressions are great ways to evaluate the relationship between two variables.

- The scattergram helps us visualize the relationship between two variables and is usually most helpful when there are relatively few observations.

- The correlation is a measure of the strength of the relationship between two variables. Here it is important to recognize that it measures the strength of a particular form of the relationship. The other examples have used a linear regression line as the form of the relationship. Although it is not covered here, it is possible for regression to have other forms of the relationship.

- The regression analysis tells us the form of the relationship. Using this line, we can estimate how much the dependent variable changes for each unit change in the independent variable.

- The standardized regression coefficient, $\beta$, measures the strength of a relationship and is identical to the correlation for bivariate regression.

- You have learned how to compute Spearman's rho for rank-order data, and you now understand its relationship to Pearson's $r$.

## 8.9   Exercises

1. Use `gss2006_chapter8.dta`. Imagine that you heard somebody say that there was no reason to provide more educational opportunities for women because so many of them just stay at home anyway. You have a variable measuring education, `educ`, and a variable measuring hours worked in the last week, `hrs1`. Do a correlation and regression of hours worked in the last week on years of education. Then do this separately for women and for men. Interpret the correlation and the slope for the overall sample and then for women and for men separately. Is there an element of truth to what you heard?

2. Use `gss2006_chapter8.dta`. What is the relationship between the hours a person works and the hours his or her spouse works? Do this for women and for men separately. Compute the correlation, the regression results, and the scattergrams. Interpret each of these. Next test if the correlation is statistically significant and interpret the results.

3. Use gss2006_chapter8.dta. Repeat figure 8.2 using your own subsample of 250 observations. Then repeat the figure using a jitter(3) option. Compare the two figures. Set your seed at 111.

4. Use gss2006_chapter8.dta. Compute the correlations between happy, hapmar, and health by using correlate and then again by using pwcorr. Why are the results slightly different? Then estimate the correlations by using pwcorr, and get the significance level and the number of observations for each case. Finally, repeat the pwcorr command so that all the $N$s are the same (that is, there is casewise/listwise deletion).

5. Use gss2002_chapter8.dta. There are two variables called happy7 and satfam7. Run the codebook command on these variables. Notice how the higher score goes with being unhappy or being dissatisfied. You always want the higher score to mean more of a variable, so generate new variables (happynew and satfamnew) that reverse these codes so that a score of 1 on happynew means very unhappy and a score of 7 means very happy. Similarly, a score of 1 on satfamnew means very dissatisfied and a score of 7 means very satisfied. Now do a regression of happiness on family satisfaction with the new variables. How correlated are these variables? Write the regression equation. Interpret the constant and the slope.

6. Use spearman.dta. Plot a scattergram, including the regression line for age and liberal, treating liberal as the dependent variable. Repeat this using the variables rankage and ranklib. Interpret this scattergram to explain why the Spearman's rho is smaller than the Pearson's correlation. Your explanation should involve the idea of one observation being an outlier.

7. Use depression.dta from the Stata Press website; that is, type

   ```
   . use http://www.stata-press.com/data/r13/depression.dta
   ```

   From this hypothetical data, you are interested in the relationship between depression (variable TotalScore) and age. (This dataset uses capitalization as an aid in reading the total score variable. This is rarely a good idea because it is hard to remember these conventions, and if you always use all lowercase, you do not need to remember when and how you used capitalization. Perhaps better options would be to label the variable totalscore or total_score.) Are older people more or less depressed?

   a. Type scatter and binscatter to describe the relationship.

   b. Interpret these results. Why is the binscatter graph easier to interpret?

8. You suspect that the relationship may be nonlinear with a gradual increase among those over about 50 years of age.

   a. Type binscatter to fit a curve.

   b. Interpret these results and compare them with the graphs created in exercise 7.

# 9 Analysis of variance

## 9.1   The logic of one-way analysis of variance

In many research situations, a one-way analysis of variance (ANOVA) is appropriate, most commonly when you have three or more groups or conditions and want to see if they differ significantly on some outcome. This procedure is an extension of the two-sample $t$ test.

You might have two new teaching methods you want to compare with a standard method. If you randomly assign students to three groups (standard method, first new method, and second new method), an ANOVA will show whether at least one of these groups has a significantly different mean. ANOVA is usually a first step. Suppose you find that the three groups differ, but you do not know how they differ. The first new method may be best, the second new method may be second best, and the standard method may be worst. Alternatively, both new methods may be equal but worse than the standard method. When you do an ANOVA and find a statistically significant result, this begs

215

a deeper look to describe exactly what the differences are. These follow-up tests often involve several specific tests (first new method versus standard method, second new method versus standard method, first new method versus second new method). When you do multiple tests like this, you need to make adjustments to the tests because they are not really independent tests, and you need to control for the type I error, $\alpha$. If you run a single test of significance that is significant at the $\alpha = 0.05$ level, everything is okay. Can you see the problem when you run a series of related significance tests? Imagine running 100 of these tests using purely random data. You would expect about 5% of them to be significant at the 0.05 level, just by chance.

ANOVA is normally used in experiments, but it can also be used with survey data. In a survey, you might want to compare Democrats, Republicans, independents, and noninvolved adults on their attitudes toward stem cell research. There is no way you could do an experimental design because you could not randomly assign people to these different party identifications. However, in a national survey, you could find many people belonging to each group. If your overall sample was random, then each of these subgroups would be a random sample as well. An ANOVA would let you compare the mean score on the value of stem cell research (your outcome variable) across the four political party identifications.

ANOVA makes a few assumptions:

- The outcome variable is quantitative (interval level).

- The errors or residuals are normally distributed. This is problematic if we have a small sample.

- The observations represent a random sample of the population.

- The outcomes are independent. (We have repeated-measures ANOVA when this assumption is violated.)

- The variance of each group is equal. You can test this assumption.

- The number of observations in each group does not vary widely.

Violating combinations of these assumptions can be especially problematic. For example, unequal $N$s for each group combined with unequal variances is far worse than unequal variances when the $N$s are equal.

## 9.2   ANOVA example

People having different political party identifications may vary in how much they support stem cell research. You might expect Democrats to be more supportive than Republicans, on average. What about people who say they are independents? What about people who are not involved in politics? Stata allows you to do a one-way ANOVA and then do multiple-comparison tests of all pairs of means to answer two questions:

- Are the means equal for all groups? This is a global question answered by ANOVA.

- Are pairs of group means different from one another? Specific tests answer this question.

There are two ways of presenting the data. Most statistics books use what Stata calls a wide format, in which the groups appear as columns and the scores on the dependent variables appear as rows under each column. An example appears in table 9.1.

Table 9.1. Hypothetical data—wide view

| | democrat | republican | independent | noninvolved | overall |
|---|---|---|---|---|---|
| | 9 | 5 | 7 | 5 | |
| | 7 | 9 | 8 | 4 | |
| | 9 | 4 | 6 | 6 | |
| | 6 | 6 | 6 | 4 | |
| | 9 | 3 | 7 | 5 | |
| | 8 | 1 | 7 | 4 | |
| | 9 | 5 | 8 | 6 | |
| | 7 | 9 | 7 | 4 | |
| | 9 | 4 | 8 | | |
| | 6 | 6 | 6 | | |
| | 9 | 3 | 6 | | |
| | 8 | 1 | 7 | | |
| | | | 7 | | |
| | | | 8 | | |
| $M$ | 8.00 | 4.67 | 7.00 | 4.75 | 6.26 |
| SD | 1.21 | 2.61 | .78 | .88 | 2.09 |

Stata can work with data arranged like this, but it is usually easier to enter data in what Stata calls a long format. We could enter the data as shown in table 9.2.

Table 9.2. Hypothetical data—long view

| stemcell | partyid |
|:---:|:---:|
| 9 | 1 |
| 6 | 1 |
| 6 | 1 |
| 9 | 1 |
| 9 | 1 |
| 7 | 1 |
| 8 | 1 |
| 7 | 1 |
| 9 | 1 |
| 8 | 1 |
| 9 | 1 |
| 9 | 1 |
| 6 | 2 |
| 5 | 2 |
| 4 | 2 |
| 1 | 2 |
| 6 | 2 |
| 4 | 2 |
| 1 | 2 |
| 3 | 2 |
| 5 | 2 |
| 9 | 2 |
| 3 | 2 |
| 9 | 2 |
| . . . | . . . |
| 5 | 4 |

There are 46 observations altogether, and table 9.2 shows 25 of them. We know the party membership of each person by looking at the `partyid` column. We have coded Democrats with a 1, Republicans with a 2, independents with a 3, and those not involved in politics with a 4. If you want to see the entire dataset, you can open up `partyid.dta` and enter the command `list, nolabel`. This method of entering data is similar to how we enter data for a two-sample $t$ test, with one important exception. This time, we have four groups rather than two groups, and the grouping variable is coded from 1 to 4 instead of from 1 to 2.

To perform a one-way ANOVA, select Statistics ▷ Linear models and related ▷ ANOVA/MANOVA ▷ One-way ANOVA, which opens the dialog box shown in figure 9.1.

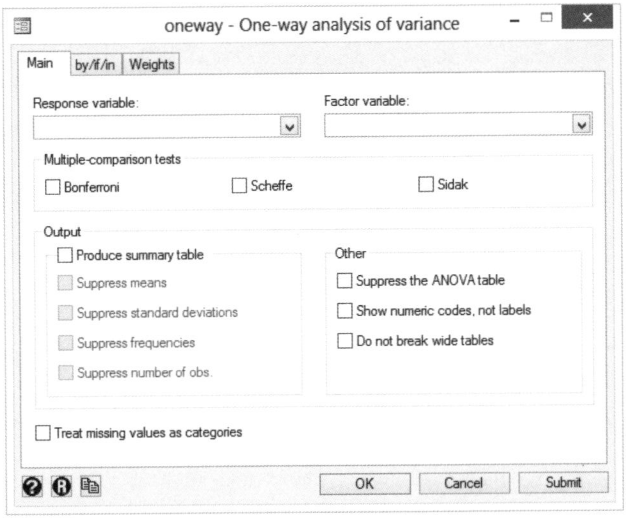

Figure 9.1. One-way analysis-of-variance dialog box

Under the Main tab, indicate that the *Response variable* is stemcell and the *Factor variable* is partyid. Many analysis-of-variance specialists call the response variable the dependent or outcome variable. They call the categorical factor variable the independent, predictor, or grouping variable. In this example, partyid is the factor that explains the response, stemcell. Stata uses the names *response variable* and *factor variable* to be consistent with the historical traditions of analysis of variance. Many sets of statistical procedures developed historically, in relative isolation, and produced their own names for the same concepts. Remember, the response or dependent variable is quantitative and the independent or factor variable is categorical.

Stata asks if we want multiple-comparison tests, and we can choose from three: *Bonferroni*, *Scheffe*, and *Sidak*. These are three multiple-comparison procedures for doing the follow-up tests that compare each pair of means. A comparison of these three approaches is beyond the purpose of this book. We will focus on the Bonferroni test, so click on *Bonferroni*. Finally, check the box to *Produce summary table* to get the mean and standard deviation on support for stem cell research for members grouped by partyid. Here are the results:

```
. oneway stemcell partyid, bonferroni tabulate
```

| party identificat ion | Summary of support for stem cell research | | |
|---|---|---|---|
| | Mean | Std. Dev. | Freq. |
| democrat | 8 | 1.2060454 | 12 |
| republica | 4.6666667 | 2.6053558 | 12 |
| independe | 7 | .78446454 | 14 |
| noninvolv | 4.75 | .88640526 | 8 |
| Total | 6.2608696 | 2.0916212 | 46 |

Analysis of Variance

| Source | SS | df | MS | F | Prob > F |
|---|---|---|---|---|---|
| Between groups | 92.7028986 | 3 | 30.9009662 | 12.46 | 0.0000 |
| Within groups | 104.166667 | 42 | 2.48015873 | | |
| Total | 196.869565 | 45 | 4.37487923 | | |

Bartlett's test for equal variances:  chi2(3) =   20.1167  Prob>chi2 = 0.000

Comparison of support for stem cell research by party identification
(Bonferroni)

| Row Mean-Col Mean | democrat | republic | independ |
|---|---|---|---|
| republic | -3.33333 0.000 | | |
| independ | -1 0.684 | 2.33333 0.003 | |
| noninvol | -3.25 0.000 | .083333 1.000 | -2.25 0.015 |

This is a lot of output, so we need to go over it carefully. Just below the command is a tabulation showing the mean, standard deviation, and frequency on support for stem cell research by people in each political party. Democrats and independents have relatively high means, $M_{\text{democrat}} = 8.00$ and $M_{\text{independent}} = 7.00$, and Republicans and those who are not involved in politics have relatively low means, $M_{\text{republican}} = 4.67$ and $M_{\text{noninvolved}} = 4.75$. (These are hypothetical data.)

Looking at the standard deviations, we can see a potential problem. In particular, the Republicans have a much larger standard deviation than that of any other group. This could be a problem because we assume that the standard deviations are equal for all groups. Finally, the table gives the frequency of each group. There are relatively fewer noninvolved people than there are people in the other groups, but the differences are not dramatic.

Next in the output is the ANOVA table. This is undoubtedly discussed in your statistics textbook, so we will go over it only briefly. The first column, Source, has two sources. There is variance between the group means, which should be substantial if the groups are really different. There is also variance within groups, which should be relatively small if the groups are different from each other, but should be homogeneous within each group. Think about this a moment. If the groups are really different, their means will be spread out, but within each group the observations will be homogeneous. The column labeled SS is the sum of squared deviations for each source, and when we divide this by the degrees of freedom (labeled df), we get the values in the column labeled MS (mean squares). This label sounds strange if you are not familiar with ANOVA, but it has a simple meaning. The between-group mean square is the estimated population variance based on differences between groups; this should be large when there are significant differences between the groups. The within-group mean square is the estimated population variance based on the distribution within each group; this should be small when most of the differences are between the groups (Agresti and Finlay 2009). The test statistic, $F$, is computed by dividing the MS(between) by the MS(within). $F$ is the ratio of two variance estimates. The $F = 30.90/2.48 = 12.46$. This ratio is evaluated using the degrees of freedom. We have 3 degrees of freedom for the numerator (30.90) and 42 degrees of freedom for the denominator (2.48). You can look this up in a table of the $F$ distribution. However, Stata gives you the probability as 0.0000. We would never report a probability as 0.0000 but instead would say $p < 0.001$.

### Can Stata give me an $F$ table?

We get the probability directly from the Stata output, so we do not need to look it up in an $F$ table. However, if you ever need to use an $F$ table, you can get one from Stata without having to find the table in a statistics textbook. Type the command search ftable, which opens a window with a link to a webpage. Go to that webpage, and click on install to install several tables. You may have already downloaded these tables when you were obtaining $z$ tests or $t$ tests. From now on, whenever you want an $F$ table, you need only enter the command ftable.

When you download these tables, you also have ttable, ztable, and chitable.

Years ago, these ANOVA tables appeared in many articles. Today, we simplify the presentation. We could summarize the information in the ANOVA table as $F(3, 42) = 12.46$, $p < 0.001$, meaning that there is a statistically significant difference between the means.

Stata computes Bartlett's test for equal variances and tells us that the $\chi^2$ with 3 degrees of freedom is 20.12, $p < 0.001$. (Do not confuse this $\chi^2$ test with the $\chi^2$ test for a frequency table. The $\chi^2$ distribution is used in many tests of significance.) One of the assumptions of ANOVA is that the variances of the outcome variable, stemcell, are equal in all four groups. The data do not meet this assumption. Some researchers discount this test because it will usually not be significant, even when there are substantial differences in variances, if the sample size is small. By contrast, it will usually be significant when there are small differences, if the sample size is large. Because unequal variances are more problematic with small samples, where the test lacks power, and less important with large samples, where the test may have too much power, the Bartlett test is often ignored. Be careful when the variances are substantially different and the $N$s are also substantially different: this is a serious problem. One thing we might do is go ahead with our ANOVA but caution readers in our reports that the Bartlett test of equal variances was statistically significant.

Given that the overall $F$ test is statistically significant, we can proceed to compare the means of the groups. This is a multiple comparison involving six tests of significance: Democrat versus Republican, Democrat versus independent, Democrat versus noninvolved, Republican versus independent, Republican versus noninvolved, and independent versus noninvolved. Stata provides three different options for this comparison, but here I consider only the Bonferroni. Multiple comparisons involve complex ideas, and we will mention only the principal issues here. The traditional $t$ test worked fine for comparing two means, but what happens when we need to do six of these tests? Stata does all the adjustments for us, and we can interpret the probabilities Stata reports in the way we always have. Thus, if the reported $p$ is less than 0.05, we can report $p < 0.05$ using the Bonferroni multiple-comparison correction.

Below the ANOVA table in our output above is the comparison table for all pairs of means. The first number in each cell is the difference in means (`Row Mean - Col Mean`). Because political independents had a mean of 7 and Democrats had a mean of 8, the table reports the difference, $7 - 8 = -1$. Independents, on average, have a lower mean score on support for stem cell research. Is this statistically significant? No; the $p = 0.684$ far exceeds a critical level of $p < 0.05$. The noninvolved people and the Republicans have similar means, and the small difference, 0.08, is not statistically significant. However, all the other comparisons are statistically significant. For example, Republican support for stem cell research has a mean that is 3.33 points lower than that of Democrats, $p < 0.001$. One way to read the table is to remember that a negative sign means that the row group mean is lower than the column group mean and that a positive sign means that the row group mean is higher than the column group mean.

How strong is the difference between means? How much of the variance in `stemcell` is explained by different political party identification? We can compute a measure of association to represent the effect size. Eta-squared ($\eta^2$) is a measure of explained variation. Some refer to this as $r^2$ because it is the ANOVA equivalent of $r^2$ for correlation and regression. Knowing the political party identification improves our ability to predict the respondent's attitude toward stem cell research. The $\eta^2$ or $r^2$ in the context of ANOVA is the ratio of the between-groups sum of squares to the total sum of squares. The `oneway` command does not compute this for you, but you can compute it by using the simple division below. Or you could obtain this by using Stata's `anova` command. For example, you could type `anova stemcell partyid` and then type the postestimation command `estat esize`.

$$\eta^2 = r^2 = \frac{\text{Between group SS}}{\text{Total SS}} = \frac{92.703}{196.870} = 0.471$$

Do not forget about Stata's calculator. You can enter `display 92.703/196.870`, and Stata will report back the answer of 0.471. This result means that 47.1% of the

variance in attitude toward stem cell research is explained by differences in party identification. Some researchers prefer a different measure that is called $\omega^2$ (pronounced
omega-squared), and you will learn how to obtain this later in the chapter.

Stata is always adding new capabilities. As of Stata 12, the `pwmean` command is a
way of doing multiple comparisons, which you might find as a useful supplement to the
straight ANOVA approach we just covered. `pwmean`'s output includes more information
than `oneway`, and it offers a wider variety of multiple-comparison procedures. Your
statistics books probably cover several of these procedures but describing them is beyond
the scope of this book. There is a brief discussion of each of these procedures in the
*Stata Base Reference Manual*. The available multiple-comparison adjustments include
the following:

- no adjustment for multiple comparisons; the default

- Bonferroni's method

- Šidák's method

- Scheffé's method

- Tukey's method

- Student–Newman–Keuls' method

- Duncan's method

- Dunnett's method

We will apply `pwmean` to our data on support for use of stem cells to see how much the
support varies across our categories of party identification using the same hypothetical
data as above. The `pwmean` command has many options and methods for adjusting the
error rate. We will use the following command:

```
. pwmean stemcell, over(partyid) effects cimeans mcompare(bonferroni)
Pairwise comparisons of means with equal variances

over          : partyid
```

| stemcell | Mean | Std. Err. | Unadjusted [95% Conf. Interval] | |
|---|---|---|---|---|
| partyid | | | | |
| democrat | 8 | .4546206 | 7.082538 | 8.917462 |
| republican | 4.666667 | .4546206 | 3.749205 | 5.584128 |
| independent | 7 | .4208969 | 6.150596 | 7.849404 |
| noninvolved | 4.75 | .5567943 | 3.626344 | 5.873656 |

| | Number of Comparisons |
|---|---|
| partyid | 6 |

| stemcell | Contrast | Std. Err. | Bonferroni t | P>\|t\| | Bonferroni [95% Conf. Interval] | |
|---|---|---|---|---|---|---|
| partyid | | | | | | |
| republican | | | | | | |
| vs | | | | | | |
| democrat | -3.333333 | .6429306 | -5.18 | 0.000 | -5.113618 | -1.553048 |
| independent | | | | | | |
| vs | | | | | | |
| democrat | -1 | .6195435 | -1.61 | 0.684 | -2.715526 | .7155256 |
| noninvolved | | | | | | |
| vs | | | | | | |
| democrat | -3.25 | .7188183 | -4.52 | 0.000 | -5.240419 | -1.259581 |
| independent | | | | | | |
| vs | | | | | | |
| republican | 2.333333 | .6195435 | 3.77 | 0.003 | .6178077 | 4.048859 |
| noninvolved | | | | | | |
| vs | | | | | | |
| republican | .0833333 | .7188183 | 0.12 | 1.000 | -1.907086 | 2.073753 |
| noninvolved | | | | | | |
| vs | | | | | | |
| independent | -2.25 | .6979785 | -3.22 | 0.015 | -4.182714 | -.3172864 |

pwmean speaks for itself. Our dependent variable is support for stem cell research. We are comparing the means over the different party identifications. We want the effects of these differences, such as Democrat versus independent, to see which individual pairs are significantly different. We have also asked for confidence intervals and means for each category by using the cimeans option. Finally, we make our mean comparisons using the Bonferroni adjustment. If we wanted these results without any adjustment for multiple comparisons, we would replace mcompare(bonferroni) with mcompare(noadjust).

The results are different from the ANOVA table in terms of the layout and what is reported. We do not get the traditional ANOVA table with the $F$ test for the overall relationship. We do get the means and standard errors for each category. We now get a confidence interval for each mean. We see that with our four categories of the party identification, we make six comparisons; these six comparisons appear in the bottom table. An advantage of this approach is that along with the contrast and standard error, we now get a Bonferroni adjusted $t$-test value reported with the two-tailed probability. Also we now get a confidence interval for each difference. For example, the `republican vs democrat` row has a difference of $-3.33$. This number is the result of the Republican mean of 4.67 minus the Democratic mean of 8.00. This difference is has a $t = -5.18$ and a $p < 0.001$ using the Bonferroni adjustment. This significance is confirmed because the confidence interval does not include 0.00.

The *Stata Base Reference Manual* has an extensive set of examples illustrating the different options and comparison criteria. `pwmean` is a nice addition to the traditional ANOVA approach.

## 9.3 ANOVA example using survey data

You will find several examples of studies like the one we just did in standard statistics textbooks. Our second example uses data from a large survey, the 2006 General Social Survey (`gss2006_chapter9.dta`), and examines occupational prestige. For our one-way ANOVA, we will see if people who are more mobile have the benefit of higher occupational prestige. We will compare three groups (the factor or independent variable must be a grouping variable). One group includes people who are living in the same town or city they lived in at the age of 16. We might think that these people have lower occupational prestige, on average, because they did not or could not take advantage of a broader labor market that extended beyond their immediate home city. The second group comprises those who live in a different city but still in the same state they lived in at the age of 16. Because they have a larger labor market, one that includes areas outside of their city of origin, they may have achieved higher prestige. The third group contains those who live in a different state than they lived in at the age of 16. By being able and willing to move this far, they have the largest labor market available to them. We will restrict the age range so that age is not a second factor; the sample is restricted to adults who are between 30 and 59 years of age.

Make sure that you open the *One-way analysis of variance* dialog box shown in figure 9.1. There is another option for analysis of variance and covariance that we will cover later in this chapter. In the Main tab, we are asked for a *Response variable*. The dependent variable is `prestg80`, so enter this as the response variable. This is a measure of occupational prestige that was developed in 1980 and applied to the 2006 sample. The dialog box asks for the *Factor variable*, so type `mobile16` (the independent variable). This variable is coded 1 for respondents who still live in the same city they lived in when they were 16; 2 if they live in a different city but still in the same state; and 3 if they live in a different state. Also click on the box by *Produce summary table* to get the means for each group on the factor variable. Finally, click on the box by *Bonferroni* in the section labeled *Multiple-comparison tests*. Because we have three groups, we can use this test to compare all possible pairs of groups, that is, same city to same state, same city to different state, and same state to different state. On the by/if/in tab, we enter the following restriction under *If: (expression)*: `age > 29 & age < 60 & wrkstat==1`. This restriction limits the sample to those who are 30–59 years old and have a `wrkstat` of 1, signifying that they work full time. Remember to use the symbol & rather than the word "and".

In our results (shown below), we first get a table of means and standard deviations. The first two means are in the direction predicted, but the third mean is not. Those who live in the same city have a mean of 44.12, those who have moved from the city but are still in the same state have a mean of 48.67, and those who have moved out of the state have a mean of 47.19. The standard deviations are similar (we can use Bartlett's test of equal variances to test this assumption). Also the $N$ for each group varies from 269 to 425. Usually, the unequal $N$s are not considered a serious problem unless they are extremely unequal and one or more groups have few observations.

```
. oneway prestg80 mobile16 if age > 29 & age < 60 & wrkstat==1, bonferroni
> tabulate
```

| GEOGRAPHIC MOBILITY SINCE AGE 16 | Summary of RS OCCUPATIONAL PRESTIGE SCORE (1980) | | |
|---|---|---|---|
| | Mean | Std. Dev. | Freq. |
| SAME CITY | 44.116162 | 13.73443 | 396 |
| SAME ST,D | 48.672862 | 13.002183 | 269 |
| DIFFERENT | 47.185882 | 13.959368 | 425 |
| Total | 46.437615 | 13.758919 | 1090 |

| Source | Analysis of Variance | | | | |
|---|---|---|---|---|---|
| | SS | df | MS | F | Prob > F |
| Between groups | 3716.07404 | 2 | 1858.03702 | 9.98 | 0.0001 |
| Within groups | 202440.184 | 1087 | 186.23752 | | |
| Total | 206156.258 | 1089 | 189.307858 | | |

Bartlett's test for equal variances:  chi2(2) =   1.6873  Prob>chi2 = 0.430

*not stat sign.*

```
                Comparison of RS OCCUPATIONAL PRESTIGE SCORE   (1980)
                       by GEOGRAPHIC MOBILITY SINCE AGE 16
                                   (Bonferroni)
```

| Row Mean-Col Mean | SAME CIT | SAME ST, |
|---|---|---|
| SAME ST, | 4.5567 0.000 | |
| DIFFEREN | 3.06972 0.004 | -1.48698 0.487 |

We can summarize the second table (the ANOVA table) by writing $F(2, 1087) = 9.98$, $p < 0.001$. Just beneath the ANOVA table is Bartlett's test for equal variances. Because $\chi^2(2) = 1.69$, this finding is not statistically significant ($p = 0.43$). This is good because equal variances is an assumption of ANOVA. The $F$ test is an overall test of significance for any differences between the group means. You can have a significant $F$ test when the means are different but in the opposite direction of what you expected.

Now let's perform multiple comparisons using the Bonferroni adjustment. Those adults who moved to a different state scored 3.07 points higher on prestige, on average, than those who stayed in their hometown ($47.19 - 44.12 = 3.07$), $p < 0.01$. Those who moved but stayed in the same state scored 4.56 points higher on the prestige scale ($48.67 - 44.12 = 4.55$), $p < 0.001$. Although these two comparisons are significant, those who moved but stayed in the same state are not significantly different from those who moved to a new state.

We can do a simple one-way bar chart of the means. The command is

```
. graph bar (mean) prestg80 if age > 29 & age < 60 & wrkstat==1, over(mobile16)
```

You can enter this command as one line in the Command window or enter it in a do-file using /// as needed for a line break. Because of the length of graph commands, it is probably a good idea to use the dialog box, as discussed in chapter 5. Select Graphics ▷ Bar chart to open the appropriate dialog box. On the Main tab, select *Graph by calculating summary statistics* and *Vertical* in the *Orientation* section. Under *Statistics to plot*, check the first box, select *Mean* from the drop-down menu and, to the right of this, type prestg80 as the variable. Next go to the Categories tab and check *Group 1*; select mobile16 as the *Grouping variable*. Finally, open the if/in tab and, under the *If: (expression)*, type age > 29 & age < 60 & wrkstat==1.

The resulting bar chart, shown in figure 9.2, provides a visual aid showing that the mean is highest for those with some geographical mobility, that is, who live in the same state but a different city. Although all three means are different in our sample, only two comparisons (staying in your hometown versus moving to a different city within your home state and staying in your hometown versus moving to a different state) reach statistical significance.

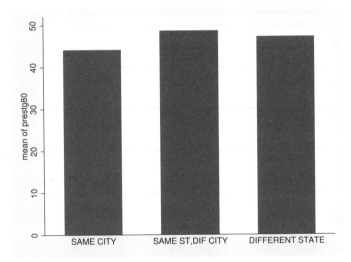

Figure 9.2. Bar graph of relationship between prestige and mobility

We can compute $\eta^2 = r^2 = 3716.074/206156.258 = 0.02$, as described in the previous section. Thus, even though the overall result is significant and two of the three comparisons are significant in the direction we predicted, the mobility variable does not explain much of the variance in prestige. Why is this? To get an answer, look at the means and standard deviations. In the previous example, the standard deviations for groups tended to be much smaller than the overall standard deviation. This is not the case in this example. The standard deviations for each group are between 13.00 and 13.96, whereas the overall standard deviation is 13.76. Although the means do differ, most of the variance still remains within each group. When the variance within the groups is not substantially smaller than the overall variance, the group differences are not explaining much variance.

## 9.4   A nonparametric alternative to ANOVA

Sometimes treating the outcome variable as a quantitative, interval-level measure is problematic. Many surveys have response options, such as agree, don't know, and disagree. In such cases, we can score these so that 1 is agree, 2 is don't know, and 3 is disagree, and then we can do an ANOVA. However, some researchers might say that the score was only an ordinal-level measure, so we should not use ANOVA. The Kruskal–Wallis rank test lets us compare the median score across the groups. If we are interested in political party identification and differences in support for stem cell research, we might use the Kruskal–Wallis rank test instead of the one-way ANOVA. If we want to use only ordinal information, it may be more appropriate to compare the median rather than the mean.

For this example, we will use `partyid.dta`. To open the dialog box for the Kruskal–Wallis rank test, select Statistics ▷ Nonparametric analysis ▷ Tests of hypotheses ▷ Kruskal-Wallis rank test. In the resulting screen, there are only two options. Type `stemcell` under *Outcome variable* and `partyid` under *Variable defining groups*. As with ANOVA, the variable defining the groups is the independent variable, and the outcome variable is the dependent variable.

```
. kwallis stemcell, by(partyid)
Kruskal-Wallis equality-of-populations rank test
```

| partyid | Obs | Rank Sum |
|---|---|---|
| democrat | 12 | 422.00 |
| republican | 12 | 174.00 |
| independent | 14 | 391.00 |
| noninvolved | 8 | 94.00 |

```
chi-squared =      22.115 with 3 d.f.
probability =       0.0001

chi-squared with ties =      22.696 with 3 d.f.
probability =       0.0001
```

This test ranks all the observations from the lowest to the highest score. With a scale that ranges from 1 to 9, there are many ties, and the command adjusts for that. If the groups were not different, the rank sum for each group would be the same, assuming an equal number of observations. Notice from the output that the Rank Sum for Democrats is 422 and for Republicans it is just 174, even though there are 12 observations in each group. This means that Democrats must have higher scores than Republicans. The output gives us two chi-squared tests. Because we have people who are tied on the outcome variable (have the same score on stemcell), use the chi-squared with ties: chi-squared$(3) = 22.696$, $p < 0.001$. Thus there is a highly significant difference between the groups in support for stem cell research.

Because we are treating the data as ordinal, it makes sense to use the median rather than the mean. Now run a tabstat to get the median. It is easiest to type the command in the Command window, but if you prefer to use the dialog box, select Statistics ▷ Summaries, tables, and tests ▷ Other tables ▷ Compact table of summary statistics.

```
. tabstat stemcell, statistics(mean median sd) by(partyid)
Summary for variables: stemcell
     by categories of: partyid (party identification)
    partyid |      mean       p50        sd
------------+------------------------------
   democrat |         8       8.5  1.206045
 republican |  4.666667       4.5  2.605356
independent |         7         7  .7844645
 noninvolved|      4.75       4.5  .8864053
------------+------------------------------
      Total |   6.26087         6  2.091621
```

We have the same pattern of results we had when we ran an ANOVA on these data. The p50 is the median. Notice that the medians are in the same relationship as the means with Democrats having the highest median, Mdn $= 8.5$; followed by the independents, Mdn $= 7.0$; and the Republicans and noninvolved having the lowest, Mdn $= 4.5$.

There are two graphs we can present to represent this information. We can do a bar chart like the one we did for the ANOVA, only this time we will do it for the median rather than for the mean. Select Graphics ▷ Bar chart to open the appropriate dialog box, or type the command

```
. graph hbar (median) stemcell, over(partyid)
> title(Median stem cell attitude score by party identification)
> ytitle(Median score on stem cell attitude)
```

which produces the graph in figure 9.3.

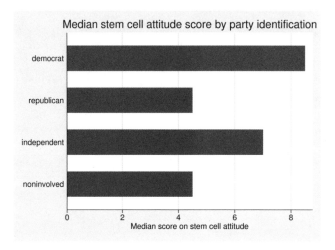

Figure 9.3. Bar graph of support for stem cell research by political party identification

Because we are working with the median, we can use a box plot. Open the appropriate dialog box by selecting Graphics ▷ Box plot. Under the Main tab, we enter `stemcell` under *Variables*. Switching to the Categories tab, we check the *Group 1* box and enter `partyid` as the variable under *Grouping variable*. Clicking on OK produces the following command and figure 9.4.

```
. graph box stemcell, over(partyid)
```

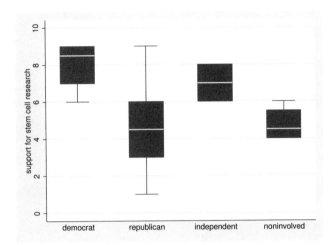

Figure 9.4. Box plot of support for stem cell research by political party identification

This is a much nicer graph than the bar chart because it not only shows the median of each group but also gives us a sense of the within-group variability. With the ANOVA, political party identification explained nearly half (0.471) of the variance in support for stem cell research. This box plot shows that not only are the medians different but also there is little variance within each group, except for the Republicans. (Again, these are hypothetical data.)

## 9.5   Analysis of covariance

We will discuss analysis of covariance (ANCOVA) as an extension of multiple regression in chapter 10. However, I will present an example of it here because it has a long history as a type of analysis of variance. ANCOVA has a categorical predictor (factor) and a quantitative dependent variable (response) like one-way ANOVA. ANCOVA adds more covariates that are quantitative variables that need to be controlled.

One of the areas in which ANCOVA developed was education, where the researchers could not assign children to classrooms randomly. Researchers needed to adjust for the fact that without random assignment, there might be systematic differences between the classrooms before the experiment started. For instance, one classroom might have students who have better math backgrounds than students in another classroom, so they might seem to do better because they were ahead of the others before the experiment started. We would need to control for initial math skills of students. People who volunteer to participate in a nutrition education course might do better than those in a control group, but this might happen because volunteers tend to be more motivated. We would need to control for the level of motivation. The idea of ANCOVA is to statistically control for group differences that might influence the result when you cannot rule out these possible differences through randomization. We might think of ANCOVA as a design substitute for randomization. Randomization is ideal, but when it is impossible to randomly assign participants to conditions, we can use ANCOVA as a fallback.

I do not mean to sound negative. ANCOVA allows us to make comparisons when we do not have randomization. Researchers often need to study topics for which randomization is out of the question. If we can include the appropriate control variables, ANCOVA does a good job of mitigating the limitations caused by the inability to use randomization.

The one-way ANOVA, with `gss2006_chapter9.dta`, showed that people who moved out of their hometown had higher prestige than people who did not. Perhaps there is a benefit to opening up to a broader labor market among people who move away from their hometown. They advance more because there are more opportunities when they cast their job search more broadly than if they never moved out of their home city. On the other hand, there might be a self-selection bias going on. People who decide to move may be people who have more experience in the first place. A person who is 30 years old may be less likely to have moved out of state than a person who is 40 years old simply because each year adds to the time in which such a move could occur. Also people who are 40 or 50 years old may have higher occupational prestige simply because they have had more time to gain experience. Because age could be related to the chances of a respondent living in the same place that he or she did at age 16 and to a respondent's prestige, we should control for age. Ideally, we would identify several other variables that need to be controlled, but for now we will just use age.

Instead of using the one-way ANOVA dialog box, we need to use the full ANOVA procedure. We have two predictors. The first is `mobile16`, which has three categories: same city, same state–different city, and different state. We are interested in differences in occupational prestige, `prestg80`, based on these fixed mobility categories. The mobility categories are not a random sample of mobility categories, but they are the specific population of categories about which we want to make a statistical inference. There are $k = 3$ groups and we should have 2 degrees of freedom ($k - 1$). The other variable is `age`. Although `age` can take on several values (30–59 in our example), we do not think of these values as categories because this is just one continuous variable, where the older you get, the higher we think your prestige will be. As a continuous variable, `age` should have 1 degree of freedom.

## What are categorical covariates and what are continuous covariates?

Whenever we have an intervention, we could think of the intensity of the intervention varying widely. We could think of the intensity as the dosage level, that is, how much of the intervention the participant received. We might be interested in the dosage of a daily aspirin, and our dependent variable could be some measure of a specific side effect. To generalize across the full range of dosage levels, we would need to randomly sample levels of the dosage. One participant might take 10 mg; another, 500 mg; a third, 50 mg; a fourth, 35 mg; a fifth, 325 mg; and so on. If we had 80 participants, each of them might get a different dosage level. Here the dosage is thought of as a continuous covariate. The dosage each participant receives would be a continuous variable. We would want to be able to generalize our findings across a wide range of possible dosage levels.

More often, we think of an intervention as a categorical covariate. One group of 20 participants would all get a daily placebo, a second group of 20 would all get a low dosage of 50 mg, a third group of 20 would all get 325 mg, and a fourth group of 20 would all get 500 mg. Although we would want to randomly assign our 80 people to each of these groups, the groups themselves represent categorical covariates of none, low level, moderate level, and high level. We want to generalize to these specific levels. We are not going to be able to generalize to other levels. We could not say what happens when the dosage is very low rather than low or when it is 400 mg—just what it is at the four categorical covariates in our study.

When we have an intervention and a control group, we think of them as categorical covariates. When we are comparing views on stem cell research of Muslims to Catholics to Protestants, we are thinking of these as three categorical covariates. We have a random sample of members of each religion, but the three religions we picked are categorical. We cannot generalize to other religions (Hinduism, Buddhism, Judaism, Taoism, etc.)—just these three. When we have a continuous covariate, we usually think of it as a continuous covariate. In our example, we do not use categorical age groups; everybody can be a different age. We are asking the question of whether prestige goes up or down with age, and if it does, we control for it in estimating the effect of the three fixed mobility categories. The `anova` command identifies a continuous variable with the prefix `c.`, as in `c.age`. If we just enter `age`, then `anova` treats this variable as a categorical covariate. As a categorical covariate, the mean of people who are 30 is compared with the overall mean, the mean of people who are 31 is compared with the overall mean, etc. To do this, treat each possible age as its own categorical covariate.

Let's use the `tabulate` command to obtain the mean prestige for each mobility category and for each year of age:

```
. tabulate mobile16 if age > 29 & age < 60 & wrkstat==1, summarize(prestg80)
```

| GEOGRAPHIC MOBILITY SINCE AGE 16 | Summary of RS OCCUPATIONAL PRESTIGE SCORE (1980) | | |
|---|---|---|---|
| | Mean | Std. Dev. | Freq. |
| SAME CITY | 44.116162 | 13.73443 | 396 |
| SAME ST,D | 48.672862 | 13.002183 | 269 |
| DIFFERENT | 47.185882 | 13.959368 | 425 |
| Total | 46.437615 | 13.758919 | 1090 |

It looks like some mobility (within the state) goes with the highest prestige and no mobility goes with the lowest prestige. These differences are not huge, though, given that the overall standard deviation is 13.76.

Let's see how prestige varies across the age range.

```
. tabulate age if age > 29 & age < 60 & wrkstat==1, summarize(prestg80)
```

| AGE OF RESPONDENT | Summary of RS OCCUPATIONAL PRESTIGE SCORE (1980) | | |
|---|---|---|---|
| | Mean | Std. Dev. | Freq. |
| 30 | 48.766667 | 15.6133 | 60 |
| 31 | 46.860465 | 13.772957 | 43 |
| 32 | 43.71875 | 13.082727 | 64 |
| 33 | 41.807692 | 12.929732 | 52 |
| 34 | 45.761905 | 13.523831 | 63 |
| 35 | 46.627451 | 13.518818 | 51 |
| 36 | 48.869565 | 14.200158 | 69 |
| *(output omitted)* | | | |
| 56 | 47.128205 | 12.825098 | 39 |
| 57 | 42.969697 | 14.08786 | 33 |
| 58 | 47 | 15.007315 | 42 |
| 59 | 45.261905 | 12.597056 | 42 |
| Total | 46.021537 | 13.863176 | 1718 |

Looking at the results for `age`, we see a mess. The prestige goes up and down but not consistently up like we speculated. When we do an ANCOVA and treat `age` as a continuous variable, we may expect that it will not be significant because prestige does not seem to go consistently up or down with `age`. Undaunted, we will still add `age` as a covariate so that we can see how this is done.

Stata identifies continuous variables when doing ANOVA by adding `c.` as a prefix to the variable. To add `age` as a continuous covariate, we must type `c.age` (with the c. telling Stata that `age` is continuous). Here are the results if we do not prefix `age` with `c.`:

```
. anova prestg80 mobile16 age if age > 29 & age < 60 & wrkstat==1
                           Number of obs =      1090     R-squared     =  0.0421
                           Root MSE      = 13.6621     Adj R-squared =  0.0140

            Source |   Partial SS    df       MS              F     Prob > F

             Model |   8676.62028    31   279.890977          1.50    0.0396

          mobile16 |   3668.50734     2   1834.25367          9.83    0.0001
               age |   4960.54623    29   171.053318          0.92    0.5943

          Residual |   197479.638  1058   186.653722

             Total |   206156.258  1089   189.307858
```

Notice that we have $k-1 = 2$ degrees of freedom for the fixed effects of `mobile16` and the effect of mobility is highly significant, $F(2, 1058) = 9.83$, $p < 0.001$. In sharp contrast, `age`, which has $k - 1 = 29$ degrees of freedom for the mobility effects is not significant, $F(29, 1058) = 0.92$, *pns*. The overall `Model`, however, is significant, $F(31, 1058) = 1.50$, $p < 0.05$. This model provides a different prediction for each of the 3 categories of `mobile16` and for each of the 30 categories of `age`.

To make Stata know that we are treating `age` as a continuous variable, we need to use `c.age`, as follows:

```
. anova prestg80 mobile16 c.age if age > 29 & age < 60 & wrkstat==1
                           Number of obs =      1090     R-squared     =  0.0183
                           Root MSE      = 13.6511     Adj R-squared =  0.0156

            Source |   Partial SS    df       MS              F     Prob > F

             Model |   3777.3325      3   1259.11083          6.76    0.0002

          mobile16 |   3750.73122     2   1875.36561         10.06    0.0000
               age |   61.2584614     1   61.2584614          0.33    0.5665

          Residual |   202378.925  1086   186.352602

             Total |   206156.258  1089   189.307858
```

The conclusion does not change in this example because `age` does not have a significant effect either way, but in other applications, the difference can be dramatic. Notice that we have just 1 degree of freedom for `age` here, where we had 29 degrees of freedom when `age` was treated as a categorical variable.

In many applications where you do not have a randomized trial, you will need to control for more than one continuous covariate. Depending on what you are studying, you may want to add age, education, ability, and motivation as additional covariates. This is done by simply adding these variables with a `c.` prefix on each of them. Typically, your primary focus is on whether the categorical variable still has a significant effect after these covariates are included. In our example, mobility categories still have a significant effect after we control for age.

Significant covariates can change the result for your categorical variable. If you have two different interventions in two classrooms and have a third classroom as a control group, you would probably adjust for possible differences in ability level of the students and their parents' education. If the difference you observe without these covariates is explainable by these covariates, then the adjusted difference might disappear. One classroom may seem to do better but does not do better when you adjust the means for the covariates. In Stata, estimating the adjusted effects is best handled using the `margins` command, which will give you the estimated mean for each level of your categorical variable for a person who was average (mean) on each of the covariates.

Immediately after running the `anova` command, you type the `margins` command. You can find the dialog box for this command by clicking on Statistics ▷ Postestimation ▷ Marginal means and predictive margins. Under the Main tab, enter `mobile16` in the box labeled *Factor terms to compute margins for.* This is the only thing to do under the Main tab. Under the At tab, we want to create a specification of the level to set our covariate, `age`. To do this, click on Create..., and in the dialog box that appears, specify the values for the covariates. Check the box to the left of the *1* for the statistics and select *Means* using the drop-down menu; for *Covariates*, type `age` (do not type `c.age` here). Clicking on OK twice produces the following command and output:

```
. margins mobile16, at((mean) age)

Adjusted predictions                              Number of obs   =       1090

Expression    : Linear prediction, predict()
at            : age             =    43.82294 (mean)
```

|  | | Delta-method | | | | |
|---|---|---|---|---|---|---|
|  | Margin | Std. Err. | z | P>\|z\| | [95% Conf. | Interval] |
| mobile16 | | | | | | |
| SAME CITY | 44.10402 | .6863205 | 64.26 | 0.000 | 42.75736 | 45.45068 |
| SAME ST,DI.. | 48.68809 | .832746 | 58.47 | 0.000 | 47.05412 | 50.32206 |
| DIFFERENT .. | 47.18756 | .6621822 | 71.26 | 0.000 | 45.88826 | 48.48686 |

The command shows that the `mobile16` variable is the categorical variable on which we are focused. We want to know the adjusted mean for each level of this categorical variable. This would be extended if we had two or more continuous covariates. The option `at((mean) age)` allows us to specify that the continuous covariate, `age`, is to be fixed at its mean.

This is a very useful result. It estimates what the mean prestige would be for each category of mobility for a person who was average on the covariate `age`. The number of observations is 1,090, which matches our `anova` results, so we know that this includes only the subsample that we analyzed in the ANOVA, namely, people between 30 and 59 who were working full time.

If we wanted to do pairwise tests of significance comparing the `prestg80` score for our three types of `mobile16`, while adjusting for age, we could add the `pwcompare` option to our `margins` command, `margins mobile16, at((mean) age) pwcompare`. This command does not show us the adjusted means, but it does show us the difference between the pairs of means.

The mean age for our subsample of people who are over 29 and under 60 and who work full time is 43.82. This is a reasonable way to adjust for a continuous covariate. A person who was average on the covariate but never left his or her hometown would have one estimated mean; a person who was average on the covariate but moved, staying within his or her state, would have a different estimated mean; and a person who was average on the covariate but moved out of state would have a different estimated mean.

What happens if you want to include a covariate that is categorical? We have seen that age was not significant when used as a covariate. We might think that gender may also be an important covariate. Perhaps men are more mobile than women and gender influences occupational prestige. First, let's check on these two possibilities. We do a cross-tabulation of mobility and gender for our subsample.

```
. tab mobile16 sex if age > 29 & age < 60 & wrkstat==1, col chi2
```

| Key |
|---|
| *frequency* |
| *column percentage* |

| GEOGRAPHIC MOBILITY SINCE AGE 16 | RESPONDENTS SEX MALE | FEMALE | Total |
|---|---|---|---|
| SAME CITY | 188 | 212 | 400 |
| | 32.36 | 40.85 | 36.36 |
| SAME ST,DIF CITY | 142 | 129 | 271 |
| | 24.44 | 24.86 | 24.64 |
| DIFFERENT STATE | 251 | 178 | 429 |
| | 43.20 | 34.30 | 39.00 |
| Total | 581 | 519 | 1,100 |
| | 100.00 | 100.00 | 100.00 |

Pearson chi2(2) =  11.0260    Pr = 0.004

This output does show a significant relationship—chi-squared(2) = 11.03, $p <$ 0.01—and notice that 43.2% of men moved to a different state compared with just 34.3% of the women, but 40.9% of the women are in the same city compared with just 32.4% of the men. Our second idea was that there would be gender differences in prestige. We could check that using a $t$ test, but we will go ahead and put it in the `anova`. To do this, we can repeat our first `anova` and simply add the variable for gender, `sex`. It is coded as 1 for male and 2 for female. We will not put a `c.` in front of `sex` because gender is a factor variable and not a continuous variable.

```
. anova prestg80 mobile16 sex c.age if age > 29 & age < 60 & wrkstat==1
```

|  | Number of obs = 1090 | R-squared | = 0.0207 |
|---|---|---|---|
|  | Root MSE    = 13.641 | Adj R-squared = | 0.0171 |

| Source | Partial SS | df | MS | F | Prob > F |
|---|---|---|---|---|---|
| Model | 4263.58121 | 4 | 1065.8953 | 5.73 | 0.0001 |
| mobile16 | 3917.43128 | 2 | 1958.71564 | 10.53 | 0.0000 |
| sex | 486.248705 | 1 | 486.248705 | 2.61 | 0.1063 |
| age | 45.250989 | 1 | 45.250989 | 0.24 | 0.6220 |
| Residual | 201892.677 | 1085 | 186.0762 | | |
| Total | 206156.258 | 1089 | 189.307858 | | |

We see that our overall model is significant, $F(4, 1085) = 5.73$, $p < 0.001$. Mobility is still significant, $F(2, 1085) = 10.53$, $p < 0.001$. However, neither sex nor age is significant. For gender, we have a significance of $F(1, 1085) = 2.61$, *pns*. For age, we have a significance of $F(1, 1085) = 0.24$, *pns*. So we can conclude that there are significant differences for prestige and these are related to mobility, but neither of the covariates we added is significant.

However, for illustrative purposes, we will show how to use the margins command to get the adjusted means. We are estimating the mean prestige for each category of mobility, adjusting for gender and age. We will open the margins dialog box again by selecting Statistics ▷ Postestimation ▷ Marginal means and predictive margins. Under the Main tab, enter mobile16 as the *Factor terms to compute margins for* (if it is not already listed). Under the At tab, confirm that *Specification 1* is selected and click on Edit, which reopens the *Specification 1* dialog box. The checkbox and text should be filled in as completed before. Because gender is categorical, we need to enter it in the bottom group for the *Fixed values*. Check the box to the left of the *1*, type sex as the *Covariate*, and type 1 2 as the *Numlist*. Our *Specification 1* dialog box will look like figure 9.5:

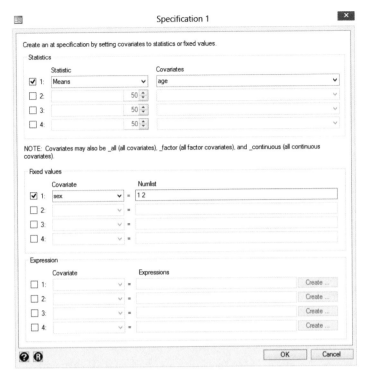

Figure 9.5. The *Specification 1* dialog box under `margins`

We click on **OK** twice. Here are the command that is generated and the results:

```
. margins mobile16, at((mean) age sex=(1 2))
Adjusted predictions                          Number of obs    =      1090
Expression    : Linear prediction, predict()
1._at         : sex             =          1
                age             =   43.82294 (mean)

2._at         : sex             =          2
                age             =   43.82294 (mean)
```

|  | Margin | Delta-method Std. Err. | z | P>|z| | [95% Conf. Interval] | |
|---|---|---|---|---|---|---|
| _at#mobile16 |  |  |  |  |  |  |
| 1#SAME CITY | 43.39188 | .8151133 | 53.23 | 0.000 | 41.7925 | 44.99126 |
| 1 # SAME ST,DI.. | 48.04545 | .9222139 | 52.10 | 0.000 | 46.23592 | 49.85497 |
| 1 # DIFFERENT .. | 46.62672 | .7471295 | 62.41 | 0.000 | 45.16074 | 48.0927 |
| 2#SAME CITY | 44.73797 | .7900194 | 56.63 | 0.000 | 43.18783 | 46.28811 |
| 2 # SAME ST,DI.. | 49.39154 | .9390421 | 52.60 | 0.000 | 47.54899 | 51.23408 |
| 2 # DIFFERENT .. | 47.97281 | .8208526 | 58.44 | 0.000 | 46.36217 | 49.58344 |

The additional specifications to the `margins` command are straightforward. We still have the `at()` option for `age` to be set at its mean, but now we have specified that `sex` be fixed at 1 and then fixed at 2. We see that we have the same 1,090 people as we had in the `anova` results. The results are a bit hard to read because we get separate estimated means when `sex` is at 1 (males) and when `sex` is at 2 (females). Both of these results have `age` set at the overall mean of 43.82.

It made sense to set the continuous variable, `age`, at its mean as a way of controlling for age. It would not make sense to set the categorical variable, `sex`, at its mean. People are either male or female. These are fixed categories. Setting `sex` at its mean for our subpopulation (the mean happens to be 1.47) would not be meaningful. What is a 1.47 sex?

The estimated means still appear under the column labeled "Margin". The first three are men (`sex` = 1) who are in the same town, 43.39; men who are in the same state, different town, 48.05; and men who are in a different state, 46.63. The bottom three rows are the corresponding values for women (`sex` = 2). It is interesting how occupational prestige for women in this age group who work full time has the same pattern but is slightly higher than the corresponding occupational prestige of men.

Using adjusted means can be very informative. Consider the effects of family structure (married, divorced, never married) on child outcomes. Generally, children living in families with married parents have better outcomes. This could be tested using a one-way ANOVA, where each outcome was a dependent variable and the three family types were a categorical independent variable. You might ask whether the differences have to do with marital status or whether they are dependent on other covariates. Never-married parents tend to have much lower income and somewhat lower maternal education than do married parents. Divorced parents tend to have much lower income than do married parents. These income and educational differences may be what is influencing child outcomes rather than just the family structure. Acock and Demo (1994) did a series of ANOVAs for different child outcomes using family structure as a categorical variable and adding income, education, and several other covariates. They found that for many of the outcomes, the advantage of the married parents structure was greatly reduced when controlling for the covariates.

In this example, some of the covariates of marital status may be intervening variables. Lower income or a change in school following a divorce may be mechanisms whereby divorce creates problems for child well-being. We could say that marital status is the driver variable and covariates such as income and school quality are variables that mediate the effect of divorce on child well-being (see section 14.4).

### Estimating the effect size and omega-squared, $\omega^2$

Some fields use a measure of effect size for each variable in an ANOVA. Three effect size measures are fairly widely used: eta-squared, $\eta^2$; omega-squared, $\omega^2$; and Cohen's $f^2$. The $\eta^2$ is the sum of squares for a variable (factor) divided by the total sum of squares. It can be interpreted as how much of the variation in the sample is explained by the predictor. Cohen (1988) suggests that a value of $\eta^2$ of 0.01 is a small effect size, 0.06 is medium, and 0.14 is large. The $\omega^2$ is an estimate of the explained variable in the population and adjusts for degrees of freedom and the error term, making it somewhat smaller than $\eta^2$. Because $\omega^2$ adjusts for chance effects, an estimated value of $\omega^2$ can be less than 0.0, even though it would be impossible to actually explain less than nothing. A negative value would be estimated only when a factor was clearly unimportant as a predictor. Cohen's $f^2$ is the ratio of the square root of the explained variance over the variance that is not explained. Cohen (1992) suggested that an $f$ of 0.10 is a small effect, 0.25 is medium, and 0.40 is large.

After running an **anova** command, we can obtain the $\eta^2$ and the $\omega^2$ using the **estat esize** postestimation command. We need to run this command twice: first, we will run the command by default to report $\eta^2$, and second, we will add the **omega** option to obtain $\omega^2$.

```
. estat esize
Effect sizes for linear models
```

| Source | Eta-Squared | df | [95% Conf. Interval] | |
|---|---|---|---|---|
| Model | .0206813 | 4 | .005142 | .0371681 |
| mobile16 | .0190342 | 2 | .0055725 | .0369944 |
| sex | .0024027 | 1 | 0 | .0116359 |
| age | .0002241 | 1 | 0 | .005375 |

```
. estat esize, omega
Effect sizes for linear models
```

| Source | Omega-Squared | df | [95% Conf. Interval] | |
|---|---|---|---|---|
| Model | .0170709 | 4 | .0014743 | .0336185 |
| mobile16 | .017226 | 2 | .0037395 | .0352193 |
| sex | .0014832 | 1 | 0 | .0107249 |
| age | 0 | 1 | 0 | .0044583 |

**Estimating the effect size and omega-squared, $\omega^2$, continued**

Philip B. Ender created a user-written command, omega2, that calculates $\omega^2$ and Cohen's $f$. If you do not have this command installed, type search omega2 and follow the link to install the command. Neither estat esize nor omega2 will run after oneway, but both commands run after anova.

```
. omega2 sex
omega squared for sex = 0.0015
fhat effect size = 0.0382


. omega2 mobile16
omega squared for mobile16 = 0.0172
fhat effect size = 0.1322


. omega2 c.age
omega squared for c.age = -0.0007
fhat effect size = 0.0000
```

These results show that the effect of our predictors was at best "small", especially so for sex and c.age. In the case of c.age, the negative $\omega^2$ indicates that the effects are not even as strong as would be expected using random numbers.

You will notice that for age, we used c.age because it is a continuous variable. From these results, we see that our one statistically significant effect, mobility, has an $\omega^2 = 0.02$ and that the effect size is 0.13. Both of these are considered small effect sizes. So along with saying that the effect of mobility is statistically significant, we should also acknowledge that it is quite weak. The other two variables, age and sex, were not statistically significant in our example, and it is not surprising that $\omega^2$ and the effect size for them is too small to be considered even a small effect.

Although omega2 does not work for the one-way ANOVA command, you can always do a one-way ANOVA by using the full anova command rather than the oneway command. Then, after doing the one-way ANOVA using the anova command, you can run the omega2 command.

## 9.6   Two-way ANOVA

Stata has enormous capacity for estimating a wide range of ANOVA models; we are only touching on a few of the ways Stata can fit ANOVA models here. One important extension of one-way ANOVA is two-way ANOVA, where we have two categorical predictors. For our example, we will use gss2006_chapter9_2way.dta to predict how many hours a person watches TV, tvhours, our dependent variable. We think this varies by marital status, married (coded 0 if a person is not married and 1 if a person is married), and work status, workfull (coded 0 if a person is not working full time and 1 if a person is working full time). We believe that people who work full time and people who are married will watch fewer hours of TV. Before doing the two-way ANOVA, let's check the mean number of hours of TV that married people watch compared with the number of hours that people who are not married watch and the corresponding mean number for people who work full time compared with the number for those who do not work. We do this with the following tabulate commands:

```
. tabulate married, summarize(tvhours)
```

| Are you married or not? | Summary of HOURS PER DAY WATCHING TV | | |
|---|---|---|---|
| | Mean | Std. Dev. | Freq. |
| Not Marri | 3.2383268 | 2.5831093 | 1028 |
| Married | 2.611691 | 1.8652487 | 958 |
| Total | 2.9360524 | 2.2864047 | 1986 |

```
. tabulate workfull, summarize(tvhours)
```

| Does R work full time or not | Summary of HOURS PER DAY WATCHING TV | | |
|---|---|---|---|
| | Mean | Std. Dev. | Freq. |
| Not Full | 3.4610778 | 2.5842414 | 1002 |
| Full-time | 2.4010152 | 1.784816 | 985 |
| Total | 2.9355813 | 2.2859255 | 1987 |

It looks like we might be on to something. Those working full time average 2.40 hours a day watching TV compared with 3.46 hours for people who do not work full time. Those who are married watch TV 2.61 hours per day compared with 3.24 hours per day for those who are not married.

To do a two-way ANOVA, we can open the anova dialog box by clicking on Statistics ▷ Linear models and related ▷ ANOVA/MANOVA ▷ Analysis of variance and covariance. In the dialog box, we enter tvhours as our *Dependent variable*. For *Model*, we simply list our two categorical variables, workfull and married. We are asking for partial sum of squares. Here are the command and results:

```
. anova tvhours workfull married
                              Number of obs =     1986    R-squared     =  0.0701
                              Root MSE      =  2.2059    Adj R-squared =  0.0692

            Source |   Partial SS     df       MS              F     Prob > F

             Model |   727.597834      2   363.798917        74.76    0.0000

          workfull |   532.878166      1   532.878166       109.51    0.0000
           married |   170.139227      1   170.139227        34.96    0.0000

          Residual |   9649.28082   1983   4.86600142

             Total |   10376.8787   1985   5.22764668
```

Both `workfull`, $F(1, 1983) = 109.51$, $p < 0.001$, and `married`, $F(1, 1983) = 34.96$, $p < 0.001$, are statistically significant. The $R^2 = 0.07$ means that we can explain just 7% of the variance. If you now use the `omega2` command, you can see that both `workfull` and `married` have fairly small effect sizes, so the relationships are significant but somewhat weak. Typing `omega2 workfull` produces an omega squared of 0.05 and an $\widehat{f}$ effect size of 0.23. Both of these results are considered small to medium effects. Typing `omega2 married` returns an omega squared of 0.02 and an $\widehat{f}$ effect size of 0.13.

Imagine that you think the effect of working full time may be different if you are not married than if you are married. You think that people who are not married watch a lot more TV if they do not work full time than if they do work full time. You think this difference will be less for people who are married because of marital obligations they have. This is the basis of what is called a statistical interaction. We can look at the combination of means to get a sense of this. We run a `tabulate` listing both of our predictors.

```
. tabulate workfull married, summarize(tvhours)
    Means, Standard Deviations and Frequencies of HOURS PER DAY WATCHING TV

      Does R |
   work full |   Are you married or
     time or |         not?
         not |  Not Marri    Married  |     Total

    Not Full |  3.8215613  3.0431034  |  3.4610778
             |  2.8497018  2.1664723  |  2.5842414
             |        538        464  |       1002

   Full-time |  2.5979592  2.2064777  |  2.4014228
             |  2.0761564  1.4163199  |  1.7856777
             |        490        494  |        984

       Total |  3.2383268   2.611691  |  2.9360524
             |  2.5831093  1.8652487  |  2.2864047
             |       1028        958  |       1986
```

It looks like you were right! An unmarried person who is not working full time watches over one more hour a day of TV than an unmarried person who is working; the means are 3.82 compared with 2.60 hours, a difference of 1.22 hours a day. By contrast, the difference is about half as much for those who are married; the means are 3.04

compared with 2.21, a difference of just 0.83 hours a day. This means that the effect of one variable, work status, on the dependent variable is different for different levels of the other variable, marital status. If you are married, the effect of work status on TV watching is weaker than if you are not married.

Two-way ANOVA lets us test this very easily. We open the anova dialog box again. This time under the Model tab, we enter workfull married workfull#married. Stata interprets workfull#married as telling it to multiply these two variables and use the product terms for testing the interaction. You might click on Examples... in the dialog box to see how you can do even more complicated models and how there are shortcuts beyond what we have used here. Here are the results:

```
. anova tvhours workfull married workfull#married

                         Number of obs =     1986    R-squared      =  0.0719
                         Root MSE      = 2.20434    Adj R-squared =  0.0705

           Source |  Partial SS    df       MS            F     Prob > F
           -------+--------------------------------------------------------
            Model |  746.133601     3    248.7112        51.18    0.0000
                  |
         workfull |   525.37818     1    525.37818      108.12    0.0000
          married |  169.421376     1   169.421376       34.87    0.0000
 workfull#married |  18.5357671     1   18.5357671        3.81    0.0509
                  |
         Residual |  9630.74505  1982   4.85910447
           -------+--------------------------------------------------------
            Total |  10376.8787  1985   5.22764668
```

These results show not only that workfull and married influence TV watching, as we saw before, but also that there is a marginally significant interaction, workfull#married, $F(1, 1982) = 3.81$, $p = 0.0509$. The tabulation we did aids our interpretation. After running this ANOVA, we probably should use a postestimation command, like margins, to estimate these means because it also provides confidence intervals for each of the means. Select Statistics ▷ Postestimation ▷ Marginal means and predictive margins. Under the Main tab, all we need to do is enter our three predictors, workfull, married, and workfull#married, in the space under *Factor terms to compute margins for*. Here are the results:

```
. margins workfull married workfull#married

Predictive margins                                    Number of obs    =      1986
Expression    : Linear prediction, predict()
```

|                      | Margin | Delta-method Std. Err. | t | P>\|t\| | [95% Conf. Interval] | |
|----------------------|--------|------------------------|-------|---------|-----------|-----------|
| **workfull**         |        |                        |       |         |           |           |
| Not Full             | 3.446051 | .0696898 | 49.45 | 0.000 | 3.309378 | 3.582724 |
| Full-time            | 2.409118 | .070326  | 34.26 | 0.000 | 2.271197 | 2.547038 |
| **married**          |        |                        |       |         |           |           |
| Not Married          | 3.215305 | .0688002 | 46.73 | 0.000 | 3.080377 | 3.350234 |
| Married              | 2.628582 | .071277  | 36.88 | 0.000 | 2.488796 | 2.768368 |
| **workfull# married** |        |                       |       |         |           |           |
| Not Full # Not Married | 3.821561 | .0950357 | 40.21 | 0.000 | 3.635181 | 4.007942 |
| Not Full # Married   | 3.043103 | .1023338 | 29.74 | 0.000 | 2.84241 | 3.243797 |
| Full-time # Not Married | 2.597959 | .0995818 | 26.09 | 0.000 | 2.402663 | 2.793255 |
| Full-time # Married  | 2.206478 | .0991778 | 22.25 | 0.000 | 2.011974 | 2.400982 |

Although the $R^2$ is small and the partial or semipartial $R^2$s for workfull, married, and workfull#married are even smaller, the means show what many would consider a substantively important effect. It is probably better to look at these means than simply rely on the measures of effect size.

If you are really good with numbers, the results of the margins command are all you need. We can see that workfull is associated with fewer hours watching TV (2.409 versus 3.446 under the Margin column). We see that those who are married also watch less TV. The interaction of working full time and marriage is hard to understand if we just look at these numbers. Fortunately, this is exactly where a graph is useful.

We have three factors in our ANOVA: workfull, married, and their interaction, workfull#married. We can produce three graphs. The first graph we create is the effect of workfull to visually display the difference in hours watching TV. We rerun the margins command, this time with only the workfull variable. (We include the quietly prefix because we do not want to see these results again.) Then we run marginsplot, which creates the graph in figure 9.6.

```
. quietly: margins workfull

. marginsplot
```

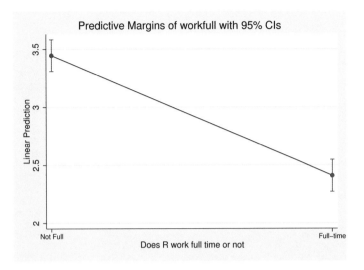

Figure 9.6. Hours of TV watching by whether the person works full time

Figure 9.6 might be improved with some judicious editing, but it is nice this way. What would you change with the Graph Editor? I would change the *y*-axis title from "Linear Prediction" to "Hours Watching TV", a more appropriate title. After examining all three graphs we will produce, I would change the range on the *y* axis to be consistent. The graph shows the difference in hours TV was watched and includes a 95% confidence interval. If we do not want this confidence interval, we simply add the `noci` option as we do in the next two graphs. To display whether the person is married, we use the following command, which creates figure 9.7:

```
. quietly: margins married
. marginsplot, noci
```

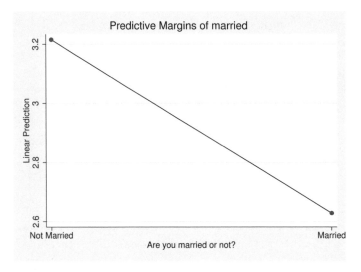

Figure 9.7. Hours of TV watching by whether the person is married

This graph makes it clear that married people watch fewer hours of TV than people who are not married. The real value of the `marginsplot` command is evident in the next example, which creates the graph in figure 9.8, when we plot the interaction between working full time and marital status:

```
. quietly: margins workfull#married
. marginsplot, noci
```

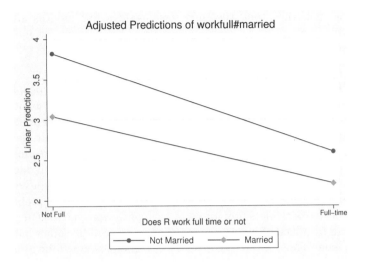

Figure 9.8. Hours of TV watching by whether the person is married and whether the person works full time

Many of your readers will find a graph like this more understandable than the table of `workfull#married` marginal means. This shows that married people watch fewer hours of TV regardless of their work status, $p < 0.001$ (from the `anova` command output). Those who work full time also watch fewer hours of TV regardless of their marital status ($p < 0.001$). The marginally significant interaction of marital status and work status, $p < 0.10$, is reflected in the lines not being parallel. Those who do not work full time have a greater gap in hours watching TV that depends on their marital status and those who work full time have less of a gap. In cases of extreme interaction, the lines might even cross over.

## 9.7  Repeated-measures design

Chapter 7 included a section on repeated-measures $t$ tests. These tests were useful when you had a before–after design in which participants were measured before an intervention and then the same participants were measured again after the intervention. We use repeated-measures ANOVA when there are more than two measurements. For example, we might want to know if the effect of the intervention was enduring. With only before and after measurements, we do not know what happens in the weeks or months following the experiment. We might have three time points, namely, pretest (before the intervention), posttest (shortly after the intervention is completed), and follow-up (another measurement a couple of months later). Students can think of many subjects where the follow-up measure would be revealing. They would have scored poorly on the final examination if they took it before taking the class. They would score much better when they took the final examination after taking the course. However, they might score poorly again if they took the same test again a few months after completing the course. This is a special problem with courses that emphasize memorization. A repeated-measures ANOVA allows us to test what happens over time.

A critical assumption of the $F$ test when doing ANOVA is that the observations are independent. This makes sense when comparing several groups; the people in one group are independent of the people in the other group. This is not a reasonable assumption when you have repeated measures on the same set of people. For example, if Alexander, Matthew, Emma, Abigail, and Jacob are each measured on their attitude toward people who are obese before an intervention, after an intervention, and three months after the intervention, each of the three measures are likely to be correlated. If Jacob is the most prejudiced at the start, the researcher hopes his prejudice will be reduced, but he is still likely to be more prejudiced than the others. If Abigail is the least prejudiced initially, she is likely to be one of the less prejudiced at each of the four measurement points. This is what we mean by a lack of independence. The scores will be somewhat homogeneous within each person. That is, three scores for the same person will be more similar than three randomly selected scores.

The way a repeated-measures ANOVA adjusts for this lack of independence is to adjust the degrees of freedom using three different strategies. The Box conservative approach is a lower bound on the adjustment. It makes it hardest to obtain statistical significance. The Greenhouse–Geisser adjustment is somewhat less conservative, but it still may have a conservative bias. The Huynh–Feldt approach is the least conservative.

There are alternatives to a repeated-measures ANOVA that go beyond the scope of this chapter. One alternative is to do a multivariate ANOVA, which Stata implements with the `manova` command. A second alternative is to do a mixed-model approach, as done with the `mixed` command. Both of these methods are beyond the scope of this chapter. We will see how to do a repeated-measures ANOVA with the caution that this does not work on a very large sample and the other approaches not covered in this book may be more appropriate.

Here are some hypothetical data for students in a small class as the data might be presented in a textbook (`wide9.dta`).

. list

|      | id | test1 | test2 | test3 |
|------|----|-------|-------|-------|
| 1.   | 1  | 55    | 85    | 80    |
| 2.   | 2  | 65    | 90    | 85    |
| 3.   | 3  | 34    | 70    | 71    |
| 4.   | 4  | 55    | 75    | 65    |
| 5.   | 5  | 61    | 59    | 65    |
| 6.   | 6  | 79    | 94    | 85    |
| 7.   | 7  | 63    | 59    | 59    |
| 8.   | 8  | 45    | 65    | 50    |
| 9.   | 9  | 54    | 70    | 60    |
| 10.  | 10 | 69    | 90    | 82    |

Stata refers to this arrangement as a *wide* layout of the data, but Stata prefers a *long* arrangement. You could make this conversion by hand with a small dataset like this, but it's easier to use the `reshape` command. Many find this command confusing, and I will illustrate only this one application of it. A detailed explanation appears at http://www.ats.ucla.edu/stat/stata/modules/reshapel.htm. For our purposes, the command to reshape the data into a long format is `reshape long test, i(id) j(time)`.

This looks strange until we analyze what is being done. The `reshape long` part lets Stata know that we want to change from the wide layout to a long layout. This is followed by a variable name, `test`, that is not a variable name in the wide dataset. However, `test` is the prefix for each of the three measures: `test1`, `test2`, and `test3`. If we had used names like `pretest`, `posttest`, and `followup`, we would have needed to rename the variables so that they would have the same stubname and then a number. We can do this by opening the Data Editor, double-clicking on the variable name we want to change, and then changing the name in the Properties pane. We could also do this by opening the Variables Manager, clicking on the variable name, changing the name in the Variable Properties pane, and clicking on the Apply button. Performing the changes in the Data Editor or the Variables Manager generates the commands `rename pretest test1`, `rename posttest test2`, and `rename followup test3`.

After the comma in the `reshape` command are two required options. (Stata calls everything after the comma an option, even if it is required for a particular purpose.) The `i(id)` option will always be the ID number for each observation. The parentheses would contain whatever variable name was used to identify the individual observations. Here we use `id`, but another dataset might use a different name, such as `ident` or `case` (if you do not have an identification variable, run the command `gen id = _n` before using the `reshape` command). The `j(time)` option refers to the time of the measurement. This option creates a new variable, `time`, that is scored a 1 if the score is for `test1`, a 2 for `test2`, or a 3 for `test3`. The choice of the name `time` is arbitrary. Here are the command and its results:

```
. reshape long test, i(id) j(time)
(note: j = 1 2 3)
Data                                wide    ->    long

Number of obs.                        10    ->        30
Number of variables                    4    ->         3
j variable (3 values)                       ->      time
xij variables:
                        test1 test2 test3   ->      test
```

```
. list

      | id    time    test |

 1.   | 1       1      55  |
 2.   | 1       2      85  |
 3.   | 1       3      80  |
 4.   | 2       1      65  |
 5.   | 2       2      90  |

 6.   | 2       3      85  |
 7.   | 3       1      34  |
 8.   | 3       2      70  |
 9.   | 3       3      71  |
10.   | 4       1      55  |

11.   | 4       2      75  |
12.   | 4       3      65  |
13.   | 5       1      61  |
14.   | 5       2      59  |
15.   | 5       3      65  |

16.   | 6       1      79  |
17.   | 6       2      94  |
18.   | 6       3      85  |
19.   | 7       1      63  |
20.   | 7       2      59  |
         (output omitted )
26.   | 9       2      70  |
27.   | 9       3      60  |
28.   | 10      1      69  |
29.   | 10      2      90  |
30.   | 10      3      82  |
```

If you are uncomfortable with the **reshape** command, you can always enter the data yourself using the long format. **id** is repeated three times for each observation, corresponding to the times 1, 2, and 3 that appear under **time**. The last column is the variable stub **test** with 55 being how person 1, id = 1, did on **test1** at **time** = 1; 85 being how person 1 did at time 2; and 80 being how person 1 did at time 3. You can see how the long format appears as a wide format to check that you have either entered the data correctly or done the **reshape** command correctly. In this example, the command is **tabdisp id time, cellvar(test)** (the results are not shown here).

With the data in the long format, we can first see how much we have violated the assumption of independence. The intraclass correlation coefficient (ICC), which is often called rho-sub-i, $\rho_I$, is a measure of the lack of independence. The variance in our three sets of test scores can be thought of as consisting of two components. There is the difference between the people because some people tend to consistently score high or consistently score low. There is also the difference within individual people, which represents the variance from one test to another for each person. If all the variability is between people, then each person would score the same at each testing and there would be complete dependence. If I know how Jacob or Abigail, or anybody else, scored at one point, I would know exactly how he or she scored at each other point. We can run a command to obtain the ICC or $\rho_I$ when the data are in the long form. Without showing the dialog box, here are the commands and output:

```
. xtset id
       panel variable:  id (balanced)

. xtreg test
Random-effects GLS regression              Number of obs      =         30
Group variable: id                         Number of groups   =         10

R-sq:  within  = 0.0000                     Obs per group: min =          3
       between = 0.0000                                    avg =        3.0
       overall = 0.0000                                    max =          3

                                            Wald chi2(0)       =          .
corr(u_i, X)       = 0 (assumed)            Prob > chi2        =          .
```

| test | Coef. | Std. Err. | z | P>\|z\| | [95% Conf. Interval] |
|---|---|---|---|---|---|
| _cons | 67.96667 | 3.510988 | 19.36 | 0.000 | 61.08526    74.84808 |

| | | |
|---|---|---|
| sigma_u | 8.8746414 | |
| sigma_e | 11.555662 | |
| rho | .37099391 | (fraction of variance due to u_i) |

The `xtreg` command is one of several xt commands. This group of Stata commands is appropriate for panel data like our example above, where the scores for each person are measured at three separate time points. Prior to running the `xtreg` command, we need to declare our data to be panel data arranged in the long form. We do this with the `xtset` command. In our example, `id` is the variable that identifies the observations (people) that have repeated measures. Each person was measured at three different times. Depending on what is repeated, the variable could have different names, for example, `xtset case`, `xtset school`, or `xtset clinic`.

The `xtreg` command estimates our intraclass correlation. The variable `test` is our dependent variable. The variable name `test` is the common stem of the three variables we had (`test1`, `test2`, and `test3`) when the data were in the wide format.

We see from the `xtreg` results that `rho` = 0.371. This is called either the ICC or the $\rho_I$, depending on your area of specialty. This would be considered a substantial lack of independence. Whenever the ICC is 0.05 or more, we need to make some adjustment. Many researchers would make an adjustment regardless of the size of the ICC because some dependency is expected as part of the research design that involves repeated measures on the same observations.

Here are the Stata command to do our repeated-measures ANOVA and the results:

```
. anova test id time, repeated(time)
                    Number of obs =        30    R-squared     =  0.8284
                    Root MSE      = 7.56233    Adj R-squared =  0.7235

       Source |  Partial SS     df       MS             F     Prob > F

        Model |  4969.56667     11   451.778788          7.90    0.0001

           id |     3328.3       9   369.811111          6.47    0.0004
         time |  1641.26667      2   820.633333         14.35    0.0002

     Residual |     1029.4      18   57.1888889

        Total |  5998.96667     29    206.86092
```

```
Between-subjects error term:  id
                   Levels:  10          (9 df)
      Lowest b.s.e. variable:  id

Repeated variable: time
                              Huynh-Feldt epsilon        =  0.7848
                              Greenhouse-Geisser epsilon =  0.6969
                              Box's conservative epsilon =  0.5000

                                     ─────────── Prob > F ───────────
       Source |    df      F     Regular    H-F      G-G        Box

         time |     2    14.35   0.0002    0.0007    0.0012    0.0043
     Residual |    18
```

In this command, we list the dependent variable, `test`; the grouping variable, `id`, because the repeated measures are repeated for each person; and the `time` variable. We included the `repeated(time)` option to make this repeated over `time`.

The top part of the results shows a traditional ANOVA table. We do not want to make much use of this information. The $R^2 = 0.828$ is misleading because this applies to the `Model`, which includes both the differences between individuals and the differences across time. We are only interested in the differences across time, so we should report the $R^2$ for `time` as the partial sum of squares for `time` divided by the total sum of squares. Hence, $R^2$ for `time` is $1641.267/5998.967 = 0.274$. If you have the `omega2` command installed, you could run `omega2`, and you would obtain $\omega^2 = 0.252$, with an effect size of 0.581. These results would indicate a substantial change over time in the test scores.

To test our significance, we need to use the table at the bottom of the results. The Huynh–Feldt epsilon will adjust the degrees of freedom by multiplying it by 0.785, the Greenhouse–Geisser will multiply it by 0.697, and Box's conservative approach would multiply it by 0.500 to estimate the probability of the $F$. Our $F = 14.35$ for our unadjusted degrees of freedom of 2 and 18. You can see that by using this degrees of freedom, our $p < 0.001$ as it is for using the Huynh–Feldt adjustment. For the

Greenhouse–Geisser adjustment, the $p < 0.01$, and this is also true by using Box's conservative adjustment. In this case, all our results are significant, but whether we report it as $p < 0.001$ or $p < 0.01$ will be determined by our choice of adjustment. Because there is no definitive best choice, it may be best to report the $p$ for all three adjustment methods. Box's approach is generally considered to be too conservative.

These results show that the scores do change significantly over time and the results are substantial in terms of $R^2$ and the effect size. However, this does not tell us exactly what the effects are. To get that, we need to run the `margins` command immediately after running the `anova` command:

```
. margins time
Predictive margins                              Number of obs    =        30
Expression   : Linear prediction, predict()
```

|      |        | Delta-method |       |       |                       |
|------|--------|--------------|-------|-------|-----------------------|
|      | Margin | Std. Err.    | t     | P>\|t\| | [95% Conf. Interval] |
| time |        |              |       |       |                       |
| 1    | 58     | 2.39142      | 24.25 | 0.000 | 52.97581    63.02419  |
| 2    | 75.7   | 2.39142      | 31.65 | 0.000 | 70.67581    80.72419  |
| 3    | 70.2   | 2.39142      | 29.35 | 0.000 | 65.17581    75.22419  |

## 9.8   Intraclass correlation—measuring agreement

The intraclass correlation coefficient, ICC or $\rho_I$, can be used as a measurement of agreement. When we have just two measures, we may be tempted to measure agreement using a simple correlation coefficient. We might have just a before and an after measure, or we might have a measure on two related people, say, a husband and his wife or a mother and her child. It is a mistake to use a simple correlation coefficient to measure agreement. Consider data reflecting how liberal a mother and her child are. If each child scored 2 points higher than the mother on a 1 to 10 scale of liberalism, then they clearly would not agree, but the correlation would be 1.0. The regression equation, `kidscore` $= 2 + 1$(`momscore`), would fit perfectly, yielding an $R^2 = 1.0$, even though each child scored 2 points higher than his or her mother. If each child scored two times the mother's score, the equation `kidscore` $= 0 + 2$(`momscore`) would fit perfectly. In either of these examples, we would say that there was agreement, and the $R^2 = 1.0$ would be inappropriate as a measure of agreement.  *NB*

The ICC will only be 1.0 if there is perfect agreement. We will illustrate it with an example from dynamics. The ICC is a better measure of agreement than the correlation, but in this example, we have three scores rather than two and agreement is across all three scores.

We randomly assign three people to each of 10 groups and have them discuss reforming Medicare. After 30 minutes of discussing the issue, each person completes a questionnaire that includes a scale measuring attitude toward welfare reform. Assume

that one member out of both groups 3 and 7 got sick, so we would have only two scores
for each group 3 and 7 on their attitudes toward reforming Medicare. You enter your
data using the long format, and the data look like this (`intraclass.dta`):

```
. list

     | medicare   group |
     |------------------|
  1. |     21        1  |
  2. |     22        1  |
  3. |     22        1  |
  4. |     17        2  |
  5. |     16        2  |
     |------------------|
  6. |     15        2  |
  7. |     15        3  |
  8. |     16        3  |
  9. |     18        4  |
 10. |     19        4  |
     |------------------|
 11. |     20        4  |
 12. |     12        5  |
 13. |     12        5  |
 14. |     14        5  |
 15. |     14        6  |
     |------------------|
 16. |     21        6  |
 17. |     24        6  |
 18. |     23        7  |
 19. |     22        7  |
 20. |     26        8  |
     (output omitted)
 26. |     35       10  |
 27. |     33       10  |
 28. |     45       10  |
     |------------------|
```

The groups, numbered 1 to 10, each appear three times, with the exception of
groups 3 and 7 (because one member in each group got sick and did not complete the
questionnaire). The three people in the first group have scores of 21, 22, and 22. The
order of members is arbitrary in this example. If the group process leads to agreement
among group members, we would expect the variance within each group to be small—all
of them would have similar scores. By contrast, we would expect there to be considerable
variance across groups. Compare the three people in group 1 with the three people in
group 2, and you can see that most of the variance is between groups. If everybody
within each group agrees with his or her group members and all the differences are
between groups, then the intraclass correlation will be 1.0.

In section 9.7, we illustrated the commands to declare data to be panel and to
produce the ICC once the data are in the long format. We can use the same statistic as
our measure of agreement. Here are the commands and results:

```
. xtset group
      panel variable:  group (unbalanced)
. xtreg medicare
Random-effects GLS regression           Number of obs      =        28
Group variable: group                   Number of groups   =        10

R-sq:  within  = 0.0000                  Obs per group: min =         2
       between = 0.0000                                 avg =       2.8
       overall = 0.0000                                 max =         3

                                         Wald chi2(0)       =         .
corr(u_i, X)   = 0 (assumed)             Prob > chi2        =         .

   medicare |     Coef.   Std. Err.      z    P>|z|     [95% Conf. Interval]

      _cons |  21.38081   2.214351     9.66   0.000     17.04077    25.72086

    sigma_u |  6.7602232
    sigma_e |  2.9907264
        rho |  .83631742   (fraction of variance due to u_i)
```

In this example, the `xtset` command specified `group` as our unit of observation. The `xtreg` command specified `medicare` as our dependent variable. Each group is one observation that has three repeated scores for the three people in each group. Notice that in section 9.8, we used `id` as the unit of observation with three repeated tests for each individual. Here we use `group` as the unit of observation with three scores for the group members.

Stata reports that the ICC is 0.836, which reflects an extremely high level of agreement. Many studies have reported correlations as measures of agreement when the ICC would be more appropriate. Sometimes researchers say a correlation is a reliable measure because when a scale is given to the same people on two occasions, we get a high correlation. This is not an appropriate measure of agreement, and the ICC should be reported instead.

# 9.9   Power analysis with ANOVA

Stata's `power` command can be very useful when you are designing a study. The `power` command has a wide range of capabilities for analyzing power in ANOVA studies, and we will illustrate only some of the basic applications. The capabilities are fully described in the *Stata Power and Sample-Size Reference Manual* (StataCorp 2013b).

## 9.9.1   One-way ANOVA

Imagine you are designing a study in which you will randomly assign women with chronic back pain to one of three conditions. Condition A uses the current, best practice treatment. Condition B uses an alternative holistic treatment. Condition C tells the women to rest and take it easy. You want to do a one-way ANOVA to see if the condition matters. How many participants will you need in your study? Let's say it is practical for you to have at least 60 participants with 20 randomly assigned to each of the three groups or conditions. However, you are writing a grant proposal to fund this project, and you could have a lot more participants if needed to have a good study. What does it mean to have a good study? One key ingredient is that your study will have the power necessary to show a significant result given that the true condition is substantively and clinically important. You need a measure of the effect size you consider important to help you decide what would be a clinically important result.

We could enter the means and standard deviations for each group, but we do not usually know them. One solution is to use a general measure of effect size used in ANOVA. We have covered $\eta^2$ and $\omega^2$. A third measure of effect size is Cohen's $f$, which is what Stata uses for its power analysis in ANOVA models. Cohen's $f$ is the square root of the ratio of the between-group to the within-group variances. This can be estimated quickly as follows:

$$f = \sqrt{\frac{\eta^2}{1 - \eta^2}}$$

An $f = 0.10$ is defined as a small effect size, an $f = 0.25$ is defined as a medium effect size, and an $f = 0.40$ is defined as a large effect size.

The cutoff values for $f$ were suggested by Cohen (1992). These reflect Cohen's own experience in the field of psychology. In some applications, you may want to have either larger or smaller cutoff values. A larger value of $f$ simply means that more of the variance is due to differences between the groups (numerator) than differences remaining within the groups (denominator). You should use cutoff values of $f$ that are appropriate to your own field.

You will need a much larger sample size for power to detect a small effect than you would need for power to detect a large effect. Let's say you want power to detect a medium effect. You decide that you need to have a power of 0.80 to detect a medium effect size of 0.25 with an alpha of 0.05. The power of 80% means that if you replicated this study an infinite number of times, 80% of the studies would obtain a statistically significant result at the $p < 0.05$ level when the true effect size was $f = 0.25$. You would still be taking a gamble, because you had the bad fortune of having one of the 20% of studies that did not obtain a significant result even though there was a true difference at the effect-size level. However, greater certainty, such as a power of 0.90, might make the study cost prohibited. Many researchers and funding agencies are satisfied to have a power of 0.80.

We will use the `power oneway` command to see what happens to the effect size we can detect with different $N$'s. We will let the total $N$ vary from 40 to 500 in increments of 20, that is, 40, 60, 80, ..., 500. We will specify a power of 0.80, an alpha of 0.05, and the three conditions or groups. We will ask for a graph and a table. Here is our command and the results:

```
. power oneway, n(40(20)500) power(0.80) alpha(0.05) ngroups(3) graph table
Performing iteration ...

Estimated effect size for one-way ANOVA
F test for group effect
Ho: delta = 0   versus   Ha: delta != 0
```

| alpha | power | N | N_per_group | delta | N_g |
|------:|------:|----:|-----------:|------:|----:|
| .05 | .8 | 40 | 13 | .5185 | 3 |
| .05 | .8 | 60 | 20 | .4115 | 3 |
| .05 | .8 | 80 | 26 | .3586 | 3 |
| .05 | .8 | 100 | 33 | .3169 | 3 |
| .05 | .8 | 120 | 40 | .287 | 3 |
| .05 | .8 | 140 | 46 | .2672 | 3 |
| .05 | .8 | 160 | 53 | .2485 | 3 |
| .05 | .8 | 180 | 60 | .2333 | 3 |
| .05 | .8 | 200 | 66 | .2223 | 3 |
| .05 | .8 | 220 | 73 | .2112 | 3 |
| .05 | .8 | 240 | 80 | .2016 | 3 |
| .05 | .8 | 260 | 86 | .1944 | 3 |
| .05 | .8 | 280 | 93 | .1868 | 3 |
| .05 | .8 | 300 | 100 | .1801 | 3 |
| .05 | .8 | 320 | 106 | .1749 | 3 |
| .05 | .8 | 340 | 113 | .1693 | 3 |
| .05 | .8 | 360 | 120 | .1643 | 3 |
| .05 | .8 | 380 | 126 | .1603 | 3 |
| .05 | .8 | 400 | 133 | .156 | 3 |
| .05 | .8 | 420 | 140 | .152 | 3 |
| .05 | .8 | 440 | 146 | .1488 | 3 |
| .05 | .8 | 460 | 153 | .1454 | 3 |
| .05 | .8 | 480 | 160 | .1421 | 3 |
| .05 | .8 | 500 | 166 | .1395 | 3 |

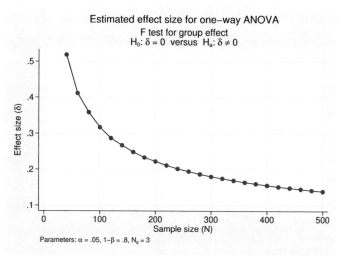

Figure 9.9. Effect size for power of 0.80, alpha of 0.05 for $N$'s from 40 to 500

If we look at the graph (the effect size is labeled $\sigma$), we can see that our original hope of doing this study with an $N = 60$ was off the mark. We now know why a funding agency would not want to fund such a study with an $N = 60$, because it would require a much greater effect size than the minimum effect size, $f = 0.25$, for which we said it would be clinically important to have a power of 0.80 with an alpha of 0.05. Based on the graph, it appears that we need about 160 observations overall with about 53 in each of our three groups. If we want a more precise way to pick our needed sample size, we can look at the tabular results. There we see that we would have a power of 0.80 to detect an effect size of 0.25 (column labeled `delta`) if we had just under 160 participants with 53 in each group. We could hone it down even more precisely by varying the total $N$ from 150 to 170 in increments of 1.

```
. power oneway, n(150(1)170) power(0.80) alpha(0.05) ngroups(3) graph table
```

Try this command and see why we would decide on $N = 159$ with 23 participants in each group.

What if we said only a large effect size would have clinical importance, $f = 0.40$? Looking at our tabular results, we see that we would need only about 60 study participants, with 20 in each group. We can also see what would happen if we insisted on detecting a small effect size. We would need a huge sample to obtain an effect size of $f = 0.10$ with an alpha of 0.05 and a power of 0.80. From the table above, we see that not even 500 participants would be enough. If we were to again type

```
. power oneway, n(500(20)1000) power(0.80) alpha(0.05) ngroups(3) graph table
```

we would see that we would need about 980 participants with 326 in each group. You can imagine that when even a small effect has important clinical implications, many research studies have samples that are too small to be effective. In that case, we do not know whether a lack of significance is because the effect is unimportant or if it is because the study design is underpowered.

### Power analysis for two-way ANOVA

Let's extend our one-way example to a two-way ANOVA. Our one-way example was restricted to women. Let's see if there is also a gender difference. With a two-way ANOVA, we have three interests: 1) whether the three conditions make a difference; 2) whether being a woman or a man makes a difference; and 3) whether there is a gender $x$ condition interaction.

For the condition difference (three columns in the two-way ANOVA), we will use the 0.80 for our power, 0.05 for our alpha, and 0.25 for our effect size. Our data now have two rows, one for women and one for men. To run the `power twoway` command, we will need to add the number of rows, `nrows(2)`, and specify that we want the results for the condition effect in the three columns, `factor(column)`. Here is our command for the condition difference and the table and graph (figure 9.10) it produces:

```
. power twoway, nrows(2) ncols(3) power(0.80) n(100(20)300) factor(column)
> graph table

Performing iteration ...

Estimated effect size for two-way ANOVA
F test for column effect
Ho: delta = 0  versus  Ha: delta != 0
```

| alpha | power | N | N_per_cell | delta | N_r | N_c |
|-------|-------|-----|------------|-------|-----|-----|
| .05 | .8 | 100 | 16 | .3221 | 2 | 3 |
| .05 | .8 | 120 | 20 | .2871 | 2 | 3 |
| .05 | .8 | 140 | 23 | .2673 | 2 | 3 |
| .05 | .8 | 160 | 26 | .251 | 2 | 3 |
| .05 | .8 | 180 | 30 | .2334 | 2 | 3 |
| .05 | .8 | 200 | 33 | .2223 | 2 | 3 |
| .05 | .8 | 220 | 36 | .2127 | 2 | 3 |
| .05 | .8 | 240 | 40 | .2017 | 2 | 3 |
| .05 | .8 | 260 | 43 | .1944 | 2 | 3 |
| .05 | .8 | 280 | 46 | .1879 | 2 | 3 |
| .05 | .8 | 300 | 50 | .1801 | 2 | 3 |

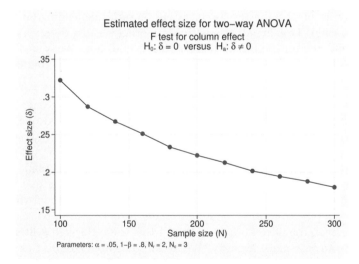

Figure 9.10. Effect size for power of 0.80 with two rows in each of the three columns for $N$'s from 100 to 300

These results are virtually the same as we had for the columns in the one-way ANOVA. In the results, the only difference is that we need 26 people in each cell.

If we were testing for gender differences, we would simply use the option `factor(row)`:

```
. power twoway, nrows(2) ncols(3) power(0.8) n(100(10)300) factor(row) graph
> table
```

Detecting a row effect of gender would require about 130 observations.

If we were testing for an interaction of condition $x$ row, we would use the option `factor(rowcol)`:

```
. power twoway, nrows(2) ncols(3) power(0.8) n(100(10)300) factor(rowcol)
> graph table
```

Detecting a medium interaction effect of $f = 0.25$ would require about 160 observations. Each of these `power` commands show that having around 160 participants in our study will work fine.

## 9.9.2   Power analysis for repeated-measures ANOVA

This time we are going to think about designing a different study. Suppose we are interested in how maternal stress varies depending on the age of a mother's oldest child. We will select three fixed ages for the oldest child. We will have one group of mothers, but we will measure their stress at four points. The first time is when the oldest child is 10. The second time is when the oldest child turns 13. The third time is when the oldest child turns 16. Finally, the fourth time is when the oldest child turns 19. Our question is whether maternal stress is consistent or varies depending on the age of the mother's oldest child. Our hypothesis is that a mother's stress does vary with the age of the oldest child and our null hypothesis is that it does not. We want to use the 0.05 alpha level of significance. We want to have a power of 0.80. How big of a sample do we need?

We know we cannot treat this as a one-way ANOVA because we do not have four independent samples. The issue is that a mother's maternal stress at one point is likely to strongly correlate with her maternal stress at another point. To allow for this lack of independence, we need to make some estimate of the correlation. Let's pick a value of $r = 0.6$. If you have no basis for estimating the correlation, you might try to repeat the analysis using different values to see how much difference that makes. You could run a simple pilot test with a few mothers to get a reasonable value for this correlation. The $r = 0.60$ was picked because we expect there to be a strong correlation between measures of the maternal stress. We might find this value through a literature review, or we may have to just make a guess. We decide that a small effect size of $f = 0.10$ is unlikely to be clinically important, but a moderate effect size of $f = 0.25$ would be important. This decision on the size of effect size can sometimes be justified by reference to literature in the field. We will be interested in how many mothers we need in our sample to have a power of 0.80 with an alpha of 0.05, four repeated measurements, and a correlation between measurements of $r = 0.60$ to detect an effect size of 0.25. Here are the command and results:

```
. power repeated, n(100(10)300) power(0.80) alpha(0.05) corr(0.60) nrepeated(4)
> ngroups(1) graph table

Performing iteration ...

Estimated within-subject variance for repeated-measures ANOVA
F test for within subject with Greenhouse-Geisser correction
Ho: delta = 0   versus   Ha: delta != 0
```

| alpha | power | N | N_per_group | delta | N_g | N_rep | corr |
|-------|-------|-----|-------------|-------|-----|-------|------|
| .05 | .8 | 100 | 100 | .3324 | 1 | 4 | .6 |
| .05 | .8 | 110 | 110 | .3167 | 1 | 4 | .6 |
| .05 | .8 | 120 | 120 | .3031 | 1 | 4 | .6 |
| .05 | .8 | 130 | 130 | .2911 | 1 | 4 | .6 |
| .05 | .8 | 140 | 140 | .2804 | 1 | 4 | .6 |
| .05 | .8 | 150 | 150 | .2708 | 1 | 4 | .6 |
| .05 | .8 | 160 | 160 | .2621 | 1 | 4 | .6 |
| .05 | .8 | 170 | 170 | .2542 | 1 | 4 | .6 |
| .05 | .8 | 180 | 180 | .247 | 1 | 4 | .6 |
| .05 | .8 | 190 | 190 | .2404 | 1 | 4 | .6 |
| .05 | .8 | 200 | 200 | .2342 | 1 | 4 | .6 |
| .05 | .8 | 210 | 210 | .2286 | 1 | 4 | .6 |
| .05 | .8 | 220 | 220 | .2233 | 1 | 4 | .6 |
| .05 | .8 | 230 | 230 | .2183 | 1 | 4 | .6 |
| .05 | .8 | 240 | 240 | .2137 | 1 | 4 | .6 |
| .05 | .8 | 250 | 250 | .2094 | 1 | 4 | .6 |
| .05 | .8 | 260 | 260 | .2053 | 1 | 4 | .6 |
| .05 | .8 | 270 | 270 | .2014 | 1 | 4 | .6 |
| .05 | .8 | 280 | 280 | .1978 | 1 | 4 | .6 |
| .05 | .8 | 290 | 290 | .1943 | 1 | 4 | .6 |
| .05 | .8 | 300 | 300 | .1911 | 1 | 4 | .6 |

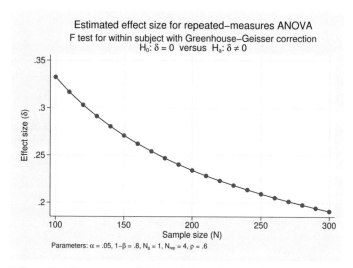

Figure 9.11.  Effect size for power of 0.80, alpha of 0.05, four repeated measurements, and a 0.60 correlation between measurements for $N$'s from 100 to 300

We decided to try sample sizes of between 100 and 300 in increments of 10. If this does not work, we could quickly try a larger or smaller set of sample sizes. We added the corr(0.60) option because our measures are not going to be independent; we indicated that we have 4 repeated measurements with the nrepeated(4) option and just one group of mothers with the ngroups(1) option.

It looks like we need about 170 mothers in our sample. We could rerun the command changing n(100(10)300) to n(160(1)180) to get a more precise estimate of the sample size required. You can see how important it is to pick the appropriate minimum effect size that would be clinically important. If you insist on being able to demonstrate a small effect size, you may need to have an impractically large sample size. If you pick a large effect size, you may find you need a very small sample, but you run a very high risk of not finding a true moderate effect to be significant.

### 9.9.3   Summary of power analysis for ANOVA

We have just scratched the surface of the capabilities of the power command when used with ANOVA. You can see how useful this is for studies involving ANOVA. When you propose to do a study and are requesting funding for your research, you need to justify your cost, and your sample size is usually a critical factor in your cost—more people means more money. A funding agency would have no reason to fund you if you requested money for a far larger sample than you need to find an appropriate effect size. At the same time, a funding agency would have no reason to fund you if you requested less money for a sample that was too small to have a good chance, say, a power of 0.80, of demonstrating a clinically worthwhile result when one existed in the population.

## 9.10   Summary

ANOVA is an extremely complex and powerful approach to analyzing data. We have only touched the surface of ANOVA here. Stata offers a powerful collection of advanced procedures that extend traditional ANOVA. In this chapter, you have learned how to

- Conduct a one-way ANOVA as an extension of the independent $t$ test to three or more groups

- Use nonparametric alternatives to ANOVA for comparing distributions or comparing medians when you have three or more groups

- Control for a covariate through ANCOVA when you do not have randomization but know how the groups differ on background variables

- Use a two-way ANOVA for situations in which you have two categorical predictors that might or might not interact with one another

- Graph interaction effects for a two-way ANOVA

- Perform a repeated-measures ANOVA as an extension of the dependent/paired $t$ test for when you have three or more measurements on each subject

- Calculate the intraclass correlation coefficient to measure agreement or homogeneity within groups

The next chapter presents multiple regression. Many of the models we have fit within an ANOVA framework can be done equally well or better within a multiple-regression format. Multiple regression is an extremely general approach to data analysis.

## 9.11 Exercises

1. Suppose that you have five people in condition A, five in condition B, and five in condition C. On your outcome variable, the five people in condition A have scores of 9, 8, 6, 9, and 5. The five in condition B have scores of 5, 9, 6, 4, and 8. The five in condition C have scores of 2, 5, 3, 4, and 6. Show how you would enter these data using a wide format. Show how you would enter them using a long format.

2. Using the data in the long format from exercise 1, do an ANOVA. Compare the means and the standard deviations. What does the $F$ test tell us? Use `pwmean` to conduct multiple-comparison tests. What do the results tell us?

3. Use `gss2002_chapter9.dta`. Does the time an adult woman watches TV (`tvhours`) vary depending on her marital status (`marital`)? Do a one-way ANOVA, including a tabulation of means; do a Bonferroni multiple-comparison test; and then present a bar chart showing the means for hours spent watching TV by marital status. Carefully interpret the results. Can you explain the pattern of means and the differences of means? Do a tabulation of your dependent variable, and see if you find a problem with the distribution.

4. Use `gss2002_chapter9.dta`. Does the time an adult watches TV (`tvhours`) depend on his or her political party identification (`partyid`)? Do a tabulation of `partyid`. Drop the 48 people who are Other party. Combine the strong Democrats, not strong Democrats, and independent near Democrats into one group and label it Democrats. Combine the strong Republicans, not strong Republicans, and independent near Republicans into one group and label it Republicans. Keep the independents as a separate group. Now do a one-way ANOVA with a tabulation and Bonferroni test to answer the question of whether how much time adults spend watching TV depends on their political party identification. Carefully interpret the results.

5. Use `gss2002_chapter9.dta`. Repeat exercise 3 using the Kruskal–Wallis rank test and a box plot. Interpret the results, and compare them with the results in exercise 3.

6. Use `nlsy97_selected_variables.dta`. A person is interested in the relationship between having fun with your family (`fun97`) and associating with peers who volunteer (`pvol97`). She thinks adolescents who frequently have fun with their families will associate with peers who volunteer. The fun-with-family variable is a count of the number of days a week an adolescent has fun with his or her family. Although this is a quantitative variable, treat this as eight categories, 0–7 days per week. Because of possible differences between younger and older adolescents, control for age (`age97`). Do an analysis of covariance, and compute adjusted means. Present a table showing both adjusted and unadjusted means. Interpret the results.

7. Use `partyid.dta`. Do a two-way ANOVA to see if there is 1) a significant difference in support for stem cell research (`stemcell`) by political party identification (`partyid`), 2) a significant difference in support by gender, and 3) a significant interaction between gender and party identification. Construct a two-way overlay graph to show the interaction and interpret it.

8. In exercise 1, you entered data for five people in each of three conditions, A, B, and C, in a wide format. Now use these same data but assume that five people were measured at three times where A is a pretest measure of the dependent variable before an intervention of some program, B is a posttest measure of the dependent variable at the conclusion of the intervention, and C is a follow-up measure 30 days after the completion of the intervention. Because you are treating this as a repeated-measures ANOVA, you need to add a variable named `id`, numbered 1–5 for the five people. You also need to label the three measures with a common stem. Let's use `time1`, `time2`, and `time3` to replace A, B, and C as labels. Use the `reshape` command to transform the data from the wide to the long format. Do a repeated-measures ANOVA and interpret the result.

9. You have 5 two-parent families, each of which has two adolescent children. You want to know if the parents agree more with each other about the risks of premarital sex or if the siblings agree more. The scores for the five sets of parents are (9, 3) (5, 5) (4, 2) (6, 7) (8, 10). The scores for the five sets of siblings are (9, 8) (7, 4) (8, 10) (3, 2) (8, 8). Enter these data in a long format, and compute the intraclass correlation for parents and the intraclass correlation for siblings. Which set of pairs has greater agreement regarding the risks of premarital sex?

# 10 Multiple regression

## 10.1   Introduction to mult

Multiple regression is an extension
a huge variety of new applications.
core statistical technique for most pub
is an exceptional tool for doing multiple regression because regression applications are
at the heart of the original conceptualization and subsequent development of Stata. In

this chapter, I introduce multiple regression and regression diagnostics. In the following chapter, I will introduce logistic regression.

## 10.2  What is multiple regression?

In bivariate regression, you had one outcome variable and one predictor. Multiple regression expands on this by allowing any number of predictors. Allowing many predictors makes sense because few outcomes have just one cause. Why do some people have more income than others? Clearly, education is an important predictor, but education is only part of the story. Some people inherit great wealth and would have substantial income whether or not they had any education. The incomes people have tend to increase as they get older, at least up to some age where their incomes may begin to decline. The careers people select will influence their incomes. A person's work ethic may also be important because those who work harder may earn more income. There are many more variables that influence a person's income, for example, race, gender, and marital status. You need more than one variable to adequately predict and understand the variation in income.

Predicting and explaining income can be even more complicated than having multiple predictors. Some of the predictors may interact such that combinations of independent variables have unique effects. For example, a physician with a strong work ethic may make much more than a physician who is lazy. By contrast, an assembly-line worker who has a strong work ethic has little advantage over an assembly-line worker who is working only at the minimum required work rate. Hence, the effect of work ethic on income can vary depending on occupation. Similarly, the effect of education on income may be different for women and men. Clearly, social science requires a statistical strategy that allows you to study many variables working simultaneously to produce an outcome; multiple regression provides that strategy.

In this chapter, we will use data from a 2004 survey of Oregon residents, called `ops2004.dta`. There is a series of 11 items with views on how concerned the person is about different environmental issues, such as pesticides, noise pollution, food contamination, air quality, and water quality. Using the alpha reliability procedures and factor analysis that are covered in chapter 12, I constructed a scale called `env_con` that has a strong alpha of 0.89. This will serve as our dependent variable; it is the outcome we want to predict.

Say that you are interested in several independent variables (predictors). These might include variables related to the person's health, education, income, and community identification. For example, we might expect people who have higher education and community identification to score higher on concern for environmental issues. Perhaps the expectation is less clear for how a person's health or income will be related to concerns for the environment, but we think they both might be important. Clearly, we have left out some important predictors, but these predictors will be sufficient to show how multiple regression analysis is done using Stata.

## 10.3 The basic multiple regression command

Select Statistics ▷ Linear models and related, and you will see a long list of regression-related commands. Stata gives us more options than I could possibly cover in this book. In this chapter, we focus on the Statistics ▷ Linear models and related ▷ Linear regression menu item. The basic dialog box for linear regression asks for the dependent variable in one box and a list of our independent variables in another box. Type env_con as the dependent variable, and then type the following independent variables: educat (years of education), inc (annual income in dollars), com3 (identification with the community), hlthprob (health problems), and epht3 (impact of the environment on the person's own health). The resulting dialog box looks like figure 10.1.

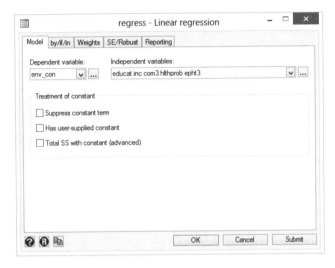

Figure 10.1. The Model tab for multiple regression

The by/if/in tab lets you fit the regression model by some grouping variable (you might estimate it separately for women and men), restrict your analysis to cases if a condition is met (you might limit it to people 18–40 years old), or apply your regression model to a subset of observations. We will go over the Weights tab and the SE/Robust tab later. The Reporting tab has several options for what is reported. Click on that tab, and check the option to show *Standardized beta coefficients*.

These dialog box options produce the simple command regress env_con educat inc com3 hlthprob epht3, beta. The syntax of this command is simple, and unless you want special options, you can just enter the command directly in the Command window. The command name, regress, is followed by the dependent variable, env_con, and then a list of the independent variables. After the comma comes the only option you are using, namely, beta. Here are the results:

```
. regress env_con educat inc com3 hlthprob epht3, beta
      Source |       SS       df       MS              Number of obs =    3769
-------------+------------------------------           F(  5,  3763) =  320.15
       Model |  647.67794       5  129.535588          Prob > F      =  0.0000
    Residual |  1522.55872    3763  .404613001         R-squared     =  0.2984
-------------+------------------------------           Adj R-squared =  0.2975
       Total |  2170.23666    3768  .575965144         Root MSE      =  .63609

----------------------------------------------------------------------------------
     env_con |      Coef.   Std. Err.      t    P>|t|                         Beta
-------------+--------------------------------------------------------------------
      educat | -.0011841     .004077    -0.29   0.772                    -.0044584
         inc | -5.51e-08    3.62e-07    -0.15   0.879                    -.0023317
        com3 |  .0503162    .0092717     5.43   0.000                     .074352
    hlthprob | -.2974035    .0248129   -11.99   0.000                    -.172927
       epht3 | -.4020741     .012687   -31.69   0.000                    -.4575999
       _cons |  3.726345    .0651735    57.18   0.000                          .
----------------------------------------------------------------------------------
```

The results have three sections. The upper left block of results is similar to what you saw in chapter 9 on analysis of variance. This time, the source called Model refers to the regression model rather than to the between-group source. The regression model consists of the set of all five predictors. The source called Residual corresponds to the error component in analysis of variance. You have 5 degrees of freedom for the model. The degrees of freedom will always be $k$, the number of predictors (here educat, inc, com3, hlthprob, and epht3) in the model.

The analysis of variance does not show the $F$ ratio like the tables in chapter 9. The $F$ appears in the block of results in the upper right section. The output says F(5, 3763) = 320.15. The $F$ of 320.15 is the ratio of the mean square for the model to the mean square for the residual. It is $129.535588/0.404613001 = 320.15$ in this example. The degrees of freedom for the numerator (Model) is 5, and for the denominator (Residual) it is 3,763. The probability of this $F$ ratio appears just below the $F$ value, Prob > $F$ = 0.0000. When this probability is less than 0.05, you can say $p < 0.05$; when it is less than 0.01, you can say $p < 0.01$; and when it is less than 0.001, you can say $p < 0.001$. In a report, you would write this as $F(5, 3763) = 320.15$, $p < 0.001$. There is a highly significant relationship between environmental concerns and the set of five predictors.

How well does the model fit the data? This question is usually answered by reporting $R^2$. The regression model explains 29.8% of the variance in environmental concerns. Stata reports this as R-squared = 0.2984. In bivariate regression, the $r^2$ measured how close the observations were to the straight line used to make the prediction. With multiple regression, you use the capital $R^2$, which measures how close the observations are to the predicted value, based on the set of predictors. What is a weak or a strong value for $R^2$ varies by the topic being explained. If you are in a fairly exploratory area, an $R^2$ near 0.3 is considered reasonably good. A rule of thumb some researchers use is that an $R^2$ less than 0.1 is weak, between 0.1 to 0.2 is moderate, and greater than 0.3 is strong. Be careful in applying this rule of thumb because some areas of research require higher values and others allow lower values. You might report our results as showing that we can explain 29.8% of the variance in environmental concern using our set of predictors, and this is a moderate-to-strong relationship.

On a small sample with several predictors, the value of $R^2$ can exaggerate the strength of the relationship. Each time you add a variable, you expect to increase $R^2$ just by chance. $R^2$ cannot get smaller as you add variables. When you have many predictors and a small sample, you may get a big $R^2$ just by chance. To offset this bias, some researchers report the adjusted $R^2$. This will be smaller than the $R^2$ because it attempts to remove the chance effects. When you have a large sample and relatively few predictors, $R^2$ and the adjusted $R^2$ will be similar, and you might report just the $R^2$. However, when there is a substantial difference between the two, you should report both values.

Across the bottom of the results is a block that has six columns containing the key regression results. A formal multiple regression equation is written as

$$\widehat{Y} = b_0 + b_1 X_1 + b_2 X_2 + \cdots + b_k X_k$$

$\widehat{Y}$ is the predicted value of the dependent variable. $b_0$ is the intercept or constant. Stata calls $b_0$ the _cons, which is an abbreviation for the constant. You can think of this as the base prediction of $Y$ when all the $X$ variables are fixed at zero. $b_1$ is the regression coefficient for the effect of $X_1$, $b_2$ is the regression coefficient for the effect of $X_2$, and $b_k$ is the regression coefficient for the last $X$ variable. You might see these coefficients called *unstandardized regression coefficients*. You can get the values for the equation from the output (see table 10.1).

Table 10.1. Regression equation and Stata output

| Name in the regression equation | Stata name | Stata results: Unstandardized coefficient |
|---|---|---|
| $b_0$ | _cons | 3.726 |
| $b_1$ | educat | −0.001 |
| $b_2$ | inc | −5.51e–08 |
| $b_3$ | com3 | 0.050 |
| $b_4$ | hlthprob | −0.297 |
| $b_5$ | epht3 | −0.402 |

You can input this information into the equation, calculating the estimated value of the dependent variable:

$$\widehat{\text{env\_con}} = 3.726 - 0.001(\text{educat}) - 0.000(\text{inc}) + 0.050(\text{com3})$$
$$- 0.297(\text{hlthprob}) - 0.402(\text{epht3})$$

The regression results next give a standard error, $t$-value, and $P > |t|$ for each regression coefficient. If you divide each regression coefficient by its standard error, you obtain the $t$-value that is used to test the significance of the coefficient. Each of the $t$

ratios has $N - k - 1$ degrees of freedom. Subtract one degree of freedom for each of the $k = 5$ predictors and another degree for the constant. From the upper right corner of the results, you can see that there are 3,769 observations, and from the analysis of variance table, you can see that there are 3,763 degrees of freedom for the residual. If you forget the $N - k - 1$ rule for degrees of freedom, you can use the number of degrees of freedom for the residual in the analysis of variance table. You might want to include the degrees of freedom in your report, but you do not need to look up the probability in a table of $t$-values because Stata gives you a two-tailed probability. For example, you might write that $b_5 = -0.402$, $p < 0.001$. In some fields, you would write $b_5 = -0.402$, $t(3763) = -31.69$, $p < 0.001$.

The unstandardized regression coefficients have a simple interpretation. They tell you how much the dependent variable changes for a unit change in the independent variable. For example, a 1-unit change in com3, identification with the community, produces a 0.05-unit change in env_con, environmental concern, holding all other variables constant. Without studying the range and distribution of each of the variables, it is hard to compare the unstandardized coefficients to see which variable is more or less important. The problem is that each variable can be measured on a different scale.

Here we measured income using dollars, and this makes it hard to interpret the $b$-value as a measure of the effect of income. After all, a $1.00 change in your annual income is something you might not even notice, and this change surely would not have much of an effect on anything. This is reflected in $b_{\text{inc}} = -5.51\text{e--}08$, which is a tiny value. (If you have not seen this format for numbers, it is a part of scientific notation, which is used for tiny numbers. The e--08 tells us to move the decimal place eight places to the left. Thus this number is $-0.0000000551$.) If we had used the natural logarithm of income or if we had measured income in 1,000s of dollars or even 10,000s of dollars, we would have obtained a different value. When you use income as a predictor, it is common to use a natural logarithm or measure income in 10,000s of dollars, where an income of 47,312 would be represented as 4.7312.

In our command above, we included the option to get beta weights ($\beta$). Beta weights are more easily compared because they are based on standardizing all variables to have a mean of 0 and a standard deviation of 1. These beta weights are interpreted similarly to how you interpret correlations in that $\beta < 0.20$ is considered a weak effect, $\beta$ between 0.2 and 0.5 is considered a moderate effect, and $\beta > 0.5$ is considered a strong effect. The beta weights tell you that education, income, and community identity all have a weak effect, and of these, only community identity is statistically significant, $p < 0.001$. Having a health problem in the family, hlthprob, has a beta weight that is still weak, $\beta = -0.173$, but it is statistically significant, $p < 0.001$. Having environmental health concerns, epht3, has $\beta = -0.458$, $p < 0.001$, which is a moderate-to-strong effect.

You might summarize this regression analysis as follows: A model including education, income, community identity, environmental health concerns, and health problems in the family explains 29.8% of the variance in environmental concerns of adults in Oregon, $F(5, 3763) = 320.15$, $p < 0.001$. Neither education nor income has a statistically significant effect. Community identity has a weak but statistically significant effect,

$\beta = 0.074$, $p < 0.001$. The two health variables are the strongest predictors in this group: having a health problem in the family has $\beta = -0.173$, $p < 0.001$, and having environmental health concerns has $\beta = -0.458$, $p < 0.001$.

If you do not want beta weights, you would simply rerun the regression without the `beta` option. Without the `beta` option, you get a confidence interval for each unstandardized regression coefficient. If you have an analysis where your primary interest is in the unstandardized coefficients, confidence intervals are a useful extension on the simple tests of significance.

## 10.4   Increment in R-squared: **Semipartial correlations**

Beta weights are widely used for measuring the effect sizes of different variables. Because it is based on standardized variables, a beta weight tells you how much of a standard deviation the dependent variable changes for each standard deviation change in the independent variable. Except for special circumstances, beta weights range from $-1$ to $+1$, with a zero meaning that there is no relationship. Sometimes the interpretation of beta weights is problematic. If you have a categorical variable like gender, the interpretation is unclear. What does it mean to say that as you go up one standard deviation on gender, you go up $\beta$ standard deviations on the dependent variable? Also, when you are comparing groups, the beta weights can be misleading. If you compare regression models predicting income for women and for men, the beta weights would be confounded with real differences in variance between women and men in the standardization of all variables.

Another approach to comparing the importance of variables is to see how much each variable increases $R^2$ if the variable is entered last. Effectively, this approach fits the model with all the variables except for the one you are interested in. You obtain an $R^2$ value for this model. Then you fit the model a second time, including the variable in which you are interested. This will have a larger $R^2$, and the difference between the two $R^2$ values will tell you how much the variable increases $R^2$. This increment is how much of the variance is uniquely explained by the independent variable you are interested in, controlling for all the other independent variables. This increment is often called simply an *increment* in $R^2$, which is a very descriptive name. Some people call it a *part-correlation square* because it measures the part that is uniquely explained by the variable. Others call it a *semipartial $R^2$*. Another way to compare variables is using what is called a *partial correlation*; we will not cover that here because it has less-desirable properties for comparing the relative importance of each variable.

Estimating the increment in $R^2$ would be a tedious process because you would have to fit the model twice for every independent variable and compute each of these differences. In this example, we would have to run 10 regression models. Fortunately, there is a Stata command, `pcorr`, that automatically does this and gives us the partial and semipartial correlations for each variable controlling for all the other variables. Select Statistics ▷ Summaries, tables, and tests ▷ Summary and descriptive statistics ▷ Partial correlations to open the `pcorr` dialog box.

In the dialog box, we enter our dependent variable in the box labeled *Display partial and semipartial correlation coefficient of variable*. Then we enter our list of independent variables in the box labeled *Against variables*. Figure 10.2 displays our updated dialog box:

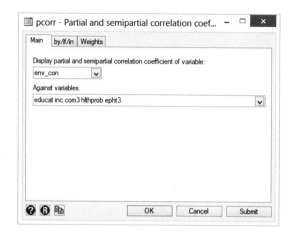

Figure 10.2. The Main tab of the pcorr dialog box

Here are the command and results:

```
. pcorr env_con educat inc com3 hlthprob epht3
(obs=3769)

Partial and semipartial correlations of env_con with
```

| Variable | Partial Corr. | Semipartial Corr. | Partial Corr.^2 | Semipartial Corr.^2 | Significance Value |
|---|---|---|---|---|---|
| educat | -0.0047 | -0.0040 | 0.0000 | 0.0000 | 0.7715 |
| inc | -0.0025 | -0.0021 | 0.0000 | 0.0000 | 0.8788 |
| com3 | 0.0881 | 0.0741 | 0.0078 | 0.0055 | 0.0000 |
| hlthprob | -0.1918 | -0.1637 | 0.0368 | 0.0268 | 0.0000 |
| epht3 | -0.4590 | -0.4327 | 0.2107 | 0.1873 | 0.0000 |

Here you are interested only in the column labeled Semipartial Corr.^2, semipartial $R^2$, which shows how much each variable contributes uniquely. The concern for environmental health problems has an increment to $R^2$ of 0.1873, $p < 0.001$. Although two other variables have statistically significant effects, this is the only variable that has a substantial increment to $R^2$. The semipartial $R^2$ is a conservative estimate of the effect of each variable because it measures only how much the $R^2$ increases when that variable is entered after all the other variables are already in the model. You might notice that the sum of the semipartial $R^2$s is just 0.220 even though the model $R^2$ is 0.298. This happens because some of the predictors are explaining the same part of the variance in the dependent variable, and whatever is shared with other predictors is not unique. The semipartial $R^2$ estimates only the unique effect of each predictor.

## 10.5   Is the dependent variable normally distributed?

Now let's examine the distribution of the dependent variable. Many applied researchers believe that multiple regression assumes the dependent variable is normally distributed. Actually, the assumption is that the residuals (error term) are normally distributed; this will be discussed in section 10.6. If the dependent variable has a highly skewed distribution or is bimodal, the residuals may not be normally distributed unless we have independent variables that predict this pattern. If we do not have independent variables that predict this pattern, the residuals may also be skewed. Because we make the assumption that the residuals are normally distributed, it would be helpful to understand the distribution of the dependent variable and consider whether we have predictors to explain any skewness and other nonnormalities. This section shows how to examine the distribution of the dependent variable.

Let's create a histogram on `env_con`:

```
. histogram env_con, frequency normal kdensity
```

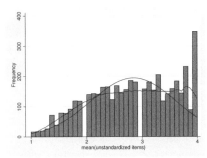

Figure 10.3. Histogram of dependent variable, `env_con`

Three distributions are represented in figure 10.3. The bars represent the actual distribution, which does not look too bad except for the big bunch of high scores (there are 348 people who have the maximum possible score of 4.0 on the environmental concerns scale). Apparently, many people in Oregon share a serious concern about the environment. The smooth, bell-shaped curve represents how the data would be distributed if they were normal.

The other curve, called a kdensity curve, is a bit hard to see in figure 10.3. The kdensity curve is an estimation of how the population data would look given our sample data. The actual data from the histogram, the normal curve, and the kdensity curve are similar up to a value of about 3.2–3.5. At the center of the distribution, the data and the kdensity curve are a little flatter than a normal curve. We have too few cases in the middle of the distribution to call it normal. The real problem, however, is on the right side of the distribution, where we have too many people who have a mean score of over about 3.5. The scale ranged from 1 to 4, and a mean score of 3.5 or more means that the person has strong environmental concern.

An alternative plot is the hanging rootogram (`ssc install hangroot`). Once the
`hangroot` command is installed, we simply enter `hangroot env_con, bar`. This produces the graph in figure 10.4.

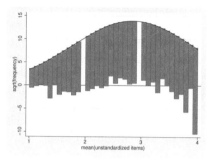

Figure 10.4. Hanging rootogram of dependent variable, `env_con`

This hanging rootogram has a smooth curve representing how the data would be distributed if it were normal. When a bar descends below the horizontal line, there are
too many observations at this value compared with a normal distribution. At a value
of 4, the bar drops far below the horizontal line, and the bars are all below the line
from a value of about 3.5 and higher. There are a few values toward the middle of the
distribution that do not drop down to the horizontal line; there are not enough values
in this area of the distribution.

The hanging rootogram is sometimes easier to read than a histogram, in part, because the vertical axis is the square root of the frequency rather than the actual frequency. This makes it much easier to see deviations from normality in the tails where
there are usually relatively few observations. John Tukey suggested this type of graph,
and Maarten L. Buis, a Stata user, wrote the `hangroot` command in 2010.

We can evaluate the normality of our dependent variable by using two additional
statistics that are known as skewness and kurtosis. We learned earlier about the `detail`
option to the `summarize` command; this option provides these statistics. Just use the
`summarize` command with the `detail` option now.

```
. summarize env_con, detail
                    mean(unstandardized items)

          Percentiles     Smallest
   1%       1.181818           1
   5%       1.545455           1
  10%            1.8           1        Obs                  4506
  25%       2.272727           1        Sum of Wgt.          4506

  50%            2.9                    Mean             2.842405
                             Largest    Std. Dev.         .7705153
  75%            3.5              4
  90%       3.818182              4      Variance         .5936939
  95%              4              4      Skewness        -.2302043
  99%              4              4      Kurtosis         2.059664
```

In the lower right column, you find skewness of $-0.23$ and kurtosis of 2.06. Skewness is a measure of whether a distribution trails off in one direction or another. For example, income is positively skewed because there are a lot of people with relatively low incomes, but there are just a few people who have extremely high incomes. By contrast, env_con is negatively skewed because there are a lot of people who have a high level of environmental concern, but there are relatively fewer people who have a low level of environmental concern. A normal distribution has skewness of 0. If the skewness is greater than 0, the distribution is positively skewed; if the skewness is less than 0 (as with env_con), the distribution is negatively skewed.

Kurtosis measures the thickness of the tails of a distribution. If you look at our histogram, you see that the tail to the left of the mean is a little too thick and the tail to the right of the mean is way too thick to be normally distributed. When a distribution has a problem with kurtosis indicated by thick tails, it will also have too few cases in the middle of the distribution. By contrast, when a distribution has a problem with kurtosis indicated by thin tails, it will also have too many cases in the middle of the distribution (peaked) for it to be normally distributed; this would happen for a variable in which most people were very close to the mean value.

A normal distribution will have a kurtosis of 3.00. A value of less than 3.00 means that the tails are too thick (hence, too flat in the middle), and a value of greater than 3.00 means that the tails are too thin (hence, too peaked in the middle). The kurtosis is 2.06 for env_con, meaning that the tails are too thick; you could tell that by looking at the histogram. Some statistical software, such as SAS or IBM SPSS Statistics, reports a value for kurtosis that is the actual value of kurtosis minus three so that a normal distribution would have a value of zero. Stata does not do this, so the correct value for a normal distribution in Stata is 3.00. You need to be careful when writing a report that will be read by people who rely on other programs that may have kurtosis of 0 for a normal distribution.

The summarize command does not give you the significance of the skewness or kurtosis coefficients. To get the significances, you need to run a command that you enter directly:

```
. sktest env_con
                   Skewness/Kurtosis tests for Normality
                                                        ──────── joint ────────
        Variable │   Obs   Pr(Skewness)   Pr(Kurtosis)  adj chi2(2)   Prob>chi2
        ─────────┼──────────────────────────────────────────────────────────────
         env_con │ 4.5e+03    0.0000         0.0000          .           0.0000
```

The **4.5e+03** under "Obs " means that there are 4,500 observations in our survey. (The **e+03** means to move the decimal place three places to the right.)

This does not report the test statistics Stata uses (the chi-squared is too big to fit in the space Stata provides for that column), but it does give us the probabilities. Combining this result with the results from the **summarize** command, we would say that skewness $= -0.23$, $p < 0.001$, and kurtosis $= 2.06$, $p < 0.001$. Both of these tested jointly have a $p < 0.001$. Hence, the distribution is not normal, the skewness tells us the distribution has a negative skew, and the kurtosis tells us the tails are too thick.

## 10.6   Are the residuals normally distributed?

An important assumption of regression is that the residuals are normally distributed. At a risk of oversimplification, this means that we are as likely to overestimate a person's score as we are to underestimate their score, and we should have relatively few cases that are extremely overestimated or underestimated.

We could examine the residual to see if it is normally distributed. Following the **regress** command that we ran on page 270, we can obtain the residual values by running the **predict** command with the **residual** option. This command generates a new variable that we have called **res**, which includes the residual value for each observation. We then run a **summarize** command to obtain the skewness and kurtosis. Here are our results:

```
. regress env_con educat inc com3 hlthprob epht3, beta
  (output omitted)

. predict res, residual
(739 missing values generated)

. summarize res, detail
                            Residuals
        ──────────────────────────────────────────────────────────
              Percentiles      Smallest
         1%    -1.425768      -2.247876
         5%    -1.05727       -1.951321
        10%    -.8467104      -1.940851       Obs               3769
        25%    -.4719241      -1.762432       Sum of Wgt.       3769

        50%     .0352801                      Mean         -4.12e-10
                                Largest       Std. Dev.     .6356698
        75%     .483786        1.849233
        90%     .7994747       1.852046       Variance      .4040761
        95%     .9625443       1.914089       Skewness     -.0997807
        99%    1.357709        1.944407       Kurtosis      2.592773
```

We see that the skewness of $-0.010$ is close to the value of $0.000$ for a normal distribution. The kurtosis of 2.592 a bit low compared with the value of 3.00 for a normal distribution (remember SAS and SPSS subtract 3.0 from the kurtosis). If we wanted to test the significance of the skewness and kurtosis, we would type

```
. sktest res
                    Skewness/Kurtosis tests for Normality
                                                   ——————— joint ———————
       Variable |   Obs   Pr(Skewness)   Pr(Kurtosis)  adj chi2(2)    Prob>chi2
                +——————————————————————————————————————————————————————————————
            res |  3.8e+03    0.0124        0.0000         41.53         0.0000
```

Both the skewness and the kurtosis are significantly different from what they would be if our residuals were normally distributed. However, we have a very large sample, and this large sample $N$ not only contributes to the tests being statistically significant, but also limits the concern we have about violating the normality of residuals assumption. You should use this test of a normal residual cautiously because it tends to result in a significant test when there is only a small departure of normality for a large sample where there is considerable robustness against violating the normality assumption. By contrast, when you have a smaller sample, say, under 100 observations, the violation of normality is more serious, but this test may have insufficient power to detect the departure from normality.

Assumptions about the distribution of the residual become more difficult when we add the fact that this distribution should be about the same for any value or combination of values for the predictors. Imagine predicting income from a person's education. With a low education, we would predict a low income. As a person goes up the scale on education, we would predict higher income. However, for those with 16 or more years of education, there is a huge range in income. Some may be in low-paying service positions, and others may be in extremely high-paying positions. If we plotted a scattergram with the regression line drawn through it, we might expect the variance around the line to increase as education increases. This effect is illustrated in figure 10.5.

This figure shows that the residual values are not normally distributed for each value of education (think of a vertical slice at a particular value of education); they are not equally likely to be above or below the predicted value for a particular value of education. Additionally, figure 10.5 shows that the variance of the residual values increases as education increases. That is, if you consider people who have less than about 12 years of education, there is little variance in the residuals. For any particular value of education up to about 12 years, a vertical slice has little variance with all the residuals closely packed together. By contrast, if you look at people who have more than 12 years of education, say, those with 16 years of education, a vertical slice of the residuals has a lot of variance. The residual values in this case range from far below the predicted value to far above it. The way the variance of the residual increases as years of education increases is called heteroskedasticity of residuals. When there is heteroskedasticity as shown in this example, we cannot predict income well for higher values of education.

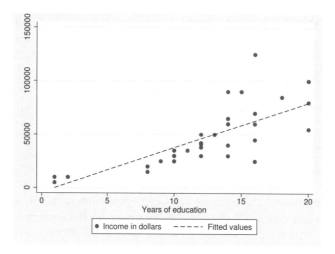

Figure 10.5. Heteroskedasticity of residuals

When we have multiple regression, it is difficult to do a graph like this because we have several predictors. The solution is to look at the distribution of residuals for different predicted values. That is, when we predict a small score on environmental concern, are the residuals distributed about the same as they are when we predict a high score?

Stata has several options to view distributions of residuals. Selecting Statistics ▷ Linear models and related ▷ Regression diagnostics ▷ Residual-versus-fitted plot opens a dialog box for the residual-versus-fitted value plot. On the Main tab, we see that we only need to click on OK. We could use the other tabs to make the graph more appealing (I used the Y axis tab to add a reference line at a value of zero), or we could just click on OK. Figure 10.6 shows the results.

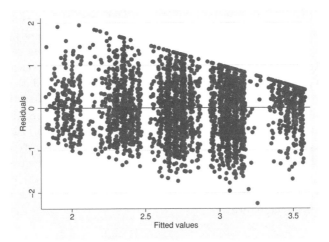

Figure 10.6. Residual-versus-fitted plot

This graph is a bit hard to read. Ideally, the observations would be normally distributed for any fitted values ($x$ axis) about the reference line (residual of 0 on $y$ axis). Where the fitted values are low, the residuals tend to be mostly positive, and where the fitted values are high, the residuals tend to be mostly negative.

A residual-versus-predicted plot, similar to the plot in figure 10.6, makes the most sense when the dependent variable takes on a larger range of values. In those situations, it may show that there is increasing or decreasing error variance as the predicted value gets larger. With our current example, where the dependent variables are between 1 and 4, it makes the problem clear but does not suggest any solution.

Sometimes it is easier to visualize how the residuals are distributed by doing a graph of the actual score of env_con on a predicted value for this score. It is easiest to show how to do this using the commands. Before doing that, however, we might want to draw a sample of the data. Because this is a very large dataset, there will be so many dots in a graph that it will be hard to read and many of the dots will be right on top of each other. To make the graph easier to read, we will include a step to sample 100 observations. Here are the commands:

```
. regress env_con educat inc com3 hlthprob epht3, beta
. predict envhat
. preserve
. set seed 111
. sample 100, count
. twoway (scatter env_con envhat) (lfit env_con envhat)
. restore
```

This series of commands does the regression first. The next line, **predict envhat**, will predict a score on **env_con** for each person based on the regression equation. Here the new variable is called **envhat**, but you can use any name because the default is to predict the score based on the regression equation. Before we take a sample of 100 observations, we enter the command `preserve`, which allows us to `restore` the dataset back to its original size. The fourth line, `set seed 111` provides a fixed start for the random sampling so that we can repeat what we did and get the same results. The command `sample 100, count` will draw a random sample of 100 cases. We are doing this to make the graph easier to read. We might want to save our data before doing this because once we drop all but 100 people, they are gone. If you save the sample of 100 observations, make sure you give the file a different name. (Using the `preserve` and `restore` commands is a simpler way to preserve the full dataset.) The next line is a `twoway` graph, as we have done before. The first section does a scattergram of `env_con` on `envhat`, and the second section does a linear regression line, regressing `env_con` on `envhat`. The final line has the command `restore`, which returns the complete dataset. The resulting graph is shown in figure 10.7, where I have used the Graph Editor to relabel some parts of the figure.

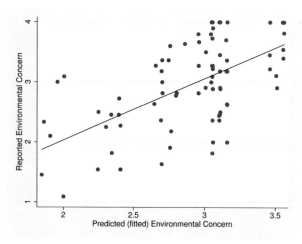

Figure 10.7. Actual value of environmental concern regressed on the predicted value

Examining this graph, we can see that in the middle, there is not much of a problem. The problem is really only at the ends, where we are predicting a very high or very low score. That is, for predictions of a high score, there are more people who actually report being lower than what we predict, and for predictions of a low score, it is just the opposite.

Is there anything we can do when we have concerns about the distribution of the residuals? One solution is to run a robust regression. A robust regression uses what is known as a sandwich estimator to estimate the standard errors. The command is

```
regress env_con educat inc com3 hlthprob epht3, vce(robust)
```

The `vce(robust)` option tells Stata to estimate the variance–covariance matrix of the errors in a way that does not assume normality. This will yield identical parameter estimates for the $b$'s, $\beta$'s, and $R^2$, but somewhat different $t$-values for testing their significance. It is important to emphasize that only the standard errors and resulting *t*-values will change, so violating normality of residuals' direct impact is on tests of significance.

Another alternative when you are concerned about the distribution of the residuals is to use a bootstrap estimation of the standard errors. This will draw several random samples with replacement (we will use 10,000 here) from your dataset. The regression will be estimated for each of these samples. Stata then examines the distribution of each parameter, for example, the *b* for `educat`, across the 1,000 results and then Stata uses the variance of that distribution to estimate a standard error. The `regress` command with bootstrap estimation of the standard errors is

```
regress env_con educat inc com3 hlthprob epht3, vce(bootstrap, reps(1000))
```

As with the robust estimator, the only thing that changes is the standard errors and hence the *t*-values. The `vce(bootstrap, reps(1000))` option may sound like it would take forever to run, but with modern computers and the efficient programming behind Stata commands, this command runs quickly.

# 10.7   Regression diagnostic statistics

Stata has a strong set of regression diagnostic tools. We will cover only a few of them. Enter the command `help regress postestimation`, and you will see a basic list of what we cover as well as several other regression diagnostic tools that are beyond the scope of this book.

## 10.7.1   Outliers and influential cases

When we are doing regression diagnostics, one major interest is finding outliers. These are cases with extreme values, and they can have an extreme effect on an analysis. One way to think of outliers is to identify those cases that the regression equation has the most trouble predicting; these are the cases that are farthest from the predicted values. These cases are of special interest for several reasons. You might want to examine them to see if there was a coding error by comparing your data with the original questionnaires. If you can contact these people, you might want to do a qualitative interview to better understand why the variables you thought would predict their scores failed. You may discover factors that you had not anticipated, and this could guide future research.

To find the cases we cannot predict, use postestimation commands. Select Statistics ▷ Postestimation ▷ Predictions, residuals, etc. to open the dialog box to predict the estimated value, `yhat`; the residual, `residual`; and the standardized residual, `rstandard`. We could also enter the following series of commands directly:

```
. use http://www.stata-press.com/data/agis4/ops2004, clear
. regress env_con educat inc com3 hlthprob epht3, beta
. predict yhat
. predict residual, residual
. predict rstandard, rstandard
. list respnum env_con yhat residual rstandard
> if abs(rstandard) > 2.58 & rstandard < .
```

The first prediction command, `predict yhat`, will predict the estimated score based on the regression. This has no option after a comma because the estimated score is the default prediction and the variable name, `yhat`, could be any name we choose. The next command, `predict residual, residual`, will predict the residual, or $Y - \widehat{Y}$, giving us a raw or unstandardized measure of how far the estimated value, `yhat`, is from the person's actual score on the environmental-concern outcome. The `residual` option tells Stata to predict this residual value. The last prediction command, `predict rstandard, rstandard`, tells Stata to estimate the standardized residual for each observation and to create a new variable called `rstandard`. This is a $z$ score that we can use to test how bad our prediction is for each case. We usually will be interested in our "bad" predictions, because these are the residual outliers. The `list` command lists all cases that have a standardized residual—that is, $z$ score—whose size is greater than 2.58 (we include the restriction that `rstandard` is also less than "."). We picked 2.58 because this corresponds to the two-tailed 0.01 level of significance. In other words, we would expect residuals this large in either direction less than 1% of the time by chance. Here are the results of this string of commands:

```
. use http://www.stata-press.com/data/agis4/ops2004, clear
. regress env_con educat inc com3 hlthprob epht3, beta
```

| Source   | SS         | df   | MS         |
|----------|------------|------|------------|
| Model    | 647.67794  | 5    | 129.535588 |
| Residual | 1522.55872 | 3763 | .404613001 |
| Total    | 2170.23666 | 3768 | .575965144 |

| | |
|---|---|
| Number of obs = | 3769 |
| F( 5, 3763) = | 320.15 |
| Prob > F = | 0.0000 |
| R-squared = | 0.2984 |
| Adj R-squared = | 0.2975 |
| Root MSE = | .63609 |

| env_con  | Coef.     | Std. Err. | t      | P>\|t\| | Beta      |
|----------|-----------|-----------|--------|-------|-----------|
| educat   | -.0011841 | .004077   | -0.29  | 0.772 | -.0044584 |
| inc      | -5.51e-08 | 3.62e-07  | -0.15  | 0.879 | -.0023317 |
| com3     | .0503162  | .0092717  | 5.43   | 0.000 | .074352   |
| hlthprob | -.2974035 | .0248129  | -11.99 | 0.000 | -.172927  |
| epht3    | -.4020741 | .012687   | -31.69 | 0.000 | -.4575999 |
| _cons    | 3.726345  | .0651735  | 57.18  | 0.000 | .         |

```
. predict yhat
(option xb assumed; fitted values)
(738 missing values generated)

. predict residual, residual
(739 missing values generated)

. predict rstandard, rstandard
(739 missing values generated)
```

```
. list respnum env_con yhat residual rstandard
> if abs(rstandard) > 2.58 & rstandard < .
```

|        | respnum | env_con  | yhat     | residual  | rstandard |
|--------|---------|----------|----------|-----------|-----------|
| 65.    | 100072  | 4        | 2.055593 | 1.944407  | 3.060185  |
| 170.   | 100189  | 4        | 2.339057 | 1.660942  | 2.616319  |
| 323.   | 100370  | 4        | 2.355076 | 1.644924  | 2.589258  |
| 539.   | 100626  | 4        | 2.353892 | 1.646108  | 2.591119  |
| 800.   | 100928  | 4        | 2.353216 | 1.646784  | 2.5918    |
| 803.   | 100931  | 1.454545 | 3.110656 | -1.656111 | -2.604731 |
| 1056.  | 101237  | 4        | 2.150767 | 1.849233  | 2.911431  |
| 1657.  | 101958  | 4        | 2.355283 | 1.644717  | 2.589034  |
| 1690.  | 101996  | 4        | 2.349444 | 1.650556  | 2.598174  |
| 2017.  | 102384  | 3.9      | 2.251481 | 1.648519  | 2.594183  |
| 2247.  | 102656  | 1.272727 | 3.213579 | -1.940851 | -3.055561 |
| 2418.  | 102866  | 4        | 2.197531 | 1.802469  | 2.836931  |
| 2608.  | 103091  | 1.818182 | 3.50535  | -1.687168 | -2.654609 |
| 3386.  | 200221  | 1.363636 | 3.058592 | -1.694955 | -2.665744 |
| 3463.  | 200325  | 1        | 2.655898 | -1.655898 | -2.604345 |
| 3655.  | 200587  | 1.111111 | 3.062432 | -1.951321 | -3.069876 |
| 3662.  | 200595  | 3.818182 | 1.904093 | 1.914089  | 3.012149  |
| 3679.  | 200621  | 4        | 2.353216 | 1.646784  | 2.5918    |
| 3736.  | 200704  | 4        | 2.356122 | 1.643878  | 2.587669  |
| 3743.  | 200712  | 4        | 2.147954 | 1.852046  | 2.918959  |
| 3745.  | 200716  | 1.3      | 3.062432 | -1.762432 | -2.772711 |
| 3762.  | 200743  | 1        | 2.666197 | -1.666197 | -2.624323 |
| 3795.  | 200785  | 1        | 3.247876 | -2.247876 | -3.542243 |
| 4117.  | 201350  | 1.363636 | 3.007862 | -1.644226 | -2.586329 |
| 4147.  | 201423  | 1        | 2.709645 | -1.709645 | -2.690599 |
| 4172.  | 201472  | 4        | 2.242008 | 1.757992  | 2.767844  |
| 4348.  | 201767  | 4        | 2.347377 | 1.652623  | 2.600933  |

In this dataset, respnum is the identification number of the participant. Participant 100072 had an actual mean score of 4 for the scale. This is the highest possible score, yet we predicted that this person would have a score of just 2.06. The residual is 1.94, indicating how much more environmentally concerned this person is than we predicted. The $z$ score is 3.06. Toward the bottom of the listing, participant 200785 had a score of 1, indicating that he or she was at the low point on our scale of environmental concern. However, we predicted that this person would have a score of 3.25, and the $z$ score is $-3.54$. We might look at the questionnaires for these two people to see if there was a miscoding. If the coding was correct, we might try to contact these participants to find out other variables we need to add to the regression equation. Unstructured interviews with participants that have extreme outliers can be helpful. We may find that those who have positive outliers are politically liberal and those who have negative outliers are politically conservative. Then we could add a variable measuring liberalism as a predictor. Correlating rstandard with other possible predictors would also be helpful.

A variable that is highly correlated with rstandard helps explain the variance in our outcome that is not already explained by the predictors in our model.

## 10.7.2    Influential observations: DFbeta

Stata offers several measures of the influence each observation has. DFbeta (the dfbeta command) is the most direct of these. It indicates the difference between each of the regression coefficients when an observation is included and when it is excluded. You could think of this as redoing the regression model, omitting just one observation at a time and seeing how much difference omitting each observation makes. A value of DFbeta $> 2/\sqrt{N}$ indicates that an observation has a large influence.

To open the dfbeta dialog box, select Statistics ▷ Linear models and related ▷ Regression diagnostics ▷ DFBETAs, but it is easier to type the one-word command, dfbeta, in the Command window:

```
. dfbeta
(739 missing values generated)
                    _dfbeta_1: dfbeta(educat)
(739 missing values generated)
                    _dfbeta_2: dfbeta(inc)
(739 missing values generated)
                    _dfbeta_3: dfbeta(com3)
(739 missing values generated)
                    _dfbeta_4: dfbeta(hlthprob)
(739 missing values generated)
                    _dfbeta_5: dfbeta(epht3)
```

These results show that Stata created five new variables. _dfbeta_1 has a DFbeta score for each person on the education variable; _dfbeta_2 does this for the income variable; _dfbeta_3 does this for the community identity variable; _dfbeta_4 does this for health problems; and _dfbeta_5 does this for environmental-specific health concerns.

We could then list cases that have relatively large values of DFbeta, using the formula DFbeta $= 2/\sqrt{3769} = 0.03$ as our cutoff because we have $N = 3769$ observations. For example, if we wanted to know problematic observations for the educat variable, the list command would be

```
. list respnum rstandard _dfbeta_1 if abs(_dfbeta_1) > 2/sqrt(3769) & _dfbeta_1 < .
```

I do not show the results of this command here. This list command, however, has a few features you might note. We use the absolute value function, abs(_dfbeta_1). We also include a simple formula, 2/sqrt(3769), where 3,769 is the sample size in our regression. This listing is just for the educat variable. dfbeta gives us information on how much each observation influences each parameter estimate. This is more specific information than that provided by the standardized residual. If there are a few observations that have a large influence on the results, we might see if they were miscoded. If possible, we might interview the individuals to see why they picked the answers they did and how they interpreted the item.

## 10.7.3   Combinations of variables may cause problems

Collinearity and multicollinearity can be a problem in multiple regression. If two independent variables are highly correlated implicitly, it is difficult to know how important each of them is as a predictor. Multicollinearity happens when a combination of variables makes one or more of the variables largely or completely redundant. Figure 10.8 shows how this can happen with just two predictors, X1 and X2, which are trying to predict a dependent variable, Y. The figure on the left shows that areas a and c represent the portion of Y that is explained by X1, and areas b and c represent the portion of Y that is explained by X2. Because X1 and X2 overlap a little bit (are correlated), the area c cannot be distributed to either X1 or X2. Still, there is a lot of variance in both predictors that is not overlapping, and we can get a good estimate of the unique effect of X1, namely, area a, and the unique effect of X2, namely, area b. Earlier in this chapter, we referred to the areas represented by a and b as the semipartial $R^2$s.

The figure on the right shows what happens when the two predictors are more correlated and, hence, overlap more. Together, X1 and X2 are explaining much of the variance of Y, but the unique effects represented by areas a and b are relatively small. You can imagine that as X1 and X2 become almost perfectly correlated, the areas a and b all but disappear while the confounded area c increases. This example illustrates what could happen with just two predictors. You can imagine what could happen in multiple regression when there are many predictors that are correlated with each other. The more correlated the predictors, the more they overlap and, hence, the more difficult it is to identify their independent effects. In such situations, you can have multicollinearity in which one or more of the predictors are virtually redundant.

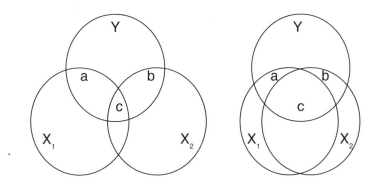

Figure 10.8. Collinearity

Stata can compute a variance inflation factor to assess the extent to which multicollinearity is a problem for each independent variable. This is computed by running the command `estat vif` after the regression. Here are the results for our example:

```
. estat vif
    Variable |       VIF       1/VIF
-------------+----------------------
      educat |      1.26    0.791143
         inc |      1.25    0.797304
       epht3 |      1.12    0.894240
    hlthprob |      1.12    0.895659
        com3 |      1.01    0.993215
-------------+----------------------
    Mean VIF |      1.15
```

This command computes both the variance inflation factor (VIF) and its reciprocal (1/VIF). Statisticians usually look at the VIF. If this is more than 10 for any variable, a multicollinearity problem may exist, and you may need to consider making an adjustment to your model. None of the variables has a problem using this criterion. If the average VIF is substantially greater than 1.00, there still could be a problem. Here the mean VIF is 1.15, and this is not a problem.

The value of 1/VIF may have a more intuitive interpretation than the VIF. If we regress educat on inc, epht3, hlthprob, and com3, we get an $R^2$ of 0.21. This means that there is an overlap of education with all the other predictors of 21% of the variance in education. Alternatively, $1 - 0.21 = 0.79$, or 79% of the variance in education is not overlapping, that is, it is not explained by the other predictors. Thus 79% of the variance in educat is available to give us an estimate of the independent effect of education on environmental concern, controlling for the other predictors. Some people call 1/VIF the tolerance. It is 1.00 minus the $R^2$ you obtain if you regress an independent variable on the set of other independent variables. It tells how much of the variance in the independent variable is available to predict the outcome variable independently. When VIF $= 10$, this means that only 10% of the variance in an independent variable is really available after you adjust for the other predictors. When VIF $> 10$ or 1/VIF $< 0.10$, there may be a multicollinearity problem.

When you have a problem with multicollinearity, consider these solutions. Dropping a variable often helps. If two variables both have a high VIF value, try dropping one of them and repeating the analysis. Dropping a variable that has a high VIF value is not as much of a problem as you might suspect; a variable that has more than 90% of its variance confounded with the other predictors probably does not explain much uniquely because it has little unique variance itself. Sometimes there are several closely related variables. For example, you might have a series of items involving concerns people have about global warming. Rather than trying to include all of these individual items, you might create a scale that combines them into one variable.

Multicollinearity sometimes shows up in what may seem like strange parameter estimates. Normally, the standardized $\beta$s should not be outside the range of $-1$ to $+1$. Yet you might have one variable that has $\beta = 2.14$ and a closely related variable that has $\beta = -1.56$. Sometimes these $\beta$s will be excessively large, but neither of them will be statistically significant. You almost certainly have a multicollinearity problem when this happens. If you drop either variable, the remaining variable may have a $\beta$ that is within the normal range and is statistically significant.

# 10.8   Weighted data

Regression allows us to weight our cases. Many large datasets will use what is known as a weighted sample. Researchers want to have an adequate number of observations of groups that otherwise would be a small subsample. One survey of a community that is mostly Caucasians may oversample Hispanics and African Americans. Another survey might oversample people who are cohabiting or people who have a disability. If you want to take a subsample of people who are disabled, cohabiting, Hispanic, or African American, this oversampling is critically important, because it means that your subsample will be large enough for meaningful analysis. However, if you want to generalize to the entire population, you will need to adjust for this oversampling.

Many surveys provide a weight that tells you how many people each respondent represents, based on the sample design. This sampling fraction is simply the population $N$ for the group, say, the number of African Americans in the United States, divided by the sample $n$ for African Americans in your sample. Other surveys provide a proportional weight so that if you oversampled African Americans by a factor of two (that is, an African American was twice as likely to be sampled as anyone else), then each African American would have a weight value of 0.5 when you tried to generalize to the entire population.

When using a weighted sample, you need to consult the documentation on the survey to know what weight variable to use. The Oregon sample we have been using has both types of weights. We can list these for the first five observations:

```
. list finalwt finalwt2 in 1/5
```

|    | finalwt  | finalwt2 |
|----|----------|----------|
| 1. | 969.7892 | 1.602893 |
| 2. | 384.8077 | .6360203 |
| 3. | 91.07467 | .1505306 |
| 4. | 5000     | 8.264132 |
| 5. | 27.0839  | .044765  |

With complex weighting, we can have different weights for each person. The first observation in the Oregon sample represents 969.79 Oregonians, and the fifth observation represents just 27.08 Oregonians, as indicated by the scores they have for the variable `finalwt`. The sum of all the `finalwt` scores will be close to the population of Oregon.

The probability weights represented by `finalwt2` accomplish the same thing for the sample. A person who is in a highly oversampled group (say, Native Americans) will have a `finalwt2` score of less than 1.00. A person who is not oversampled because we know we will have plenty of them (say, Caucasian males) in our sample will have a `finalwt2` score of more than 1.00. The sum of all the `finalwt2` scores will approximate the total sample size.

Weighting is complicated and beyond the scope of this book. Consult the documentation for the dataset you are using to find out the right weight variable to use.

Stata can adjust for different kinds of weights, but here we will just use what Stata calls pweights. In the dialog box for multiple regression, you will find a tab called **Weights**. Click on this tab, and select that you want to use *Sampling weights* (this is the name Stata's interface uses for what Stata calls pweights). Then all you need to do is enter the name of the weight variable; here use finalwt. The command and results are as follows:

```
. regress env_con educat inc com3 hlthprob epht3 [pweight = finalwt], beta
(sum of wgt is    2.3167e+06)
```

| Linear regression |  |  |  |  | Number of obs = | 3769 |
|---|---|---|---|---|---|---|
|  |  |  |  |  | F( 5,  3763) = | 120.73 |
|  |  |  |  |  | Prob > F    = | 0.0000 |
|  |  |  |  |  | R-squared   = | 0.3073 |
|  |  |  |  |  | Root MSE    = | .60771 |

| env_con | Coef. | Robust Std. Err. | t | P>\|t\| | Beta |
|---|---|---|---|---|---|
| educat | -.0077949 | .0071749 | -1.09 | 0.277 | -.0303352 |
| inc | -1.64e-07 | 5.96e-07 | -0.28 | 0.783 | -.0076229 |
| com3 | .0467495 | .0181525 | 2.58 | 0.010 | .0694138 |
| hlthprob | -.2687917 | .0449038 | -5.99 | 0.000 | -.1620468 |
| epht3 | -.4099326 | .0226743 | -18.08 | 0.000 | -.4702973 |
| _cons | 3.797229 | .1182412 | 32.11 | 0.000 | . |

When you do a weighted regression this way, Stata automatically uses the robust regression—whether you ask for it or not—because weighted data require robust standard errors. The results of this robust regression have important differences from the original regression. Because it is robust regression, we do not get the ANOVA table and cannot get an adjusted $R^2$. Because we used a weighted sample, we get a line saying that the sum of wgt is 2.3167e+06. (Remember that when Stata encounters a number that is extremely small or extremely large, it uses this notation. The e+06 means to move the decimal place six places to the right.) Expanded, the sum of the weights is 2,316,700. This is the approximate total adult population of Oregon at the time of this survey. It is good to check this value, which should be the total population size if you use this type of weight or the total sample size if you use the proportional weight.

The number of observations is 3,769, and this is the actual size of the sample used for tests of significance. We should check this to make sure that we did the weighting correctly. If we had used the other weight, finalwt2, the line would read sum of wgt is 3.8291e+03. Moving the decimal place three spaces to the right, this is 3,829, which is close to the actual sample size we have of 3,769. The Oregon Survey includes 4,508 observations. Because regression uses casewise deletion, we have only 3,769 observations with valid scores on all the variables used in the regression. Because the sum of the weights for the observations with incomplete data is not equal to the actual number of observations with incomplete data, the sum of the weights for the complete data will not be identical to 3,769.

You may have observed that many of the results are different from the unweighted results. For example, the $F$ value is 120.73 compared with 320.15 for the unweighted sample. This difference is partly due to the weighting and partly due to using robust regression. The parameter estimates of the coefficients and the $\beta$s are different, as are the standard errors and $t$ tests. When the weighted sample gives different results, this means that the weighting is important. Imagine that the authors of the study decided they needed many Native Americans, so they oversampled Native Americans by a factor of 10. This means that each Native American would count 10 times as much in the unweighted sample as he or she would if you wanted to generalize the overall population. If the relationship between the variables were somehow different for Native Americans than for other groups, these differences would be overrepresented by a factor of 10. Weighting makes every case accurately represent its proportion of the population.

# 10.9    Categorical predictors and hierarchical regression

Smoking by adolescents is a major health problem. We will examine this problem in this section by using `nlsy97_selected_variables.dta` as a way of illustrating what can be done with categorical predictors. Let's pick a fairly simple model to illustrate the use of categorical variables. Say that we think smoking behavior depends on the adolescent's age (a continuous variable), gender (a dichotomous variable), peer influence (a continuous variable), and race/ethnicity (a multicategory nominal variable). Let's also say that we think of these variables as being hierarchical. Start with just age and gender, and see how much they explain. Then add peer influence and see how much it adds. Finally, add race/ethnicity to see if it has a unique effect above and beyond what is explained by age, gender, and peer influence.

When we are working with categorical predictors, it is important to distinguish between dichotomous predictors, such as gender, and multicategorical predictors, such as race, for which there are more than two categories. When we have a dichotomous predictor, such as gender, we can create an indicator variable coded 0 for one gender, say, females, and 1 for the other gender, males. We can then enter this as a predictor in a regression model. We can do this by using the `recode` command; select Data ▷ Create or change data ▷ Other variable-transformation commands ▷ Recode categorical variable. In the *Variables* box, type `gender97` because this is the name of the variable for gender in our dataset. Because males are coded as 1 and females as 2, we need to recode. An indicator or dummy variable is a 0/1 variable, where a 1 indicates the presence of the characteristic and a 0 indicates its absence. Click on the Options tab, and check *Generate new variables*. Name the variable `male` because that is the category we will code as 1. If we had named the variable `female`, we would code females as 1. The choice is arbitrary and only changes the sign of the regression coefficient. Returning to the Main tab, enter the *Required* rule: (1 = 1 Male) (2 = 0 Female). If you want to run the command directly from the command line, enter

```
. recode gender97 (1 = 1 Male) (2 = 0 Female), generate(male)
```

## Names for categorical variables

The conventions for coding categorical variables precede the widespread use of multiple regression. Conventions are slow to change, and we usually need to recode categorical variables before doing regressions. Most surveys code dichotomous response items using codes of 1 and 2 for `yes/no`, `male/female`, and `agree/disagree` items. We need to convert these to codes of 0 and 1. With more than two category response options for categorical variables (for example, religion or marital status), we need to recode the variable into a series of dichotomous variables, each of which is coded as 0 or 1. I explain how to do this in the text, but there is some inconsistency in names used for these recoded variables.

A common name for a 0/1 variable is *dummy variable*. Others call these variables *indicator variables*, and still others call them *binary variables*. You should use whatever naming convention is standard for your content area. Here I will use these names interchangeably.

When we pick a name for one of these variables, it is useful to pick a name representing the category coded as 1. If we code women with 1 and men with 0, we would call the variable `female`. Calling the variable `gender` would be less clear when we interpret results. If whites are coded 1 and nonwhites are coded 0, it would make sense to call the variable `white`. Which category is coded 1 and which is coded 0 is arbitrary and affects only the sign of the coefficient. A positive sign signifies that the category coded 1 is higher on the dependent variable, and a negative sign signifies that it is lower on the dependent variable.

We must always check our changes, so enter a cross-tabulation, `tab2 gender97 male, missing`, to verify that we did not make a mistake. Once we have an indicator variable, we can enter it in the regression as a predictor just like any other variable. The unstandardized regression coefficient tells us the change in the outcome for a 1-unit change in the predictor. Because a 1-unit change in the predictor, `male`, constitutes being a male rather than being a female, the regression coefficient is simply the difference in means on the dependent variable for males and females, controlling for other variables. Although this regression coefficient is easy to interpret, many researchers also report the $\beta$s. These make a lot of sense for a continuous variable; that is, a 1-standard-deviation change in the independent variable produces a $\beta$-standard-deviation change in the dependent variable. However, the $\beta$s are of dubious value for an indicator variable. It does not make sense for an indicator variable to vary by degree. Going from 0 for female to 1 for male makes sense, but going up or down one standard deviation on gender does not make much sense.

Using dummy variables is more complex when there are more than two possible scores. Consider race: We might want a variable to represent differences between Caucasians, African Americans, Hispanic Americans, and others. We might code a variable `race` as 1 for Caucasian, 2 for African American, 3 for Hispanic American, and 4 for other. This variable, `race`, is a nominal-level categorical variable, and it makes no sense to think of a code of 4 being higher or lower than a code of 1 or 2 or 3. These are simply nominal categories and the order is meaningless. To represent these four racial or ethnic groups, we need three indicator variables. In general, when there are $k$ categories, we need $k - 1$ indicator variables.

How does this work? First, choose one category that serves as a reference group. It makes sense to pick the group we want to compare with other groups. In this example, it would make sense to pick Caucasian as our reference group. This is a large enough group to give us a good estimate of its value, and we will probably often want to compare other groups with Caucasians. It would make little sense to pick the "other" group as the reference category because this is a combination of several different ethnicities and races (for example, Pacific Islander and Native American). Also, there are relatively few observations in the "other" category.

Next generate three dummy or indicator variables (`aa`, `hispanic`, and `other`), allowing us to uniquely identify each person's race/ethnicity. A score of 1 on `other` means that the person is in the "other" category. A score of 1 on `hispanic` means that the person is Hispanic. A score of 1 on `aa` means that the person is African American. You may be wondering how we know that a person is Caucasian. A Caucasian person will have a score of 0 on `aa`, 0 on `hispanic`, and 0 on `other`.

The dataset used two questions to measure race/ethnicity. This is done to more accurately represent Hispanics, who may be of any race. The first item asks respondents if they are Hispanic. If they say no, they are then asked their race. The coding is fairly complicated, and here are the Stata commands used.

```
. generate race=race97
. replace race=1 if race97==1 & ethnic97==0
. replace race=2 if race97==2 & ethnic97==0
. replace race=3 if ethnic97==1
. replace race=4 if (race97==4 | race97==5) & ethnic97==0
. tab2 race race97 ethnic97
. recode race (2 = 1 African_American) (1 3/4 = 0 Other), generate(aa)
. recode race (3 = 1 Hispanic) (1/2 4 = 0 Other), generate(hispanic)
. recode race (4 = 1 Other_race) (1/3 = 0 W_AA_H), generate(other)
. tab1 aa hispanic other
```

The first line creates a new variable, `race`, which is equal to the old variable, `race97`. We never want to change the original variable. We then change the code of `race` based on the codes the person had on two variables, his or her original `race97` variable and his or her ethnicity, `ethnic97`. You can type the command `codebook race97 ethnic97` to see how these were coded. For example, we make `race=1` if the person had a code of 1 on `race97` and a code of 0 on `ethnic97`. We use the command `tab2 race race97 ethnic97` to check that we have done this correctly. We then use the `recode` command to generate the three dummy variables. Finally, we do frequency tabulations for the three dummy variables. Here only African Americans will have a code of 1 on `aa`, only Hispanics will have a code of 1 on `hispanic`, only others will have a code of 1 on `other`, and only Caucasians will have a code of 0 on all three indicator variables.

Now we are ready to do the multiple regression. In this example, we will enter the variables in blocks. The first block is estimated by using

```
. regress smday97 age97 male
> if !missing(smday97, age97, male, psmoke97, aa, hispanic, other), beta
```

| Source | SS | df | MS | | |
|---|---|---|---|---|---|
| Model | 22864.9705 | 2 | 11432.4853 | | |
| Residual | 418279.523 | 3464 | 120.75044 | | |
| Total | 441144.493 | 3466 | 127.277696 | | |

Number of obs = 3467
F( 2, 3464) = 94.68
Prob > F = 0.0000
R-squared = 0.0518
Adj R-squared = 0.0513
Root MSE = 10.989

| smday97 | Coef. | Std. Err. | t | P>\|t\| | Beta |
|---|---|---|---|---|---|
| age97 | 1.838514 | .1337569 | 13.75 | 0.000 | .2274106 |
| male | .2707643 | .3735982 | 0.72 | 0.469 | .0119908 |
| _cons | -20.33657 | 1.995249 | -10.19 | 0.000 | . |

We have a special qualification on who is included in this regression with the `if` qualifier. By inserting `if !missing(smday97, age97, male, psmoke97, aa, hispanic, other)`, we exclude people who did not answer all the items. If we did not exclude these people, the number of observations for each regression might differ. When you put this qualification with a command, there are a couple of things to remember: 1) `!missing` means "not missing"; programmers like to use the exclamation mark to signify "not". 2) You must insert the commas between variable names in this particular function; this is inconsistent with how Stata normally lists variables (without commas) but is necessary because `missing()` is a Stata function, not a command.

The $R^2 = 0.05$, $F(2, 3464) = 94.68$, $p < 0.001$. Although age and gender explain just a little bit of the variance in adolescent smoking behavior, their joint effect is statistically significant. How important is age as a predictor? For each year an adolescent gets older, he or she smokes an expected 1.84 more days per month. This is highly significant: $t(3464) = 13.75$, $p < 0.001$. This seems like a substantial increase in smoking behavior. However, $\beta = 0.23$ suggests that this is not a strong effect. What about gender—is it important? Because we coded males as 1 and females as 0, the coefficient 0.27 for `male` indicates that males smoke an average of 0.27 more days per month than females do, controlling for age. This difference does not seem great, and it is not significant. The $\beta$ is very weak, $\beta = 0.01$, but we will not try to interpret $\beta$ because gender is a categorical variable.

The next regression equation adds peer influence. Repeat the command, but add the variable `psmoke97`, which represents the percentage of peers who smoke. The command and results are

```
. regress smday97 age97 male psmoke97
> if !missing(smday97, age97, male, psmoke97, aa, hispanic, other), beta
```

| Source | SS | df | MS | | | |
|---|---|---|---|---|---|---|
| | | | | Number of obs = | 3467 | |
| | | | | F( 3, 3463) = | 134.24 | |
| Model | 45956.7038 | 3 | 15318.9013 | Prob > F = | 0.0000 | |
| Residual | 395187.789 | 3463 | 114.117179 | R-squared = | 0.1042 | |
| | | | | Adj R-squared = | 0.1034 | |
| Total | 441144.493 | 3466 | 127.277696 | Root MSE = | 10.683 | |

| smday97 | Coef. | Std. Err. | t | P>\|t\| | Beta |
|---|---|---|---|---|---|
| age97 | 1.286699 | .1356942 | 9.48 | 0.000 | .1591551 |
| male | .9523383 | .3663385 | 2.60 | 0.009 | .0421741 |
| psmoke97 | 2.250031 | .1581743 | 14.23 | 0.000 | .2406996 |
| _cons | -19.47606 | 1.940615 | -10.04 | 0.000 | . |

This new model explains 10% of the variance in smoking behavior: $R^2 = 0.10$, $F(3, 3463) = 134.24$, $p < 0.001$. Does `psmoke97` make a unique contribution? There are two ways of getting an answer. First, we can see that the regression coefficient of 2.25 is significant: $t(3463) = 14.23$, $p < 0.001$, and $\beta = 0.24$.

A second way of testing the unique effect of adding `psmoke97` is to use the `pcorr` command, as follows:

```
. pcorr smday97 age97 male psmoke97 if !missing(smday97, age97, male, psmoke97,
> aa, hispanic, other)
(obs=3467)
```

Partial and semipartial correlations of smday97 with

| Variable | Partial Corr. | Semipartial Corr. | Partial Corr.^2 | Semipartial Corr.^2 | Significance Value |
|---|---|---|---|---|---|
| age97 | 0.1591 | 0.1525 | 0.0253 | 0.0233 | 0.0000 |
| male | 0.0441 | 0.0418 | 0.0019 | 0.0017 | 0.0094 |
| psmoke97 | 0.2350 | 0.2288 | 0.0552 | 0.0523 | 0.0000 |

So we can say that the influence of peer smoking adds 5% to the explained variance because the semipartial $R^2 = 0.05$, $p < 0.001$. The significance level is reported as a probability without any test statistic. There are two ways of obtaining the test statistic. Most statistics books show how to do an $F$ test for adding a single variable (or set of variables). More simply, because we are interested in the contribution of a single added variable, the $t$ test Stata reports for `psmoke97` will be the same as the square root of the $F$ test. We could say that the semipartial $R^2 = 0.05$, $t(3463) = 14.23$, $p < 0.001$. This $t$-value was reported in the previous regression command.

The next model we want to fit includes race/ethnicity. We can fit this by simply adding the variables `aa`, `hispanic`, and `other` to the regression command. Here are the command and its results:

```
. regress smday97 age97 male psmoke97 aa hispanic other
> if !missing(smday97, age97, male, psmoke97, aa, hispanic, other), beta
```

| Source | SS | df | MS |
|---|---|---|---|
| Model | 59879.4994 | 6 | 9979.91657 |
| Residual | 381264.994 | 3460 | 110.192195 |
| Total | 441144.493 | 3466 | 127.277696 |

|  |  |
|---|---|
| Number of obs = | 3467 |
| F( 6, 3460) = | 90.57 |
| Prob > F = | 0.0000 |
| R-squared = | 0.1357 |
| Adj R-squared = | 0.1342 |
| Root MSE = | 10.497 |

| smday97 | Coef. | Std. Err. | t | P>|t| | Beta |
|---|---|---|---|---|---|
| age97 | 1.308304 | .1333711 | 9.81 | 0.000 | .1618275 |
| male | 1.058956 | .3601243 | 2.94 | 0.003 | .0468957 |
| psmoke97 | 2.173683 | .1557058 | 13.96 | 0.000 | .2325322 |
| aa | -4.676358 | .4592876 | -10.18 | 0.000 | -.1674716 |
| hispanic | -3.223299 | .4638056 | -6.95 | 0.000 | -.1144606 |
| other | .3825068 | 1.143897 | 0.33 | 0.738 | .0053334 |
| _cons | -18.01775 | 1.911984 | -9.42 | 0.000 | . |

When we add the set of three indicator variables that collectively represent race and ethnicity, we increase our $R^2$ to 0.14, $F(6, 3460) = 90.57$, $p < 0.001$. This represents a $0.1357 - 0.1042 = 0.0315$, or 3.2%, increase in $R^2$ when we add the set of three indicator variables that represent race/ethnicity.

The $t$ tests for **aa** and **hispanic** are statistically significant, and the $t$ test for **other** is not. This shows that both African Americans and Hispanics smoke fewer days per month than do Caucasians. If we want to know whether the set of three indicators is statistically significant (this refers to the three variables simultaneously and tests the significance of the combined race/ethnicity variable), we can type the following **test** command immediately after the regression and before doing another multiple regression (because it uses the most recent results):

```
. test aa hispanic other
 ( 1)   aa = 0
 ( 2)   hispanic = 0
 ( 3)   other = 0
       F(  3,  3460) =   42.12
           Prob > F =   0.0000
```

This **test** command is both simple and powerful. It does a test that three null hypotheses are all true. It is testing that the effect of **aa** is zero, of **hispanic** is zero, and of **other** is zero. It gives us $F(3, 3460) = 42.12$, $p < 0.001$. Thus we can say that the effect of race/ethnicity is weak because it adds just 3.2% to the explained variance, but it is statistically significant. If you paid careful attention to the unstandardized regression coefficients, you may object to the statement that the effect is weak. Uniquely explaining 3.2% of the variance sounds weak, which is why we described it this way. However, $b = -4.68$ for African Americans, and this means that they are expected to smoke almost five fewer days per month than Caucasians. If you think this is a substantial difference, you might want to focus on this difference rather than the explained variance.

### More on testing a set of parameter estimates

We have seen one important use for the **test** command, namely, testing a set of indicator variables (**aa**, **hispanic**, **other**) that collectively define another variable (race/ethnicity). There are many other uses for this command. You may want to test whether a set of control variables is significant. When you have interaction terms, you may want to test a set of them. The test is always done in the context of the multiple regression most recently estimated. In this example, if we ran the command **test age97 male**, we would get a different result from when we entered these two variables by themselves. This is because we are testing whether these two variables are simultaneously significant, controlling for all the other variables that are in the model.

There is a useful command called **nestreg**. The procedures we have been using here involve nested regressions. The regressions are nested in the sense that the first regression is nested in the second regression because all the predictors in the first regression are included in the second. Likewise, the second regression is nested in the third regression, and so on. Some people call this hierarchical regression. If you call it hierarchical regression, you should not confuse it with hierarchical linear modeling, which is the name of a program that does multilevel analysis or mixed regression, which is related to the command in Stata called **mixed**. Nested regression is used where we have blocks of variables that we want to enter in a sequence, each step adding another block. Our method, using a series of regressions followed by the **test** command, is extremely powerful and flexible. However, the **nestreg** command was created to automate this process.

To use the **nestreg** command, you add the **nestreg:** prefix (the colon must be included) to the regression command. You then write a **regress** command, but put each block of predictors in parentheses. Here we have three blocks of predictors: (**age97 male**), (**psmoke97**), and (**aa hispanic other**). The usual options for the **regress** command are available. We selected the option to produce the beta weights.

```
. nestreg: regress smday97 (age97 male) (psmoke97) (aa hispanic other), beta
Block  1: age97 male
```

| Source | SS | df | MS | | Number of obs | = | 3467 |
|---|---|---|---|---|---|---|---|
| | | | | | F( 2, 3464) | = | 94.68 |
| Model | 22864.9705 | 2 | 11432.4853 | | Prob > F | = | 0.0000 |
| Residual | 418279.523 | 3464 | 120.75044 | | R-squared | = | 0.0518 |
| | | | | | Adj R-squared | = | 0.0513 |
| Total | 441144.493 | 3466 | 127.277696 | | Root MSE | = | 10.989 |

| smday97 | Coef. | Std. Err. | t | P>|t| | Beta |
|---|---|---|---|---|---|
| age97 | 1.838514 | .1337569 | 13.75 | 0.000 | .2274106 |
| male | .2707643 | .3735982 | 0.72 | 0.469 | .0119908 |
| _cons | -20.33657 | 1.995249 | -10.19 | 0.000 | . |

```
Block  2: psmoke97
        Source |       SS       df       MS              Number of obs =    3467
---------------+------------------------------           F(  3,  3463) =  134.24
         Model | 45956.7038       3  15318.9013          Prob > F      =  0.0000
      Residual | 395187.789    3463  114.117179          R-squared     =  0.1042
---------------+------------------------------           Adj R-squared =  0.1034
         Total | 441144.493    3466  127.277696          Root MSE      =  10.683

       smday97 |    Coef.    Std. Err.      t     P>|t|                      Beta
---------------+------------------------------------------           ------------------
         age97 |  1.286699   .1356942     9.48   0.000                  .1591551
          male |  .9523383   .3663385     2.60   0.009                  .0421741
       psmoke97|  2.250031   .1581743    14.23   0.000                  .2406996
         _cons | -19.47606   1.940615   -10.04   0.000                         .
```

```
Block  3: aa hispanic other
        Source |       SS       df       MS              Number of obs =    3467
---------------+------------------------------           F(  6,  3460) =   90.57
         Model | 59879.4994       6  9979.91657          Prob > F      =  0.0000
      Residual | 381264.994    3460  110.192195          R-squared     =  0.1357
---------------+------------------------------           Adj R-squared =  0.1342
         Total | 441144.493    3466  127.277696          Root MSE      =  10.497

       smday97 |    Coef.    Std. Err.      t     P>|t|                      Beta
---------------+------------------------------------------           ------------------
         age97 |  1.308304   .1333711     9.81   0.000                  .1618275
          male |  1.058956   .3601243     2.94   0.003                  .0468957
       psmoke97|  2.173683   .1557058    13.96   0.000                  .2325322
            aa | -4.676358   .4592876   -10.18   0.000                 -.1674716
      hispanic | -3.223299   .4638056    -6.95   0.000                 -.1144606
         other |  .3825068   1.143897     0.33   0.738                  .0053334
         _cons | -18.01775   1.911984    -9.42   0.000                         .
```

```
              |         Block  Residual                         Change
        Block |      F     df         df    Pr > F        R2     in R2
--------------+---------------------------------------------------------
            1 |  94.68      2       3464    0.0000    0.0518
            2 | 202.35      1       3463    0.0000    0.1042    0.0523
            3 |  42.12      3       3460    0.0000    0.1357    0.0316
```

This command will run the three regressions, and report increments in $R^2$ and whether these are statistically significant. A special strength of this command is that it works with many types of nested models, as you can see by typing help nestreg. The nestreg command is hard to find under the menu system. You find it at Statistics ▷ Other ▷ Nested model statistics. In addition to working with our multiple regression example, the dialog box provides a way of evaluating nested models by using 18 specialized extensions of multiple regression, such as logistic regression.

The results show the three regressions we ran earlier—calling them Block 1, Block 2, and Block 3—and list the variables added with each block. After the last block is entered, there is a summary table showing the change in $R^2$ for the block along with its significance. For example, the second block added the variable psmoke97. This block increased the $R^2$ by 0.0523, from 0.0518 to 0.1042. This increment in $R^2$ is significant, $F(1, 3463) = 202.35$, $p < 0.001$. The significance of $R^2 = 0.104$ at this stage is taken from the `Block 2` regression, $F(3, 3463) = 134.24$, $p < 0.001$. Avoid confusing the $F$ test for the increment in $R^2$ from the summary table and the $F$ test for the $R^2$ itself from the regression for the block. Adding the race/ethnicity variables to the model in `Block 3` increased $R^2$ by 0.032, and this is significant, $F(3, 3460) = 42.12$, $p < 0.001$. So, too, is the final $R^2 = 0.136$, $F(6, 3460) = 90.57$, $p < 0.001$.

**Tabular presentation of hierarchical regression models**

You may find it helpful to have a table showing the results of your nested (hierarchical) regressions. The table should show $b$'s, standard errors, and standardized betas for each model you fit. You should also include $R^2$ for each model and the $F$ test of the change in the $R^2$. In many fields, it is conventional to show the significance of regression coefficients by using asterisks. One asterisk is used for parameter estimates that are significant at the 0.05 level, two asterisks for parameters significant at the 0.01 level, and three asterisks for parameters significant at the 0.001 level. Some fields—for example, psychology through the American Psychological Association—have recommended tables that only report the coefficients for the variables that are added at each step. This approach ignores the fact that the coefficients for the variables from previous blocks will change when the new block of variables is added. This can become misleading because a variable in block 1 may be significant at that step, but it may become insignificant when subsequent blocks are added. A detailed example of a summary table format for nested regression is shown at http://oregonstate.edu/~acock/tables/regression.pdf.

Nested regression is commonly used to estimate the effect of personality characteristics or individual beliefs after you have controlled for background variables. If you want to know whether motivation was important to career advancement, you would do a multiple regression entering background variables first and then adding motivation as a final step. For example, you might enter gender, race, and parents' education in the first step as background variables over which the individual has no control; then enter achievement variables, such as education, in a second step; and finally, enter motivation. If motivation makes a significant increase in $R^2$ in the final step, then you have a much better test of the importance of motivation than if you had just done a correlation of career achievement and motivation without entering the control variables first.

# 10.10   A shortcut for working with a categorical variable

In the previous analysis, we did a regression of `smday97` on `age97`, `male`, `psmoke97`, `aa`, `hispanic`, and `other`. The last three are our race/ethnicity variables, where we used Caucasian as our reference group and created dummy variables for African Americans, Hispanics, and others. There is a much easier way of doing this, but you need to understand how it works.

Stata treats categorical variables, regardless of the number of categories, as factor variables. Gender is a two-level factor variable. Our race/ethnicity variable is a four-level factor variable. Stata then has an elegant way of producing the dummy variables for us and entering them in a regression analysis. Using this feature makes the most sense when the first category is a good reference category and when you want to use all the categories. If you put i. as a stub in front of a categorical variable, Stata will make the first category the reference category and then generate a dummy variable for each of the remaining categories. Here are two regression results. The first uses the dummy variables as we named them before and the second uses the factor-variable feature.

```
. regress smday97 age97 male psmoke97 aa hispanic other
```

| Source   | SS         | df   | MS         |   | Number of obs = | 3467   |
|----------|------------|------|------------|---|-----------------|--------|
|          |            |      |            |   | F( 6, 3460) =   | 90.57  |
| Model    | 59879.4994 | 6    | 9979.91657 |   | Prob > F    =   | 0.0000 |
| Residual | 381264.994 | 3460 | 110.192195 |   | R-squared   =   | 0.1357 |
|          |            |      |            |   | Adj R-squared = | 0.1342 |
| Total    | 441144.493 | 3466 | 127.277696 |   | Root MSE    =   | 10.497 |

| smday97   | Coef.     | Std. Err. | t      | P>|t| | [95% Conf. | Interval]  |
|-----------|-----------|-----------|--------|-------|------------|------------|
| age97     | 1.308304  | .1333711  | 9.81   | 0.000 | 1.04681    | 1.569798   |
| male      | 1.058956  | .3601243  | 2.94   | 0.003 | .3528779   | 1.765033   |
| psmoke97  | 2.173683  | .1557058  | 13.96  | 0.000 | 1.868399   | 2.478968   |
| aa        | -4.676358 | .4592876  | -10.18 | 0.000 | -5.57686   | -3.775856  |
| hispanic  | -3.223299 | .4638056  | -6.95  | 0.000 | -4.132659  | -2.313938  |
| other     | .3825068  | 1.143897  | 0.33   | 0.738 | -1.860274  | 2.625288   |
| _cons     | -18.01775 | 1.911984  | -9.42  | 0.000 | -21.76648  | -14.26902  |

```
. regress smday97 age97 male psmoke97 i.race
```

| Source   | SS         | df   | MS         |   | Number of obs = | 3467   |
|----------|------------|------|------------|---|-----------------|--------|
|          |            |      |            |   | F( 6, 3460) =   | 90.57  |
| Model    | 59879.4994 | 6    | 9979.91657 |   | Prob > F    =   | 0.0000 |
| Residual | 381264.994 | 3460 | 110.192195 |   | R-squared   =   | 0.1357 |
|          |            |      |            |   | Adj R-squared = | 0.1342 |
| Total    | 441144.493 | 3466 | 127.277696 |   | Root MSE    =   | 10.497 |

| smday97   | Coef.     | Std. Err. | t      | P>|t| | [95% Conf. | Interval]  |
|-----------|-----------|-----------|--------|-------|------------|------------|
| age97     | 1.308304  | .1333711  | 9.81   | 0.000 | 1.04681    | 1.569798   |
| male      | 1.058956  | .3601243  | 2.94   | 0.003 | .3528779   | 1.765033   |
| psmoke97  | 2.173683  | .1557058  | 13.96  | 0.000 | 1.868399   | 2.478968   |
|           |           |           |        |       |            |            |
| race      |           |           |        |       |            |            |
| 2         | -4.676358 | .4592876  | -10.18 | 0.000 | -5.57686   | -3.775856  |
| 3         | -3.223299 | .4638056  | -6.95  | 0.000 | -4.132659  | -2.313938  |
| 4         | .3825068  | 1.143897  | 0.33   | 0.738 | -1.860274  | 2.625288   |
|           |           |           |        |       |            |            |
| _cons     | -18.01775 | 1.911984  | -9.42  | 0.000 | -21.76648  | -14.26902  |

Notice that under `race` in the second result, we have three categories that are labeled 2, 3, and 4. Category 2 is the second category and corresponds to the `aa` category in the first results. Similarly, category 3 matches the `hispanic` category, and category 4 matches the `other` category.

This method saves us the work of having to generate the dummy variables, but it has the possible disadvantage of not having clear labels for the resulting categories. It is easy to remember that `aa` is African American, `hispanic` is Hispanic, and `other` is other races. It is harder to remember which category coincides with 2, 3, or 4. Still, if you have several categorical variables, you can save a lot of recoding time by using the factor-variable feature. It is possible to change the reference category or what Stata refers to as the `baselevel`. If we want to make Hispanics the reference category, we could type

```
. regress smday97 age97 male psmoke97 ib3.race
```

where the `ib3.` makes category 3 the base. If we wanted African Americans to be the reference category, we would use `ib2.race`. Rarely would we want the last category to be the reference category, but if we did, we would use `ib(last).race`.

# 10.11  Fundamentals of interaction

In many situations, the effect of one variable depends on where you are on another variable. For example, there is a well-established relationship between education and income. The more education you have, the more income you can expect, on average. There is also an established relationship between gender and income. Men, on average, make more money than do women. With this in mind, we could do a regression of income on both education and gender to see how much of a unique effect each predictor has. There is, however, an important relationship that we could miss by doing this. In addition to the possibility of men making more money than women, the payoff men get for each additional year of education may be greater than the payoff women receive. Thus men are advantaged in two ways. First, they make more money on average, and second, they get a greater return on their education. This second issue is what we mean by interaction. That is, the effect of education on income depends on your gender—stronger effect for men, weaker effect for women. How can we test this relationship using multiple regression?

As an example here, we will fit two models. The first model includes just education and gender as predictors, or main effects, to distinguish them from interaction effects. The second model adds the interaction between education and gender. Let's use `c10interaction.dta`, a dataset with hypothetical data. The first model is fit using `regress inc educ male`, where `inc` is income measured in thousands of dollars, `educ` is education measured in years, and `male` represents gender, with men coded 1 and women coded 0. Here are the command and results:

```
. regress inc educ male, beta

      Source |       SS       df       MS              Number of obs =     120
-------------+------------------------------           F(  2,   117) =   37.19
       Model | 100464.105        2 50232.0527          Prob > F      =  0.0000
    Residual | 158015.895      117 1350.5632           R-squared     =  0.3887
-------------+------------------------------           Adj R-squared =  0.3782
       Total |    258480       119 2172.10084          Root MSE      =   36.75

-------------+----------------------------------------------------------------
         inc |      Coef.   Std. Err.       t    P>|t|                    Beta
-------------+----------------------------------------------------------------
        educ |   8.045694   1.008586     7.98    0.000                .5775017
        male |   19.04991   6.719787     2.83    0.005                .2052297
       _cons |  -42.54411    14.2919    -2.98    0.004                       .
-------------+----------------------------------------------------------------
```

We can explain 38.87% of the variance in income with education and gender, $F(2,117) = 37.19$, $p < 0.001$, and both education and gender are significant in the ways we anticipated. Each additional year of education yields an expected increase in income of \$8,046, and men expect to make \$19,050 more than women. (Remember, these are hypothetical data!)

We can make a graph of this relationship. First, generate a predicted income for women as a new variable and a predicted income for men as a new variable. Just after the regression, type the following commands:

```
. predict incfnoi if male==0
. predict incmnoi if male==1
```

The first command generates a predicted value for income for women (`male==0`), and the second command does the same for men. Then do an overlaid two-way graph by using the dialog box we discussed earlier in the book. The graph command used to produce figure 10.9 is long:

```
. twoway (connected incmnoi educ if male == 1, lcolor(black) lpattern(dot)
> msymbol(diamond) msize(large)) (connected incfno educ if male == 0,
> lcolor(black) lpattern(solid) msymbol(circle) msize(large)),
> ytitle(Income in thousands) xtitle(Education) legend(order(1 "Men" 2 "Women"))
> scheme(s2manual)
```

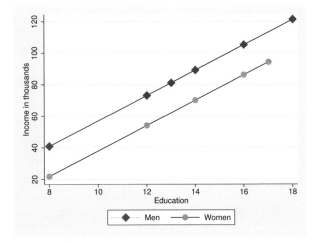

Figure 10.9. Education and gender predicting income, no interaction

There is one line for the education-to-income relationship for men (the higher line) and another line for women. This shows a strong positive connection between education and income, as well as a big gender gap. We use this model that includes just the main effects and has both lines parallel because it is the best model, as long as we assume that the gender gap is constant. If the size of the gender gap increases with increasing age, this model would not be able to show that.

Testing for interaction takes us a huge step further. When we test for interaction, we want to see if the slopes are different. In other words, we want to see if the slope between education and income is different and perhaps steeper for men than it is for women. You may be thinking that the graph shows the lines to be parallel, but this is because we set up the equation without an interaction term; without the interaction term, Stata gives us the best parallel lines. To see if the lines really should not have this restriction, we add an interaction term. We get the interaction term as the product of gender and education.

There are two ways to include an interaction term as the product of gender and education. One way is to use the **generate** command, for example,

```
. generate ed_male = educ*male
```

where **ed_male** is the interaction term. Then we would refit the model by adding the interaction term:

```
. regress inc educ male ed_male, beta
```

A better way to include an interaction term is to use factor-variable notation. This produces identical parameter estimates, but has two advantages. First, you do not need to generate the interaction or product term, `ed_male`, before running the regression. This is a small advantage in this example, but in more complex models, it can save considerable time. The second, and more important advantage, is that some postestimation commands (for example, `margins` and `marginsplot`) require that you used the factor-variable notation. For a categorical variable like gender, we insert `i.` at the front of the variable name, `i.male`. For a continuous variable, like education, we insert `c.` at the front of the name of the variable, `c.educ`. Stata uses the `##` notation to indicate a product of variables. Thus we can write a command that includes both the main effects of gender and education and the interaction between gender and education as follows:

```
. regress inc i.male##c.educ, beta
```

With the `##` notation, Stata knows to include both `male` and `educ` along with their product. Thus this command is equivalent to the command in the above paragraph, `regress inc educ male ed_male, beta`. Here are the results:

```
. regress inc i.male##c.educ, beta
```

| Source   | SS         | df  | MS         |
|----------|------------|-----|------------|
| Model    | 122604.719 | 3   | 40868.2397 |
| Residual | 135875.281 | 116 | 1171.33863 |
| Total    | 258480     | 119 | 2172.10084 |

Number of obs = 120
F( 3, 116) = 34.89
Prob > F = 0.0000
R-squared = 0.4743
Adj R-squared = 0.4607
Root MSE = 34.225

| inc         | Coef.     | Std. Err. | t     | P>\|t\| | Beta      |
|-------------|-----------|-----------|-------|---------|-----------|
| 1.male      | -91.88539 | 26.27242  | -3.50 | 0.001   | -.9899052 |
| educ        | 3.602369  | 1.388076  | 2.60  | 0.011   | .2585699  |
| male#c.educ |           |           |       |         |           |
| 1           | 8.196446  | 1.885263  | 4.35  | 0.000   | 1.287508  |
| _cons       | 16.84834  | 19.07279  | 0.88  | 0.379   | .         |

These results show that we can explain 47.43% of the variance, $F(3, 116) = 34.89$, $p < 0.001$, in income after we add the effect of the interaction of education and gender. This is an increase of 8.56% in the explained variance we had when we did not include the interaction. Is the interaction significant? In the result, we see that `male#c.educ`, the interaction, has a $t = 4.35$, $p < 0.001$ so we know the interaction is significant. (The output does not include the `i.` at the start of the categorical variable, but it is implied.) We could also get this increase in explained variance increase by using `pcorr` for the semipartial $R^2$ or by using `nestreg: regress inc (educ male) (ed_male), beta`.

### Centering quantitative predictors before computing interaction terms

Although we have not done it here, some researchers choose to center quantitative independent variables, such as education, before computing the interaction terms. The interaction term is the product of the centered variables, but the interaction term will not be centered itself. Centering involves subtracting the mean from each observation. If the mean of education were 12 years, centered education would be `generate c_educ = educ - 12`. A better way to do this is to summarize the variable first. Stata then saves the mean to many decimal places in memory as a variable called `r(mean)`. We would use two commands: `summarize educ` and then `generate educ_c = educ - r(mean)`. If you plan to center many variables, you may want to install the user-written command `center` (type `search center` or `ssc install center`). Ben Jann wrote this command in 2004, with subsequent revisions. Once it is installed, you can type `center educ` to automatically create a new variable named `c_educ`.

If you center these variables, the mean has a value of zero. Because the intercept is the estimated value when the predictors are zero, the intercept is the estimated value when the centered variable is at its mean. This often makes more sense, especially when a value of zero for the uncentered variable is a rare event, like it is in the example of years of education. The centered education variable becomes a measure of how far a person is above or below average on years of education rather than how many years of education they have had. This centering can help you interpret the intercept.

Interpreting the interaction term (`i.male#c.educ`) and the main effects (`educ`, `male`) is tricky when the interaction is significant. Just glancing at them seems to suggest ridiculous relationships. The $-91.89$ for the `male` variable might look like men make significantly less income than do women, but we cannot interpret it this way in the presence of interaction. We need to make a separate equation for each level of the categorical variable. Thus we need to make an equation for men and a separate equation for women. This is not hard to do; just remember that women have a score of zero on two variables: `male` and `ed_male` (educ $\times$ 0 = 0). For men, by substituting a value of 1 for the `male` variable, the equation simplifies to

$$\widehat{\text{inc}} = (16.85 - 91.89) + (3.60 + 8.20)\texttt{educ}$$
$$\widehat{\text{inc}} = -75.04 + 11.80(\texttt{educ})$$

For women, by substituting a value of 0 for the `male` variable, the equation simplifies to

$$\widehat{\text{inc}} = 16.85 + 3.60(\texttt{educ})$$

Here the payoff of one additional year of education for men is \$11,800 compared with just \$3,600 for women. The adjusted constant or intercept for men of $-\$75,040$ does not make much sense. In the data, the lowest education is 8 years, and the constant refers to a person who has no education. Do not interpret an intercept that is out of the range of the data.

A nice way to see how the interaction works is with a graph. In this example, we are interested in how the relationship between income and education varies by gender. We want to see if the numerical results shown above really show a substantively important difference in this relationship. We will use two commands to do this. The `margins` command is run first to find out what the estimated income is for specific years of education for women and for men. Let's do this by specifying that education years will have values of 8 years, 10 years, 12 years, 14 years, 16 years, and 18 years. The `margins` has the categorical variable, `male`, listed immediately following the command. Then, after the command, we specify the specific values for the continuous variable, `educ`. Here are the command and results:

```
. margins male, at(educ=(8 10 12 14 16 18))
Adjusted predictions                          Number of obs    =         120
Model VCE       : OLS

Expression      : Linear prediction, predict()

1._at           : educ            =           8

2._at           : educ            =          10

3._at           : educ            =          12

4._at           : educ            =          14

5._at           : educ            =          16

6._at           : educ            =          18
```

|           |        | Delta-method |        |       |                      |
|-----------|--------|--------------|--------|-------|----------------------|
|           | Margin | Std. Err.    | t      | P>\|t\| | [95% Conf. Interval] |
| _at#male  |        |              |        |       |                      |
| 1 0       | 45.66729 | 8.66112    | 5.27   | 0.000 | 28.51285    62.82173 |
| 1 1       | 19.35346 | 8.545103   | 2.26   | 0.025 | 2.42881     36.27812 |
| 2 0       | 52.87203 | 6.431252   | 8.22   | 0.000 | 40.13412    65.60993 |
| 2 1       | 42.95109 | 6.496577   | 6.61   | 0.000 | 30.0838     55.81838 |
| 3 0       | 60.07676 | 4.808436   | 12.49  | 0.000 | 50.55305    69.60048 |
| 3 1       | 66.54872 | 4.940843   | 13.47  | 0.000 | 56.76276    76.33468 |
| 4 0       | 67.2815  | 4.505014   | 14.93  | 0.000 | 58.35875    76.20425 |
| 4 1       | 90.14635 | 4.431483   | 20.34  | 0.000 | 81.36924    98.92346 |
| 5 0       | 74.48624 | 5.734395   | 12.99  | 0.000 | 63.12855    85.84393 |
| 5 1       | 113.744  | 5.280516   | 21.54  | 0.000 | 103.2853    124.2027 |
| 6 0       | 81.69097 | 7.802914   | 10.47  | 0.000 | 66.23632    97.14563 |
| 6 1       | 137.3416 | 7.01066    | 19.59  | 0.000 | 123.4561    151.2271 |

The first part of the results show that we picked 6 values for education, where the first value represents 8 years and the sixth value represents 18 years. The second part shows the estimated incomes for each value when `male` equals a value of 0 (woman) and when it equals a value of 1 (man).

Next we want to plot the estimated values of income on education separately for women and men. We simply type `marginsplot, noci`. We added the `noci` option to the `marginsplot` command because we did not want confidence intervals. If you retype this command without the option, you will see that the confidence interval is widest at the extremes (8 years of education and 18 years of education) in the resulting graph; see figure 10.10. (Note that I have made some minor enhancements to the figure with the Graph Editor.)

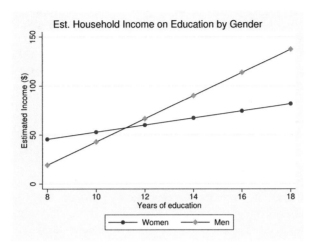

Figure 10.10. Education and gender predicting income, with interaction term

Figure 10.10 shows that at low levels of education, men are estimated (using a linear model) to make less than women, but at higher levels of education, the men make much more than women. It shows that the rate of increase in income for additional years of education is much steeper for men than it is for women. Remember that we are using hypothetical data.

**Do not compare correlations across populations**

We have found a significant interaction showing that women are relatively disadvantaged in the relationship between education and income. We did this by examining the unstandardized slopes, which showed that the payoff for an additional year of education was $11,800 for men, compared with $3,600 for women. If these were actual data, this would be compelling evidence of a gross inequity. Some researchers compare standardized beta weights or correlations to make this argument, but this is a serious mistake. The standardized beta weights depend on the form of the relationship (shown with the unstandardized slopes in figure 10.10) and the differences in variance. Unless the men and women have identical variances, comparing the correlations or beta weights can yield misleading results. Correlation is a measure of fit or the clustering of the observations around the regression line. You can have a steep slope like that for men, but the observations may be widely distributed around this, leading to a small correlation. Similarly, you can have a much flatter slope, like that for women, but the observations may be closely packed around the slope, leading to a high correlation. Whether the correlation between education and income is higher or lower for women than it is for men measures how closely the observations are to the estimated values. How steep the slopes are, measured by the unstandardized regression coefficients as we have done, measures the form of the relationship.

## 10.12   Nonlinear relations

We have been assuming that the relationship between each independent variable and our dependent variable is linear and thereby described by a straight line. This is often an adequate assumption, even when it is not the best possible model. Many relationships have a curve. You could think of one example as a decreasing marginal utility. If the income of the average college student goes up or down $10,000 per year, this represents a big change in their resources and will influence many outcome variables. What about the average billionaire? As their income goes up or down $10,000 per year, they are unlikely to notice the change and the change will not influence any outcomes. Figure 10.11(A) shows such a relationship. Note the following equation:

$$\widehat{y} = 0.10 \; + \; 0.122x \; - \; 0.006x^2$$
$$\phantom{\widehat{y} = 0.10 \; + \;} \text{linear} \quad\;\; \text{quadratic}$$

We fit this by adding a quadratic term, $x^2$. The $0.122x$ is the linear component and has a positive slope, but the $-0.006x^2$ has a negative slope. The positive component is pushing the estimated value of $y$ up and the negative quadratic is pushing the estimated value of $y$ down. Because the $-0.006$ is a much smaller value than the $0.122$, the positive linear push dominates initially. Although the $-0.006$ is a small value, it is multiplied by $x^2$ rather than by $x$. Because $x$ changes arithmetically, $x^2$ changes geometrically and the quadratic component will eventually bring the estimated value of $y$ down. We can illustrate this using different values of $x$. Because $x$ increases from 0 to 10, the estimated value of $y$ increases from 0.91 to 1.53; after that, additional increases in $x$ actually lead to a lowering of the estimated value of $y$.

| Value of $x$ | Value of $x^2$ | Estimated value of $y$ using $y = 0.910 + 0.122x - 0.006x^2$ |
|:---:|:---:|:---:|
| 0 | 0 | 0.91 |
| 10 | 100 | 1.53 |
| 20 | 400 | 0.95 |
| 30 | 900 | $-0.83$ |
| 40 | 1600 | $-3.81$ |
| 50 | 2500 | $-7.99$ |

Sometimes we need a model that allows a positive relationship for part of the range of the independent variable and a negative relationship for the rest of the range. How often do a husband and wife have conflict? If they have conflict all the time, they will be unhappy. However, one theory argues that couples that never or rarely have conflict are also unhappy and those who have a moderate degree of conflict are the happiest. Figure 10.11(B) shows such a relationship. Notice that the positive linear component is pushing the estimated value of $y$, happiness, up and that the negative quadratic component is pushing the estimated value of $y$ down. Within the range of the data, the value of $x^2$ becomes so great that the initially small quadratic component effect reverses the effect of the linear component.

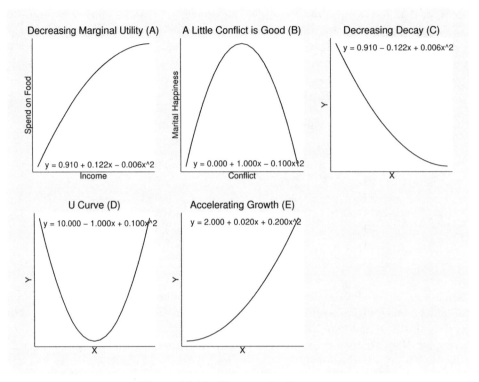

Figure 10.11. Five quadratic curves

Figure 10.11 illustrates some of the models that can be fit by adding a quadratic term to our regression model. We have discussed the first two models, which are examples of a positive linear component and a negative quadratic component. The `regress` command provides the best estimates of the linear and quadratic parameters for all of these curves. Figure 10.11(C) is labeled decreasing decay. Here a negative effect begins being quite strong but then levels out for high values of $x$. Notice figure 10.11(C) is obtained by having a negative value for the linear component, $x$, and a very small positive value for the quadratic component, $x^2$. Can you think of an example where this would apply? Suppose you were studying infant mortality ($y$) at the nation level and your independent variable is the gross national product (GNP). Countries with the lowest GNP would have a high rate of infant mortality, but this would drop quickly in countries with somewhat higher GNP's. However, at some point, a further increase in GNP would not lead to a substantial reduction in infant mortality. There would be a leveling off of infant mortality for nations that have higher GNP's.

When using a quadratic, you never want to make an inference that goes beyond your range of the independent variable. What would happen if the $x$ axis were extended farther to the right? The small positive quadratic component would push up more and more, and you would get something like figure 10.11(D), a U curve. Imaging a group of patients who have headache pain ($y$), you administered varying dosages of a pain medicine ($x$). Initially, you have a large drop in the pain, but as you increase the dosage, the benefit diminishes. You know you have a model like that in figure 10.11(C) or 10.11(D) when you have a negative linear coefficient and a positive quadratic component. The distinction between models like those in figures 10.11(C) and 10.11(D) is often difficult to distinguish without creating a graph.

Figure 10.11(E) shows an example of accelerating growth. Both the linear and the quadratic components are positive. The linear component is fairly small meaning that at the lower range of the $x$ variable, the estimate $y$ value is increasing only slightly. The positive quadratic term is being added to this effect as an extra push to a higher estimated value of $y$.

## 10.12.1 Fitting a quadratic model

Consider the relationship between the log of your wages and your total years of experience. We believe that the more years of experience you have, the greater the log of your wages. Let's use data that come from Stata:

```
. use http://www.stata-press.com/data/r13/regsmpl.dta
(NLS Women 14-26 in 1968)
```

We run a linear regression and obtain the following results:

```
. regress ln_wage ttl_exp, beta
```

| Source | SS | df | MS | | |
|---|---|---|---|---|---|
| Model | 1150.37005 | 1 | 1150.37005 | | |
| Residual | 5371.51384 | 28532 | .188262787 | | |
| Total | 6521.88388 | 28533 | .228573367 | | |

Number of obs = 28534
F( 1, 28532) = 6110.45
Prob > F = 0.0000
R-squared = 0.1764
Adj R-squared = 0.1764
Root MSE = .43389

| ln_wage | Coef. | Std. Err. | t | P>\|t\| | Beta |
|---|---|---|---|---|---|
| ttl_exp | .0431613 | .0005522 | 78.17 | 0.000 | .4199835 |
| _cons | 1.406646 | .0042866 | 328.15 | 0.000 | . |

We can represent these results as a graph (figure 10.12) by typing

```
. twoway lfit ln_wage ttl_exp
```

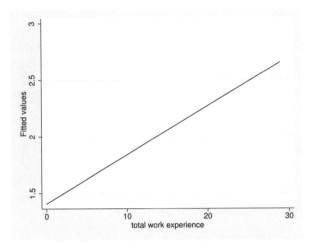

Figure 10.12. Graph of quadratic model

Figure 10.12 shows a moderately strong relationship between the total years of work experience and the estimated log wage. Do you see a problem? With many occupational positions, there is a top wage, so after several years, your wage rate may not increase or may only increase very gradually. We expect that wages will increase fairly rapidly the first several years and then level off in latter years. We can get a feel of whether this is correct using the **binscatter** command that we discussed in section 8.4.

```
. binscatter ln_wage ttl_exp
```

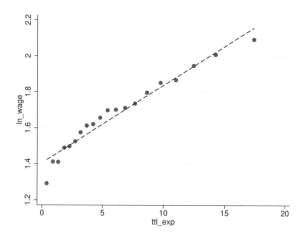

Figure 10.13. `binscatter` representation of nonlinear relationship between the log of wages and total years of experience

How does adding a quadratic component represent this? We expect the linear component to be positive because initially there is a regular increase in your wage rate, but we expect diminishing returns for additional years, especially after you have worked for many years. Therefore, we expect the linear component to be positive and the quadratic component to have a small, but significant, negative effect. We can handle the quadratic term the same way we could handle the interaction term. We could generate a new variable by multiplying the variables together, for example,

```
. generate ttl_sq = ttl_exp*ttl_exp
```

and then adding the new term to the regression model:

```
. regress ln_wage ttl_exp ttl_sq, beta
```

As with the interaction term, we have a better alternative. We can use the factor-variable notation for the continuous variable. Stata lets us use this special notation to indicate that a variable is continuous. We will use the `##` notation like we did with an interaction term, `c.ttl_exp##c.ttl_exp`. This notation keeps the main effect of total experience, `ttl_exp`, and it adds the product term to the model, `ttl_exp*ttl_exp`. This method is better than generating the squared term in a separate step because postestimation commands will behave properly. The double hash marks, `##`, along with the `c.` prefixes (for continuous independent variables) tell Stata to include both the linear component of `ttl_exp` and the quadratic component `ttl_exp2`.

Here are the command and results:

```
. regress ln_wage c.ttl_exp##c.ttl_exp, beta

      Source |       SS           df       MS            Number of obs =    28534
-------------+----------------------------------          F(  2, 28531) =  3166.13
       Model |  1184.57867          2  592.289333         Prob > F      =   0.0000
    Residual |  5337.30522      28531  .187070387         R-squared     =   0.1816
-------------+----------------------------------          Adj R-squared =   0.1816
       Total |  6521.88388      28533  .228573367         Root MSE      =   .43252

      ln_wage |      Coef.   Std. Err.      t    P>|t|                        Beta
-------------+----------------------------------------------------------------------
      ttl_exp |   .0660334   .0017787    37.12   0.000                    .6425416

  c.ttl_exp#
  c.ttl_exp |  -.0013929    .000103   -13.52   0.000                   -.2340455

        _cons |   1.348441   .0060651   222.33   0.000                           .
```

These results are exactly the same as we would have obtained by using the **generate** command to create the **ttl_sq** variable.

Just like with interaction terms, we can use the **margins** and **marginsplot** commands to graph these nonlinear results. This time we just have one predictor so we must set representative values for it. Running the **fre ttl_exp** (or **tab ttl_exp**) command, we find that experience ranges from 0 years to 28.885 years. Let's specify values from 0 to 28 years in increments of 2.

```
. margins, at(ttl_exp = (0(2)28))
Adjusted predictions                                Number of obs   =      28534
Model VCE     : OLS

Expression    : Linear prediction, predict()
1._at         : ttl_exp         =           0
2._at         : ttl_exp         =           2
3._at         : ttl_exp         =           4
4._at         : ttl_exp         =           6
5._at         : ttl_exp         =           8
6._at         : ttl_exp         =          10
7._at         : ttl_exp         =          12
8._at         : ttl_exp         =          14
9._at         : ttl_exp         =          16
10._at        : ttl_exp         =          18
11._at        : ttl_exp         =          20
12._at        : ttl_exp         =          22
13._at        : ttl_exp         =          24
14._at        : ttl_exp         =          26
15._at        : ttl_exp         =          28
```

|      | Margin | Delta-method Std. Err. | t | P>\|t\| | [95% Conf. Interval] | |
|------|--------|------------------------|------|--------|-----------|-----------|
| _at  |        |          |        |       |          |          |
| 1    | 1.348441 | .0060651 | 222.33 | 0.000 | 1.336553 | 1.360329 |
| 2    | 1.474936 | .0037037 | 398.24 | 0.000 | 1.467676 | 1.482195 |
| 3    | 1.590288 | .0029503 | 539.03 | 0.000 | 1.584505 | 1.59607 |
| 4    | 1.694497 | .0033365 | 507.87 | 0.000 | 1.687957 | 1.701036 |
| 5    | 1.787563 | .0038029 | 470.06 | 0.000 | 1.780109 | 1.795016 |
| 6    | 1.869485 | .0040283 | 464.08 | 0.000 | 1.861589 | 1.877381 |
| 7    | 1.940265 | .0042471 | 456.84 | 0.000 | 1.93194 | 1.948589 |
| 8    | 1.999901 | .0050573 | 395.45 | 0.000 | 1.989989 | 2.009814 |
| 9    | 2.048394 | .0069713 | 293.83 | 0.000 | 2.03473 | 2.062059 |
| 10   | 2.085745 | .0100468 | 207.60 | 0.000 | 2.066052 | 2.105437 |
| 11   | 2.111952 | .0141597 | 149.15 | 0.000 | 2.084198 | 2.139705 |
| 12   | 2.127016 | .0192161 | 110.69 | 0.000 | 2.089351 | 2.16468 |
| 13   | 2.130936 | .0251649 | 84.68 | 0.000 | 2.081612 | 2.180261 |
| 14   | 2.123714 | .0319792 | 66.41 | 0.000 | 2.061033 | 2.186395 |
| 15   | 2.105348 | .039644 | 53.11 | 0.000 | 2.027644 | 2.183053 |

In this `margins` command, we use a shortcut, 0(2)28, meaning to start at a value of 0 and increment it by 2 until you get to 28 (that is, 0 2 4 6 8 ... 28). This will give us an estimate value of the log of income for 15 different values on total years of experience.

We now run the `marginsplot` command, which produces figure 10.14.

    . marginsplot

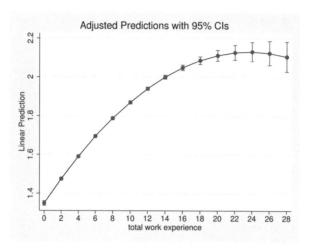

Figure 10.14.  Quadratic model of relationship between total experience and log of income

The linear component is 0.066, and it is significant, $t = 37.12$, $p < 0.000$. The quadratic component, $-0.001$, is significant, $t = -13.52$, $p < 0.001$. We can interpret these results in terms of their significance and the fact that they are in the predicted direction, that is, positive for the linear component and negative for the quadratic component. Caution is needed for interpreting the coefficients beyond noting their direction and significance. Converting to standardized $\beta$ values will not simplify the interpretation.

There is no single slope coefficient that describes the relationship like we had without the quadratic term. We cannot make a meaningful statement about the effect of a one-unit difference in experience on wages using the linear component or the quadratic component individually. Examining figure 10.14, we see that going from 0 to 1 year of work experience has a steep increase in the wages, going from 22 to 23 years has almost no increase in estimated wages, and going from 26 to 28 years of experience actually results in a slight decrease in estimated wages. We should be cautious about the 26 to 28 difference because this is at the upper end of the range of our independent variable and we have relatively few observations at that part of the distribution. You can see this in the confidence interval Stata constructed for each estimated value.

It can be shown that the linear slope, $B = 0.066$, is actually the tangent of the curve at a value of 0.0 on total years of experience. This may or may not be of much interest, but it is critical to not represent this as the difference in $y$ for a unit difference in $x$ because this only applies at a value of zero on the $x$ variable. It is important to show a figure so the reader can see the big picture.

There are interpretations you can make based on the coefficients that are useful. The intercept (_cons) is simply the estimated value of $y$ when $x$ is zero, and this is the same whether we have a linear or a quadratic model. The linear slope is the tangent of the curve when $x = 0$. In our example, the tangent of the curve for the ln_wage variable is 0.066 when you have zero total work experience. What else can we say? The curve is flat (the tangent to the curve being 0) at a value of $-B_1/2B_2$. For our example, the curve is flat at $-0.0660334/\{2(-0.0013929)\} = 23.704$ total years of experience. (It is important to keep all the decimal places you have in calculating this value and to use the unstandardized coefficients.) You can verify that the curve is flat in this area by observing where the tangent of the curve approaches zero in figure 10.14.

## 10.12.2   Centering when using a quadratic term

Often, it is useful to center the independent variable to provide a more reasonable value for the linear component. We center a variable by subtracting the mean of the variable from each score. We use this centered variable to generate the squared value. We covered centering a bit more when we introduced interaction models. Here is what we do for now by replacing the uncentered total experience variable, ttl_exp, with its centered value, cttl_exp:

```
. summarize ttl_exp
    Variable |        Obs        Mean    Std. Dev.       Min        Max
-------------+--------------------------------------------------------
     ttl_exp |      28534    6.215316    4.652117         0   28.88461
. generate cttl_exp = ttl_exp - 6.215
. regress ln_wage c.cttl_exp##c.cttl_exp, beta
      Source |       SS       df       MS              Number of obs =   28534
-------------+------------------------------           F(  2, 28531) = 3166.13
       Model | 1184.57867        2  592.289333         Prob > F      =  0.0000
    Residual | 5337.30522    28531  .187070387         R-squared     =  0.1816
-------------+--------------------------------         Adj R-squared =  0.1816
       Total | 6521.88388    28533  .228573367         Root MSE      =  .43252

--------------------------------------------------------------------------------
     ln_wage |      Coef.   Std. Err.      t    P>|t|                      Beta
-------------+------------------------------------------------------------------
    cttl_exp |   .0487197   .0006869    70.92   0.000                  .4740703
             |
 c.cttl_exp# |
  c.cttl_exp |  -.0013929    .000103   -13.52   0.000                 -.0903913
             |
       _cons |   1.705036   .0033948   502.25   0.000                         .
--------------------------------------------------------------------------------
```

The intercept, _cons $= 1.705$, is the estimated log income when cttl_exp $= 0$ that is at the mean on experience, which is 6.215 years. At this point, with 6.215 years of experience, the tangent is 0.0487, notably less steep than when we did not center experience. We will generate our figure to show the centered experience results using the following commands:

```
. margins, at(cttl_exp = (-6(2)22))
. marginsplot
```

The figure corresponding to the centered solution appears in figure 10.15. This is the same shape as in figure 10.14, where the total experience was not centered except for the shift in the $x$ variable due to subtracting the mean (6.215) from each person's score on ttl_exp. Thus instead of ranging from 0 years of experience to 28 years, it ranges from $-6$ years of experience to 22 years. The only added value of this approach, which is the linear component—now 0.049—is the instantaneous rate of change for a person with an average amount of experience, that is, with 6.215 years of experience.

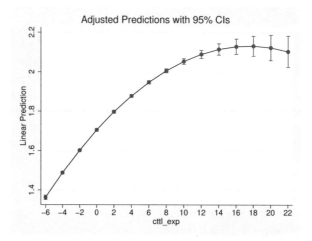

Figure 10.15. Quadratic model relating log of income to total experienced where experience is centered

Centering is important for independent variables where a value of zero may not be meaningful. For example, we would not be interested in a person who had zero years of education compared with a person who has an average amount of education. We would not be interested in the instantaneous rate of change for a person who had a graduate record examination of zero because that is not even a possible value. Other examples where a zero value is probably not meaningful and you should center include body mass index, weight, depression measured on a 30–75 point scale, and age in subsample of retired people. In each instance, a value of zero is neither useful nor meaningful. Nobody has a body mass index of zero or weight zero. No retired person has an age of zero and nobody has a score of zero on a scale that ranges from 30–75. With centering the independent variable, the size of the linear component still needs to be interpreted with care, but it is often more useful than using uncentered data. With a centered independent variable, the linear component represents the instantaneous rate of change for the person who has an average score on the independent variable.

## 10.12.3   Do we need to add a quadratic component?

The possible answers to this question are yes, no, or maybe. Consider our example of predicting a person's wage based on their total years of work experience. Figure 10.16 shows the result for both the linear model and the quadratic model.

```
. twoway (lfit ln_wage ttl_exp) (qfit ln_wage ttl_exp)
```

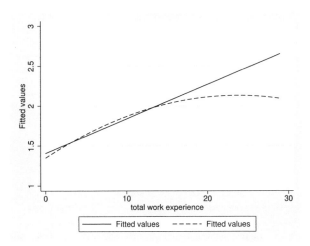

Figure 10.16. Comparison of linear and quadratic models

Examining figure 10.16, if you are interested in people with fewer than about 15 years of work experience, the linear model and the quadratic model are nearly indistinguishable. The linear model, being simpler, might be preferred. For people with more than 15 years of total work experience, the alternative models are different. The downward trend in the quadratic model indicates that it may be important to think of the wage benefit of increasing experience as leveling off. Sometimes the linear component works nearly as well as a quadratic model or at least does so over a range of the $x$ variable. Science always favors the simplest solution that works, so the linear model may be preferred at least for a partial range of the independent variable.

We can use the **nestreg** command to determine if adding the quadratic term strengthens $R^2$ and whether the improvement is statistically significant. However, it is important to note that this command only works if we do not use the factor-variable notation. The **nestreg** command requires you to generate the squared value of the independent variable; that is, first, you would type

```
. generate ttl_sq = ttl_exp*ttl_exp
```

We can then open the dialog box for the **nestreg** command by selecting Statistics ▷ Other ▷ Nested model statistics. In the *Dependent variable* box, type **ln_wage**. In the *Command* box, select **regress**. Check the *Block 1* and type the linear component, ttl_exp. Check the *Block 2* and type the quadratic component, ttl_sq. Alternatively, you could type the command directly. Here are the command and results:

```
. nestreg : regress ln_wage (ttl_exp) (ttl_sq)
Block  1: ttl_exp
```

| Source | SS | df | MS |
|---|---|---|---|
| Model | 1150.37005 | 1 | 1150.37005 |
| Residual | 5371.51384 | 28532 | .188262787 |
| Total | 6521.88388 | 28533 | .228573367 |

Number of obs =    28534
F( 1, 28532) = 6110.45
Prob > F      =  0.0000
R-squared     =  0.1764
Adj R-squared =  0.1764
Root MSE      =  .43389

| ln_wage | Coef. | Std. Err. | t | P>\|t\| | [95% Conf. Interval] | |
|---|---|---|---|---|---|---|
| ttl_exp | .0431613 | .0005522 | 78.17 | 0.000 | .042079 | .0442435 |
| _cons | 1.406646 | .0042866 | 328.15 | 0.000 | 1.398244 | 1.415048 |

```
Block  2: ttl_sq
```

| Source | SS | df | MS |
|---|---|---|---|
| Model | 1184.57867 | 2 | 592.289333 |
| Residual | 5337.30522 | 28531 | .187070387 |
| Total | 6521.88388 | 28533 | .228573367 |

Number of obs =    28534
F( 2, 28531) = 3166.13
Prob > F      =  0.0000
R-squared     =  0.1816
Adj R-squared =  0.1816
Root MSE      =  .43252

| ln_wage | Coef. | Std. Err. | t | P>\|t\| | [95% Conf. Interval] | |
|---|---|---|---|---|---|---|
| ttl_exp | .0660334 | .0017787 | 37.12 | 0.000 | .0625471 | .0695197 |
| ttl_sq | -.0013929 | .000103 | -13.52 | 0.000 | -.0015948 | -.001191 |
| _cons | 1.348441 | .0060651 | 222.33 | 0.000 | 1.336553 | 1.360329 |

| Block | F | Block df | Residual df | Pr > F | R2 | Change in R2 |
|---|---|---|---|---|---|---|
| 1 | 6110.45 | 1 | 28532 | 0.0000 | 0.1764 | |
| 2 | 182.86 | 1 | 28531 | 0.0000 | 0.1816 | 0.0052 |

The "Block 1" result replicates our linear regression solution. The "Block 2" result replicates our results for the quadratic model that includes both ttl_exp and ttl_sq. The key results are in the bottom table. "Block 1" has an $F(1, 28532) = 6110.45$, $p < 0.001$, and an $R^2 = 0.176$. "Block 2", for which we add the quadratic component, increases the $R^2$ by just 0.005 to a total $R^2 = 0.182$. The increase in $R^2$ of 0.005 is small, but it is statistically significant, $F(1, 28531) = 182.86$, $p < 0.001$.

Is our answer to the question posed in this section "yes", "no", or "maybe"? Based on the test of significance, the answer is yes. The quadratic model significantly improves our explanatory power. However, we need considerable caution with this result and should acknowledge that the improvement, while it should not be attributed to chance, is quite small. We should examine our graphs to further note the quadratic effect that applies mostly latter in a person's career. Although the curvature is interesting, the tiny increase in the $R^2$ is likely because only a small portion (less than 5%) of the total sample have scores on `ttl_exp` of more than 20 years.

## 10.13   Power analysis in multiple regression

There are two types of research questions for which we need to do a power analysis. The first type of power analysis estimates the power we have to show that our set of variables is significant. This is comparable to the overall $F$ test in multiple regression. Do age, gender, and education explain a significant portion of the variance in income? The null hypothesis is the $R^2 = 0.00$. To do a power analysis, we need to decide how much explained variance is substantively important. Do we care if race/ethnicity, gender, and education explain 5% of the variance; how about 10%, 20%, or 30%? If we are only interested in our results so that we can explain at least 20% of the variance, we will need far fewer people than if we wanted to have the power to obtain statistical significance of the true explained variance in our population for just 5%. Let's start with an alternative hypothesis that $R^2 = 0.20$.

Philip Ender and Xiao Chen, at the UCLA Academic Technology Services Center, wrote a useful command, `powerreg`, that we will use for this power analysis. To install this command, type `search powerreg` and follow the online instructions. To do a power analysis for our first type of question, we specify that $H_0$: $R^2 = 0.00$; $H_a$: $R^2 = 0.20$, alpha $= 0.05$, power $= 0.90$; and the number of variables is three. The command needs this information but uses a slightly different notation than what we have been using. $H_0$ is `r2f(.20)`, $H_a$ is `r2r(0)`, alpha is `alpha(.05)`, power is `power(.90)`, the number of variables is `nvar(3)`, and the number of variables we are testing is `ntest(3)`.

```
. powerreg, r2f(.2) r2r(.0) nvar(3) ntest(3) alpha(.05) power(.90)
Linear regression power analysis
alpha=.05  nvar=3  ntest=3
R2-full=.2  R2-reduced=0  R2-change=0.2000

nominal      actual
 power       power           n
 0.9000      0.8957         60
```

These results indicate that we only need $N = 60$ observations to have a power of 0.90 with alpha of 0.05 when we use three predictors and are interested in a significant result only if the true $R^2$ in the population is at least 0.20.

What if we want to be able to detect a small effect, say, a population $R^2$ of just 0.05? This is a much more sensitive test. It is going to be harder, meaning that we need more observations, to show that a small effect is statistically significant than to show that a relatively larger effect is statistically significant.

```
. powerreg, r2f(.05) r2r(.0) nvar(3) ntest(3) alpha(.05) power(.90)
Linear regression power analysis
alpha=.05  nvar=3  ntest=3
R2-full=.05  R2-reduced=0  R2-change=0.0500

nominal     actual
power       power          n
0.9000      0.8962        270
```

You can see why it is important to make a good judgment of how big an $R^2$ or change in an $R^2$ needs to be for it to be considered important. When we specified that and $R^2 = 0.20$ was the minimum we would consider important, we needed only 60 observations, but when we said we wanted to be able to show statistical significance for an $R^2 = 0.05$, we needed 270 observations.

It is often difficult to decide on a specific effect size, that is, $R^2$, to use. You can look to published research in your area of interest. If you cannot find a standard for your field, there is a general guideline you can use. Many researchers rely on a statistic called $f^2$ (Cohen 1988) that is simply the ratio of how much variance you explain to how much you do not explain.

$$f^2 = \frac{R^2}{1 - R^2}$$

Dattalo (2008) presents a table (10.2) showing what is generally recognized as a small, moderate, and large effect size. This table shows both the $f^2$ and the $R^2$ values. Remember, these are general recommendations and your area may use different values.

Table 10.2. Effect size of $f^2$ and $R^2$

| Effect size | $f^2$ | $R^2$ |
|---|---|---|
| small | 0.02 | 0.02 |
| medium | 0.15 | 0.13 |
| large | 0.35 | 0.26 |

These numbers may sound small, but remember that they are squared values. The corresponding $R$s would be 0.14, 0.36, and 0.51.

The second type of power analysis involves the ability to test the effect of adding one variable or a set of variables to an existing regression equation. Our previous research showed that age and education explained 35% of the variance in income, $R^2 = 0.35$. We want to see if race/ethnicity using our four categories (Caucasian, African American, Hispanic, and other) can explain an additional 10% of the variance. We would be adding three dummy variables using Caucasian as our reference group. The change in $R^2$ for the new "full" model is 0.45. We will assume that our alpha is 0.05 and that we want a power of 0.90.

```
. powerreg, r2f(.45) r2r(.35) nvar(5) ntest(3) alpha(.05) power(.90)
Linear regression power analysis
alpha=.05  nvar=5  ntest=3
R2-full=.45  R2-reduced=.35  R2-change=0.1000

 nominal     actual
  power       power           n
  0.9000      0.8997          82
```

These results show that adding three variables (`ntest()`) to the two we already had means that we have a total of five variables (`nvar()`). Using an alpha of 0.05 and a power of 0.90, we will need 82 observations.

Instead of race/ethnicity, what would happen if we wanted to add one variable, say, `motivation`? We will keep the other parameter estimates the same. Our command would be as follows:

```
. powerreg, r2f(.45) r2r(.35) nvar(3) ntest(1) alpha(.05) power(.90)
Linear regression power analysis
alpha=.05  nvar=3  ntest=1
R2-full=.45  R2-reduced=.35  R2-change=0.1000

 nominal     actual
  power       power           n
  0.9000      0.9007          60
```

All of these examples of power analysis are the type you conduct before you begin your study. The power analysis is done to make sure that you have a large enough sample to be able to show a result you feel is statistically significant and that you are not using a larger sample than is necessary to show this.

It is possible to reverse the process using `powerreg` to see what the power was in a study that has already been conducted. Imagine reading a study that argued that gender does not have a significant effect on income after you control for education, age, and uninterrupted time at the current employer. The researchers report that education, age, and uninterrupted time explained 25% of the variance in income and that this is significant. However, when gender is added, it explains only 5% and that increase is not statistically significant. Reading more carefully, you notice that they had only 50 people in their study. You decide that increasing the variance by 5% sounds important enough that you do not want to dismiss gender as an irrelevant variable. You do an a posteriori power analysis. How much power did they have with this size of sample? The `powerreg` command is done the same way as before except that this time the unknown is the power and the known is the sample size, $N$. Thus, instead of putting in a value for power, we put in the sample size, that is, `n(50)`.

```
. powerreg, r2f(.30) r2r(.25) nvar(4) ntest(1) alpha(.05) n(50)
Linear regression power analysis
alpha=.05  nvar=4  ntest=1
R2-full=.3  R2-reduced=.25  R2-change=.05

n = 50    power = 0.4561
```

It seems clear that the researchers did not have sufficient power with a power analysis estimate of just 0.46 for an $N = 50$. When a study has inadequate power, we can be skeptical of any results that are reported as insignificant. We do not know if the results are insignificant because they are unimportant or simply because their sample was too small to have a meaningful test.

## 10.14  Summary

Multiple regression is an extremely powerful and general strategy for analyzing data. We have covered several applications of multiple regression and diagnostic strategies. These provide you with a strong base, and Stata has almost unlimited ways of extending what you have learned. Here are the main topics we covered in chapter 10:

- What multiple regression can do, and how it is an extension of the bivariate correlation and regression we covered in chapter 8

- The basic command for doing multiple regression, including the option for standardized beta weights

- How we measure the unique effects of a variable, controlling for a set of other variables, which included a discussion of semipartial correlation and the increment in $R^2$

- Diagnostics of the dependent variable and how we test for normality of the error term

- Diagnostics for the distribution of residuals or errors in prediction

- Diagnostics to find outlier observations that have too much influence and may involve a coding error

- Collinearity and multicollinearity, along with how to evaluate the seriousness of the problem and what to do about it

- How to use weighted samples where different observations have different probabilities of being included, such as when there is oversampling of certain groups and we want to generalize to the entire population

- How to work with categorical predictors with just two categories and with more than two categories

- How to work with nested (hierarchical) regression in which we enter blocks of variables at a time

- How to work with interaction between a continuous variable and a categorical variable

If you are a Stata beginner, you may be amazed by the number of things you can do with regression and by what we have covered in this chapter. If you are an experienced user, I have only whetted your appetite. The next chapter, on logistic regression, includes several references that you should pursue if you want to know more about Stata's capabilities for multiple regression. Stata has special techniques for working with different types of outcome variables, for working with panel data, and for working with multilevel data where subgroups of observations share common values on some variables.

The next chapter is the first extension of multiple regression. I will cover the case in which there is a binary outcome because this happens often. You will have data with which you want to predict whether a couple will get divorced, whether a business will survive its first year, whether a patient recovers, whether an adolescent reports having considered suicide, whether a bill will become a law, or something similar. In each case, the outcome will be either a success or a failure. Then we will learn how to predict a success or a failure.

# 10.15   Exercises

1. Use `gss2002_chapter10.dta`. You are interested in how many hours a person works per week. You think that men work more hours, older people work fewer hours, and people who are self-employed work the most hours. The GSS has the needed measures, `sex`, `age`, `wrkslf`, and `hrs1` (hours worked last week). Recode `sex` and `wrkslf` into dummy variables, and then do a regression of hours worked on the predictors. Write out a prediction equation showing the $\beta$. What are the predicted hours worked for a female who is 20 years old and works for herself? How much variance is explained and is this statistically significant?

2. Use `census.dta`. The variable `tworace` is the percentage of people in each state and the District of Columbia who report being a combination of two or more races. Predict this by using the percentage of the state that is white (`white`), the median household income (`hhinc`), and the percentage with a BA degree or more education (`ba`). Predict the estimated value, calling it `yhat`. Predict the standardized residual, calling it `rstandard`. Plot a scattergram like the one in figure 10.5. Do a listing of states that have a standardized residual of more than 1.96. Interpret the graph and the standardized residual list.

3. Use `census.dta`. Repeat the regression in exercise 2. Compute the DFbeta score for each predictor, and list the states that have relatively large DFbeta values on any of the predictors. Explain why the problematic states are problematic.

4. Use `gss2002_chapter10.dta`. Suppose that you want to know if a man's socioeconomic status depends more on his father's or his mother's socioeconomic status. Run a regression (use `sei`, `pasei`, and `masei`), and control for how many hours a week the man works (`hrs1`). Do a test to see if `pasei` and `masei` are equal.

Create a graph similar to that in figure 10.7, and carefully interpret it for the distribution of residuals.

5. Use `gss2002_chapter10.dta`. You are interested in predicting the socioeconomic status of adults. You are interested in whether conservative political views predict higher socioeconomic status uniquely, after you have included background and achievement variables. The socioeconomic variable is `sei`. Do a nested regression. In the first block, enter gender (you will need to recode the variable `sex` into a dummy variable, `male`), mother's education (`maeduc`), father's education (`paeduc`), mother's socioeconomic index (`masei`), and father's socioeconomic index (`pasei`). In the second block, enter education (`educ`). In the third block, add conservative political views (`polviews`). Carefully interpret the results. Create a table summarizing the results.

6. Use `c10interaction.dta`. Repeat the interaction analysis presented in the text, but first center education and generate the interaction term to be male times the centered score on education. Centering involves subtracting the mean from each observation. You can either use the `summarize` command to get the mean for education and then subtract this from each score or type `search center` and install the ado-file called `center`. Type `center educ` to generate a new variable called `c_educ`. Then generate the interaction term with the command `generate educc_male = c_educ*male`. After running the regression (`regress inc male c_educ educc_male`) and creating a two-way overlay graph, compare the results with those in the text. Why is the intercept so different? Hint: The intercept is when the predictors are zero—think about what a value of zero on a centered variable is.

7. Repeat exercise 6 using the `i.`, `c.`, and `#` factor-variable operators. You do not need to do the graph, just type the commands, and show how the results match the results in exercise 6.

8. You want to do a study of sleep problems of college students. You have a scale measuring sleep problems. You think the following are all predictors: 1) use of social drugs (yes/no), 2) frequency of use of social drugs (0–30 times a month), 3) hours worked for pay, 4) stress about academic performance, and 5) stress about social relationships. How big of a sample do you need if you are interested in showing that these variables have a moderate effect on sleep problems? You need to justify the $R^2$ value you use along with the alpha and power.

9. Suppose that the study described in exercise 8 was conducted and the $R^2 = 0.20$. You think these predictors are important, but another predictor should have been financial stress. You think financial stress would explain at least an additional 5% of variance. You are interested in showing that such an increase would be statistically significant. Setting your alpha and power, what sample size will you need?

10. You are going to do the study described in exercise 9, but just as you are starting, you find an article that already did it. Using a sample of 75 people, the researchers

found that financial stress did increase the $R^2$ by 6%, but it was not a significant increase. How much power did they have? Why is this a problem?

11. In section 8.4, we used `nlsw88.dta`, which is installed with Stata. We again load this dataset by typing `sysuse nlsw88`. We will also again drop people who are under 35 or over 45 and keep only blacks and whites by keeping only people who had a code of less than 3 on the variable `race`. We will use the `binscatter` command. If you did not install this command in section 8.4, type `ssc install binscatter`.

    a. Type `binscatter wage tenure`, and then type `binscatter wage tenure, line(qfit)`.

    b. Does it look like a quadratic component is needed?

    c. Use the `nestreg` prefix command to test whether adding the quadratic term improves the fit.

12. Repeat the previous exercise using `nlsw88.dta` but this time center `tenure`.

    a. What is different?

    b. What is the same?

    c. Carefully interpret the linear effect.

# 11   Logistic regression

## 11.1   Introduction to logistic regression

The regression models that we covered in chapter 10 focused on quantitative outcome variables that had an underlying continuum. There are important applications of regression in which the outcome is binary—something either does or does not happen and we want to know why. Here are a few examples:

- A woman is diagnosed as having breast cancer

- A person is hired

- A married couple get divorced

- A new faculty member earns tenure

- A participant in a study drops out of the program

- A candidate is elected

We could quickly generate many more examples where something either does or does not happen. Binary outcomes were fine as predictors in chapter 10, but now we are interested in them as dependent variables. Logistic regression is a special type of regression that is used for binary-outcome variables.

## 11.2   An example

One of the best predictors of marital stability is the amount of positive feedback a spouse gives her or his partner.

- You observe 20 couples in a decision-making task for 10 minutes.

- You do this 1 week prior to their marriage and record the number of positive responses each of them makes about the other. An example would be "you have a good idea" or "you really understand my goals".

- You look just at the positive comments made by the husband and call these `positives`.

- You wait 5 years and see whether they get divorced. You call this `divorce`.

- You assign a code of 1 to those couples who get divorced and a code of 0 to those who do not.

- The resulting dataset has just two variables and is called `divorce.dta`.

The following is a list of the data in `divorce.dta`:

```
. list divorce positives
```

|      | divorce | positi~s |
|------|---------|----------|
| 1.   | 0       | 10       |
| 2.   | 0       | 8        |
| 3.   | 0       | 9        |
| 4.   | 0       | 7        |
| 5.   | 0       | 8        |
| 6.   | 0       | 5        |
| 7.   | 0       | 9        |
| 8.   | 0       | 6        |
| 9.   | 0       | 8        |
| 10.  | 0       | 7        |
| 11.  | 1       | 1        |
| 12.  | 1       | 1        |
| 13.  | 1       | 3        |
| 14.  | 1       | 1        |
| 15.  | 1       | 4        |
| 16.  | 1       | 5        |
| 17.  | 1       | 6        |
| 18.  | 1       | 3        |
| 19.  | 1       | 2        |
| 20.  | 1       | 0        |

If we graph this relationship (see figure 11.1), it looks strange because the outcome variable, `divorce`, has just two possible values. (The graph has been enhanced by using the Graph Editor, but the basic command is `scatter divorce positives`.)

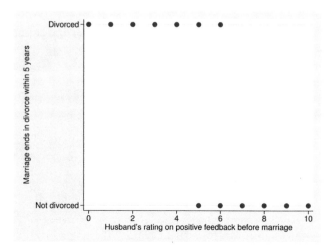

Figure 11.1. Positive feedback and divorce

A couple with a very low rating, say, 0–4 on `positives`, is almost certain to have a score of 1 on `divorce`. With a score of 7–10, a couple is almost certain to not get divorced, that is, to score 0 on `divorce`. A couple somewhere in the middle on `positives`, with a score of 5 or 6, may or may not get divorced. This is the zone of transition. It would not make sense to use a straight line as a prediction rule like we did with bivariate regression.

If we use logistic regression to predict the probability of a divorce based on the husband's rating on giving positive feedback prior to the marriage (we will soon learn how to do this), we get a predicted probability of divorce based on the score a husband has for `positives`. This conforms to what is sometimes called an *S*-curve (figure 11.2).

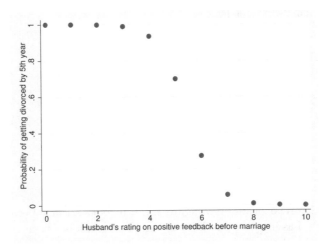

Figure 11.2. Predicted probability of positive feedback and divorce

Now this graph makes sense; it is a nonlinear relationship. The probability of divorce is nonlinearly related to the number of positive responses of the husband. If he gives relatively few positive responses, the probability of a divorce is very high; if he gives a lot of positive feedback, the probability of divorce is very low. Those husbands who are somewhere in the middle on positive feedback are the hardest to predict a divorce outcome for. This is exactly what we would expect and what our figure illustrates.

To fit a model like this, we do not estimate the probability directly. Instead, we estimate something called a logit. When the logit is linearly related to the predictor, the probability conforms to an *S*-curve like the one in figure 11.2. The reason we go through the trouble of computing a logit for each observation and using that in the logistic regression as the dependent variable is that the logit can be predicted by using a linear model. Notice that the probability of getting a divorce varies from 0 to 1, but, in fact, the logit ranges from about −10 to about 10 in this example. Although the probability of divorce is not linearly related to `positives`, the logit is. The relationship between the positive feedback the husband gives and the logit of the `divorce` variable is shown in figure 11.3. From a statistical point of view, figure 11.3 predicting the logit of divorce is equivalent to figure 11.2 predicting the probability of divorce.

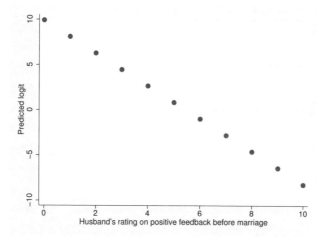

Figure 11.3. Predicted probability of positive feedback and logit of divorce

Before logistic regression was available, people did ordinary least-squares (OLS) regression, as described in chapter 10, when there was a binary-outcome variable. One problem with this is that the probability of the outcome being a 1 is much more likely to follow an $S$-curve than it is to follow a straight line. A second problem is that there is an absolute lower limit of zero and an absolute upper limit of one for a binary variable. If you predict 1 using OLS regression, you are saying that the probability is 1.0. If you predict 0.5, you are saying that the probability is 0.5. What if you are predicting $-0.2$ or 1.5? These are impossible values, but they can happen using OLS regression. Figure 11.4 shows what would happen using OLS regression because OLS regression has no way of knowing that your outcome cannot be below 0 or above 1.

```
. regress divorce positives
```

| Source | SS | df | MS | | |
|---|---|---|---|---|---|
| Model | 3.52343538 | 1 | 3.52343538 | | |
| Residual | 1.47656462 | 18 | .082031368 | | |
| Total | 5 | 19 | .263157895 | | |

| | | | | | |
|---|---|---|---|---|---|
| Number of obs = | 20 |
| F( 1, 18) = | 42.95 |
| Prob > F = | 0.0000 |
| R-squared = | 0.7047 |
| Adj R-squared = | 0.6883 |
| Root MSE = | .28641 |

| divorce | Coef. | Std. Err. | t | P>\|t\| | [95% Conf. Interval] |
|---|---|---|---|---|---|
| positives | -.1381739 | .021083 | -6.55 | 0.000 | -.1824677   -.0938801 |
| _cons | 1.211596 | .1260582 | 9.61 | 0.000 | .9467574   1.476434 |

```
. predict divols
(option xb assumed; fitted values)
```

```
. twoway (scatter divorce positives) (lfit divols positives)
```

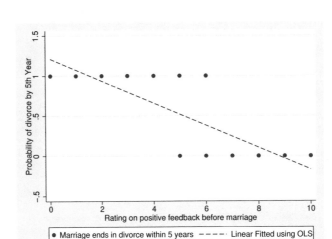

Figure 11.4. Positive feedback and divorce using OLS regression

You can see the problem with using a straight line with OLS regression. A couple
with a score of 10 on `positives` would have a probability of getting divorced within 5
years of less than zero. A couple with a score of 0 on `positives` would have a probability
of about 1.25. Both of these are impossible values because a probability must not be
greater than 1.0 nor less than 0.0. Comparing figure 11.2 and figure 11.4, you can see
that a logistic regression approach makes sense for binary-outcome variables.

## 11.3   What is an odds ratio and a logit?

I will explain odds ratios and logits by using a simple example with hypothetical data,
`environ.dta`. Suppose that you are interested in the relationship between environmental support and support for a liberal candidate in a community election. Your survey
asks items that you can use to divide your sample into two categories of environmental
concern, namely, high or low environmental concern. You also ask participants whether
they support a particular liberal candidate, and they either support this candidate or
they do not. We have two variables that we will call `environ` (1 means high environmental concern; 0 means low environmental concern) and `libcand` (1 means that they
support this liberal candidate; 0 means that they do not). Here is how the relationship
between these variables looks:

```
. tab2 environ libcand, row

-> tabulation of environ by libcand
```

| Key |
| --- |
| *frequency* |
| *row percentage* |

| Environmental concern: 1 high, 0 low | support liberal candidate: 1 yes, 0 no | | Total |
| --- | --- | --- | --- |
| | 0 | 1 | |
| 0 | 6 | 4 | 10 |
| | 60.00 | 40.00 | 100.00 |
| 1 | 3 | 7 | 10 |
| | 30.00 | 70.00 | 100.00 |
| Total | 9 | 11 | 20 |
| | 45.00 | 55.00 | 100.00 |

- The probability of supporting a liberal candidate (1 on support liberal) is 11/20, or 0.55.

- The probability of supporting a liberal candidate if you have low environmental concern (1 on support liberal if 0 on environmental concern) is 4/10, or 0.4.

- The probability of supporting a liberal candidate if you have high environmental concern (1 on support liberal if 1 on environmental concern) is 7/10, or 0.7.

From these last two probabilities, we can see that there is a relationship. With only 4 of the 10 people who have a low environmental concern supporting the liberal candidate compared with 7 of the 10 people who have high environmental concern, this looks like a fairly strong difference.

What are the odds of supporting a liberal candidate? The odds are the ratio of those who support the candidate to those who do not.

- The odds of supporting a liberal candidate are $11/9 = 1.22$. This means that there are 1.22 people supporting the liberal candidate for each person who opposes the liberal candidate.

- The odds of supporting a liberal candidate if you have low environmental concern are $4/6 = 0.67$. This means that among those with low environmental concern there are just 0.67 people supporting the liberal candidate for each person who opposes the liberal candidate.

- The odds of supporting a liberal candidate if you have high environmental concern are $7/3 = 2.33$. This indicates that among those with high environmental concern, there are 2.33 people supporting the liberal candidate for each person who opposes the liberal candidate.

## 11.3.1   The odds ratio

We need a way to combine these two odds, 2.33 and 0.67, into one number. We can take the ratio of the odds. The odds ratio of supporting the liberal candidate is $2.33/0.67 = 3.48$. The odds of supporting a liberal candidate are 3.48 times as great if you have high environmental concern as they are if you have low environmental concern.

The odds ratio gives us useful information for understanding the relationship between environmental concern and support for the liberal candidate. Environmental concern is a strong predictor of this support because the odds of a person high on environmental concern supporting the liberal are 3.48 times as great as the odds of a person with low environmental concern supporting the liberal.

## 11.3.2   The logit transformation

Odds ratios make a lot of sense for interpreting data, but for our purposes, they have some problems as a score on the dependent variable. The distribution of the odds ratio is far from normal.

An odds ratio of 1.0 means that the odds are equally likely and that the predictor makes no difference. Thus an odds ratio of 1.0 is equivalent to a beta weight of 0.0. If those with and without environmental concerns were equally likely to support the liberal candidate, the odds ratio would be 1.0.

An odds ratio can go from 1.0 to infinity for situations where the odds are greater than 1.0. By contrast, the odds ratio can go from 1.0 to just 0.0 for situations where the odds are less than 1.0. This makes the distribution extremely asymmetrical, which could create estimation problems. On the other hand, if we take the natural logarithm of the odds ratio, it will not have this distributional problem. This logarithm of the odds ratio is the logit:

$$\text{Logit} = \ln(\text{odds ratio})$$

Although the probability of something happening makes a lot of sense to most people and the odds ratio also makes sense to most of us, a lot of people have trouble understanding the score for a logit. Because the logistic regression is predicting a logit score, the values of the parameter estimates are difficult to interpret. To get around this problem, we can reverse the transformation process and get coefficients for the odds ratio that are easier to interpret.

# 11.4 Data used in the rest of the chapter

The National Longitudinal Survey of Youth, 1997 (`nlsy97_chapter11.dta`) asked a series of questions about drinking behavior among adolescents between the ages of 12 and 16. One question asked how many days, if any, each youth drank alcohol in the past month, `drday97`. We have dichotomized this variable into a binary variable called `drank30` that is coded 1 if the youth reported having had a drink in the past month or 0 if the youth reported not having had a drink in the past month.

Say that we want to see if peer behavior is important and if having regular meals together with the family is important. Our contention is that the greater the percentage of your peers who drink, the more likely you are to drink. By contrast, the more often you have meals together with your family, the less likely you are to drink. Admittedly, having meals together with your family is a limited indicator of parental influence. Because older youth are more likely to drink, we will control for age. We will also control for gender.

### Predicting a count variable

We have a variable that is a count of the number of days the youth reports drinking in the last month. We have dichotomized this as none or some. This makes sense if we are interested in predicting whether a youth drinks or not. What if we were interested in predicting how often a youth drinks rather than simply whether they did? Certainly, there is a difference between an adolescent who drank once in the last month and an adolescent who drank 30 days in the last month.

Stata has commands for estimating a dependent variable that is a count, although these commands are beyond the scope of this book. Using OLS regression with a count variable can be problematic when the count is of a rare event such as the number of days a youth drank in the past month. The distribution of rare events is skewed and sometimes has a substantial percentage of participants reporting a count of zero.

Here are two things we could do. One is to use what is called *Poisson regression of the count*. Poisson regression is useful for counts that are skewed, with most people doing the behavior rarely or just a few times. A second thing Stata can do is called *zero-inflated Poisson regression*, in which there is not only a skewed distribution but there are more zeros than you would expect with a Poisson distribution. If you asked how many times a female adolescent had a fist fight in the past 30 days, most would say zero times. Zero-inflated Poisson regression simultaneously estimates two regressions, one for the binary result of whether the count is zero and the other for how often the outcome occurred. Different factors may predict the onset of the outcome than would predict the count of the outcome. Long and Freese (2006) have an excellent book that discusses these extensions to what we cover in this chapter.

Interpreting a logistic regression depends on understanding something of the distribution of the variables. First, do a summary of the variables so that we know their means and distributions. If we need to eliminate any observations that did not answer all the items, we drop the cases that have missing values on any of the variables. It would also be important to do a tabulation, although those results are not shown here. The summary of the variables is

```
. summarize drank30 age97 pdrink97 dinner97 male
> if !missing(drank30, age97, pdrink97, dinner97, male)
    Variable |      Obs       Mean    Std. Dev.      Min        Max

     drank30 |     1654    .382104    .4860487         0          1
       age97 |     1654   13.67352    .9371347        12         16
    pdrink97 |     1654   2.108222    1.214858         1          5
    dinner97 |     1654   4.699516    2.349352         0          7
        male |     1654   .5405079    .4985071         0          1
```

We see that the mean for drank30 is 0.382. This means that 38.2% of the students reported drinking at least once a month. Logistic regression also works well even when the outcome is rare, such as whether a person dies after surgery. When the mean is around 0.50, the OLS regression and logistic regression produce consistent results, but when the probability is close to 0 or 1, the logistic regression is especially important.

## 11.5  Logistic regression

There are two commands for logistic regression: logit and logistic. The logit command gives the regression coefficients to estimate the logit score. The logistic command gives us the odds ratios we need to interpret the effect size of the predictors. Select Statistics ▷ Binary outcomes. From here, we can select Logistic regression or Logistic regression (reporting odds ratios). Selecting the latter produces the dialog box in figure 11.5.

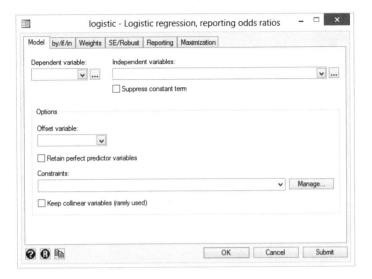

Figure 11.5. Dialog box for doing logistic regression

This dialog box is pretty straightforward. Type **drank30** as the *Dependent variable* and **age97 male pdrink97 dinner97** as the *Independent variables*. The *Offset variable* is for special purposes and is not discussed here. The option to *Retain perfect predictor variables* should not be checked in most applications. In a rare case where there is a category that has no variation on the outcome variable, Stata drops the predictor. Checking the *Retain perfect predictor variables* box forces Stata to include the variable, which can lead to instabilities in the estimates. The other tabs have the usual set of options to restrict our sample, weight the observations, or use different estimators.

The only difference between the dialog boxes for logistic regression and for logistic regression with odds ratios is that the default for logistic regression is the estimated regression coefficients and the default for the logistic regression with odds ratios is the odds ratios. Regardless of which dialog box you choose, on the Reporting tab, we can specify whether we want the regression coefficients or the odds ratios. The following commands and output show the results of both dialogs. The first results are for the command **logistic**, where we get the odds ratios, and the second results are for the command **logit**, where we get the regression coefficients.

```
. logistic drank30 age97 male pdrink97 dinner97
Logistic regression                              Number of obs   =      1654
                                                 LR chi2(4)      =     78.01
                                                 Prob > chi2     =    0.0000
Log likelihood = -1061.0474                      Pseudo R2       =    0.0355
```

| drank30 | Odds Ratio | Std. Err. | z | P>\|z\| | [95% Conf. Interval] | |
|---|---|---|---|---|---|---|
| age97 | 1.169241 | .0684191 | 2.67 | 0.008 | 1.042546 | 1.311332 |
| male | .9794922 | .1046935 | -0.19 | 0.846 | .7943646 | 1.207764 |
| pdrink97 | 1.329275 | .0598174 | 6.33 | 0.000 | 1.217056 | 1.451841 |
| dinner97 | .942086 | .0208682 | -2.69 | 0.007 | .9020603 | .9838878 |
| _cons | .0524677 | .0415938 | -3.72 | 0.000 | .0110944 | .2481314 |

```
. logit drank30 age97 male pdrink97 dinner97
Iteration 0:    log likelihood = -1100.0502
Iteration 1:    log likelihood =  -1061.142
Iteration 2:    log likelihood = -1061.0474
Iteration 3:    log likelihood = -1061.0474
Logistic regression                              Number of obs   =      1654
                                                 LR chi2(4)      =     78.01
                                                 Prob > chi2     =    0.0000
Log likelihood = -1061.0474                      Pseudo R2       =    0.0355
```

| drank30 | Coef. | Std. Err. | z | P>\|z\| | [95% Conf. Interval] | |
|---|---|---|---|---|---|---|
| age97 | .1563548 | .0585158 | 2.67 | 0.008 | .0416659 | .2710437 |
| male | -.020721 | .1068855 | -0.19 | 0.846 | -.2302128 | .1887708 |
| pdrink97 | .2846336 | .0450001 | 6.33 | 0.000 | .1964351 | .3728321 |
| dinner97 | -.0596587 | .022151 | -2.69 | 0.007 | -.1030739 | -.0162434 |
| _cons | -2.947557 | .7927494 | -3.72 | 0.000 | -4.501317 | -1.393797 |

Both commands give the same results, except that `logit` gives the coefficients for estimating the logit score and `logistic` gives the odds ratios. The results of the `logit` command show the iterations Stata went through in obtaining its results. Logistic regression relies on maximum likelihood estimation rather than ordinary least squares; this is an iterative approach where various solutions are estimated until the best solution of having the maximum likelihood is found. The results show the number of observations followed by a likelihood-ratio (LR) chi-squared test and something called the pseudo-$R^2$. In this example, the likelihood-ratio chi-squared(4) = 78.01, $p < 0.001$. There are several coefficients that are called pseudo-$R^2$. The one reported by Stata is the McFadden pseudo-$R^2$. This is often a small value, and it should not be confused with $R^2$ for OLS regression. Many researchers do not report this measure.

It is difficult to interpret the regression coefficients for the `logit` command. We can see that both the percentage of peers who drink and the number of dinners a youth has with his or her parents are statistically significant. The peer variable has a positive coefficient, meaning that the more of the youth's peers who drink, the higher the logit for the youth's own drinking. The `dinner97` variable has a negative coefficient, and this is as we expected; that is, having more dinners with your family lowers the logit for drinking. We can also see that age is significant and in the expected direction. Sex, by

contrast, is not statistically significant, so there is no statistically significant evidence
that male versus female adolescents are more likely to drink.

The first output is for the `logistic` command, which gives us the odds ratios. We
need to spend a bit of time interpreting these odds ratios. The variable `age97` has an
odds ratio of 1.17, $p < 0.01$. This means that the odds of drinking are multiplied by 1.17
for each additional year of age. When the odds ratio is more than 1, the interpretation
can be simplified by subtracting 1 and then multiplying by 100: $(1.17 - 1.00) \times 100 =$
17%. This means that for each increase of 1 year, there is a 17% increase in the odds of
drinking. This is an intuitive interpretation that can be understood by a lay audience.
You have probably seen this in the papers without realizing the way it was estimated.

For example, you may have seen that there is a 50% increase in the risk of getting
lung cancer if you smoke. People sometimes use the word *risk* instead of saying that
there is a 50% increase in the odds of getting cancer. What would you say to a lay
audience? Each year older an adolescent gets, the odds of drinking increases by 17%.
You need to be careful when you interpret the odds ratio. There is a separate coefficient
called the relative-risk ratio, which is discussed in the box a few pages from here about
the odds ratio versus the relative-risk ratio. Lay audiences often act as if both of these
were the same thing, but they are not and they can be different when the outcome is
not a rare event. Odds ratios tell us what happens to the odds of an outcome, whereas
risk ratios tell us what happens to their probability.

What happens if you compare a 12-year-old with a 15-year-old? The 15-year-old is
3 years older, so you might be tempted to say that the risk factor is $3 \times 17\% = 51\%$
greater. However, this underestimates the effect. If you are familiar with compound
interest, you have already guessed the problem. Each additional year builds on the
previous year's compounded rate. To compare a 15-year-old with a 12-year-old, you
would first cube the odds ratio, $1.17^3 = 1.60$ (to compare a 16-year-old with a 12-year-
old, you would compute $1.17^4 = 1.87$). Thus the odds of drinking for a 15-year-old is
$(1.60 - 1.00) \times 100 = 60\%$ greater than that for a 12-year-old. The odds for a 16-year-old
is $(1.87 - 1.00) \times 100 = 87\%$ greater than that for a 12-year-old.

When an odds ratio is less than 1.00, you need to change the calculation just a
bit. We like to talk of the decrease in the odds ratio, so we need to subtract the
odds ratio from 1.00 to find this decrease. Thus for each extra day per week that
an adolescent has dinner with his or her family, the odds of drinking are reduced by
$(1.00 - 0.94) \times 100 = 6\%$. Consider two adolescents. One has no family dinners, and the
other has them every day of the week. Once again, you could be tempted to say that
the odds of drinking are now $(7 \times 6\%) = 42\%$ lower. This is just as wrong for odds ratios
smaller than one as it is for odds ratios larger than one. Just as before, we must raise the
original odds ratio to a power: the odds of drinking for the adolescent who eats at home
is $0.94^7 = 0.65$ times as large as the odds of drinking for the adolescent who never eats
with his or her family. Again we want to express this as a percentage change. Because
the multiplier is less than 1.00, we compute the change as $(1.00 - 0.65) \times 100 = 35\%$,
meaning that a youth who has dinner with the family every night of the week has 35%
lower odds of drinking than a youth who has no meals with his or her family.

### Using Stata as a calculator

> When doing logistic regression, we often need to calculate something to help us interpret the results. We just said that $1.17^4 = 1.87$. To get this result, in the Stata Command window, enter `display 1.17^4`, which results in 1.87.
>
> The odds ratios are a transformation of the regression coefficients. The coefficient for `pdrink97` is 0.2846. The odds ratio is defined as $\exp(b)$, which for this example is $\exp(0.2846)$. This means that we exponentiate the coefficient. More simply, we raise the mathematical constant $e$ to the $b$ power, $e^b$ or $e^{0.2846}$ for the `pdrink97` variable. This would be a hard task without a calculator. To do this within Stata, in the Command window, type `display exp(0.2846)`, and 1.33 is displayed as the odds ratio.

It is hard to compare the odds ratio for one variable with the odds ratio for another variable when they are measured on different scales. The `male` variable is binary, going from 0 to 1, and the variable `dinner97` goes from 0 to 7. For binary predictor variables, you can interpret the odds ratios and percentages directly. For variables that are not binary, you need to have some other standard. One solution is to compare specific examples, such as having no dinners with the family versus having seven dinners with them each week. Another solution is to evaluate the effect of a 1-standard-deviation change for variables that are not binary. This way, you could compare a 1-standard-deviation change in dinners per week with a 1-standard-deviation change in peers who drink. When both variables are on the same scale, the comparison makes more sense, and this is exactly what we can do when we use the standard of a 1-standard-deviation change.

Using the 1-standard-deviation change as the basis for interpreting odds ratios and percentage change is a bit tedious if you need to do the calculations by hand. There is a command, `listcoef`, that makes this easy. Although this command is not part of standard Stata, you can do a `search spost` command and install the most recent version of `spost`, which will include `listcoef`. There are several items that appear when you enter `search spost9`. Toward the end of this list of items is a section called `packages found`. Under this heading, go to `spost9_ado`, and click on the link to install it. When you install `spost9_ado`, you will install a series of commands that are useful for interpreting logistic regression. Be sure to install the latest version of this package of commands because versions for earlier Stata releases, for example, Stata 7, are still posted. To run this command, after doing the logistic regression, enter the command `listcoef, help`. The `help` option gives a brief description of the various values the command estimates. Here are the results:

```
. listcoef, help

logit (N=1654): Factor Change in Odds

  Odds of: 1 vs 0
```

| drank30 | b | z | P>\|z\| | e^b | e^bStdX | SDofX |
|---------|-----|-----|-------|-----|---------|-------|
| age97 | 0.15635 | 2.672 | 0.008 | 1.1692 | 1.1578 | 0.9371 |
| male | -0.02072 | -0.194 | 0.846 | 0.9795 | 0.9897 | 0.4985 |
| pdrink97 | 0.28463 | 6.325 | 0.000 | 1.3293 | 1.4131 | 1.2149 |
| dinner97 | -0.05966 | -2.693 | 0.007 | 0.9421 | 0.8692 | 2.3494 |

```
        b = raw coefficient
        z = z-score for test of b=0
   P>|z| = p-value for z-test
      e^b = exp(b) = factor change in odds for unit increase in X
e^bStdX = exp(b*SD of X) = change in odds for SD increase in X
    SDofX = standard deviation of X
```

This gives the values of the coefficients (called $b$), their $z$ scores, and the probabilities. These match what we obtained with the `logit` command. The next column is labeled `e^b` and contains the odds ratios for each predictor. The label may look strange if you are not used to working with logarithms. If you raise the mathematical constant $e$, which is about 2.718, to the power of the coefficient $b$, you get the odds ratio. Thus $e^{0.15635} = 1.1692$.

We can interpret the odds ratio for `male` directly because `male` is a binary variable. The other variables are not binary, so we use the next column, which is labeled `e^bStdX` to interpret their odds ratios. This column displays the odds ratio for a 1-standard-deviation change in the predictor. For example, the odds ratio for a 1-standard-deviation change in age is 1.16, and this is substantially smaller than the odds ratio for a 1-standard-deviation change in the percentage of peers who drink, 1.41.

If you prefer to use percentages, you can use the command

```
. listcoef, help percent

logit (N=1654): Percentage Change in Odds

  Odds of: 1 vs 0
```

| drank30 | b | z | P>\|z\| | % | %StdX | SDofX |
|---------|-----|-----|-------|-----|-------|-------|
| age97 | 0.15635 | 2.672 | 0.008 | 16.9 | 15.8 | 0.9371 |
| male | -0.02072 | -0.194 | 0.846 | -2.1 | -1.0 | 0.4985 |
| pdrink97 | 0.28463 | 6.325 | 0.000 | 32.9 | 41.3 | 1.2149 |
| dinner97 | -0.05966 | -2.693 | 0.007 | -5.8 | -13.1 | 2.3494 |

```
        b = raw coefficient
        z = z-score for test of b=0
   P>|z| = p-value for z-test
        % = percent change in odds for unit increase in X
    %StdX = percent change in odds for SD increase in X
    SDofX = standard deviation of X
```

For communicating with a lay audience, this last result is remarkably useful. The odds of drinking are just 2.1% lower for males than for females, and this is not statistically significant. Having a 1-standard-deviation-higher percentage of peers who drink increases a youth's odds of drinking by 41.3%, and having dinner with his or her family 1 standard deviation more often reduces the odds by 13.1%. Both of these influences are statistically significant.

It is often helpful to create a graph showing the percentage change in the odds ratio associated with each of the predictors. This could be done with any graphics package, such as PowerPoint. To do this using Stata, we need to create a new dataset, c11barchart.dta. This dataset has just four variables, which we will name Age, male, peers, and dinners. There is just one observation, and this is the appropriate percentage change in the odds ratio. For Age, peers, and dinners, we enter 15.8, 41.3, and −13.1, respectively, because we are interested in the effects of a 1-standard-deviation change in each of these variables. For male, we enter −2.1 because we are interested in a 1-unit change for this dichotomous variable (female versus male). Then we construct a bar chart. You can use Graphics ▷ Bar chart to create this graph. The command is

```
. graph bar (asis) Age male peers dinners, bargap(10)
> blabel(name, position(outside)) ytitle(Percentage Change in Odds)
> title(Percentage Change in Odds of Drinking by)
> subtitle("Age, Gender, Percent of Peers Drinking, Meals with Family")
> legend(off) scheme(s2manual)
```

Note that if your title or subtitle includes commas, as shown in the subtitle() option above, you must enclose the title in quotes. We had Stata put the variable labels just above or below each bar; we must have very short labels for this to work. Also, on the Bar tab, we put a *Bar gap* of 10 that so the bars would be separated a little bit. The resulting bar chart is shown in figure 11.6.

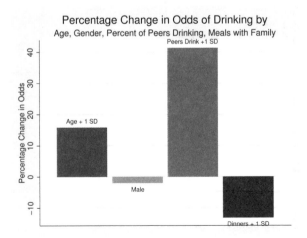

Figure 11.6. Risk factors associated with teen drinking

This bar graph shows that as adolescents get older, there is a substantial increase in the odds of their drinking; that drinking risks are roughly comparable for males and females; that having peers who drink is a substantial risk factor; and that the more the adolescent shares dinners with his or her family, the lower the odds of drinking.

**Odds ratio versus relative-risk ratio**

In section 11.3, we learned how to compute the odds ratio. There we noted that among the 10 people who favored environmental protection, seven voted liberal and three voted conservative, creating odds of voting liberal of $7/3 = 2.33$. By contrast, the 10 people who did not favor environmental protection had odds of voting liberal of $4/6 = 0.67$. The resulting odds ratio was $2.33/0.67 = 3.48$.

The analogous measure for percentages or probabilities is the relative-risk ratio. There are 7 of 10 (70% or 0.70) pro-environmentalists supporting a liberal candidate. There are 4 of 10 (40% or 0.40) people who are not pro-environment supporting a liberal candidate. The relative risk of supporting a liberal candidate for people who are pro-environment is $0.70/0.40 = 1.75$. People who are not used to odds ratios and popular media such as newspapers often misinterpret the odds ratio as if it were the relative risk. Generally, the odds ratio will be larger than the relative-risk ratio. The difference is small with a rare outcome, but with an outcome like this, there is a substantial difference. Being pro-environment makes the odds of voting for a liberal 3.48 times as great. Being pro-environment increases the risk of voting for a liberal candidate 1.75 times. In this example, there is a 248% increase in the odds of voting liberal if you are pro-environment, but there is only a 75% increase in the risk of voting liberal.

Stata's `logit` and `logistic` commands do not estimate the relative-risk ratio (sometimes called incidence relative risk). However, Joseph Hilbe, a Stata user, wrote a command, `oddsrisk`, in 2008 that converts odds ratios to risk ratios. You can install this command by entering `ssc install oddsrisk`.

The relative risk is appealing, but it should not be used in a study that controls the number of people in each category. For example, you should not use the relative risk if you are comparing 100 stroke victims to 100 people who did not have a stroke on their participation in a rigorous physical activity. By controlling the number of people in each group, you have artificially (nonrandomly) inflated the number of stroke victims because 50% of your sample have had a stroke—a far greater percentage than would be obtained in a representative sample. For more on relative risk, visit http://www.emrevision.co.uk/resources/Exams/FCEM-Exam/Academic/Stats_-Odds-ratio-versus-relative-risk.pdf, which was written by Steve Simon.

# 11.6    Hypothesis testing

There are many types of hypothesis testing that can be done with logistic regression. I
will cover some of them here. As with multiple regression, we have an overall test of the
model, tests of each parameter estimate, and the ability to test hypotheses involving
combinations of parameters. As noted above, with logistic regression there is a chi-
squared test that has $k$ degrees of freedom, where $k$ is the number of predictors. For
our current example, the $\chi^2(4) = 78.01$, $p < 0.001$. This tells us only that the overall
model has at least one significant predictor.

## 11.6.1    Testing individual coefficients

There are two tests we can use to test individual parameters, and they often yield
slightly different probabilities. The most common is a Wald test, commonly reported in
journal articles. The $z$ test in the Stata output is actually the square root of the Wald
chi-squared test. Some other statistical packages test individual parameters by using
the Wald chi-squared test (evaluated with 1 degree of freedom), but this is identical to
the square of the $z$ test in Stata.

Many statisticians prefer the likelihood-ratio chi-squared test to the Wald chi-squared
test or the equivalent $z$ test reported by Stata. The likelihood-ratio chi-squared test
for each parameter estimate is based on comparing two logistic models, one with the
individual variable we want to test included and one without it. The likelihood-ratio
test is the difference in the likelihood-ratio chi-squared values for these two models (this
appears as `LR chi2(1)` near the upper right corner of the output). The difference between the two likelihood-ratio chi-squared values is 1 degree of freedom. In standard
Stata, the likelihood-ratio test is rather tedious to get. If you want the best possible test
of the significance of the effect of age on drinking, with `nlsy97_chapter11.dta`, you
would need to type the command—`logistic drank30 male dinner97 pdrink97`—
and record the chi-squared. Then you would type the command—`logistic drank30
age97 male dinner97 pdrink97`—and record the chi-squared. The difference between
the chi-squared values with 1 degree of freedom would be the best test of the effect of
age on drinking.

Stata lets us simplify this process a bit with the following set of commands:

```
. logistic drank30 male dinner97 pdrink97
. estimates store a
. logistic drank30 age97 male dinner97 pdrink97
. lrtest a
```

The first line runs the logistic regression without including `age97`. The second line
saves all the estimates and stores them in an estimation set labeled `a`. This estimation
set includes the `LR chi2(1)` value. The third line runs a new logistic regression, this
time including `age97`. The last line, `lrtest a`, subtracts the chi-squared values and
estimates the probability of the chi-squared difference. Although this is not too difficult,
doing this process for each of the four predictors would be a bit tedious. An automated
approach is available with a command written by user Zhiqiang Wang in 1999 called

`lrdrop1`, which gives the likelihood-ratio test for each parameter estimate, along with some other information we will not discuss (`lrdrop1` works with the `logit`, `logistic`, and `poisson` regression commands). Type the command `ssc install lrdrop1` to install this command. Immediately after running the full model (with all four predictors), simply type the following command:

```
. lrdrop1
Likelihood Ratio Tests: drop 1 term
logistic regression
number of obs = 1654
-----------------------------------------------------------------------
drank30      Df     Chi2    P>Chi2    -2*log ll   Res. Df   AIC
-----------------------------------------------------------------------
Original Model                         2122.09      1649    2132.09
   -age97       1     7.18    0.0074    2129.27      1648    2137.27
   -male        1     0.04    0.8463    2122.13      1648    2130.13
-dinner97       1     7.23    0.0072    2129.33      1648    2137.33
-pdrink97       1    40.62    0.0000    2162.72      1648    2170.72
-----------------------------------------------------------------------

Terms dropped one at a time in turn.
```

Here we can see that the likelihood-ratio chi-squared for the `age97` variable is $\chi^2(1) = 7.18$, $p < 0.01$. If we take the square root of 7.18, we obtain 2.68, which is close to the square root of the Wald chi-squared, that is, the $z$ test provided by Stata of 2.67. The likelihood-ratio chi-squared and the Wald statistic are not always this close, but the two approaches rarely produce different conclusions. Although the likelihood-ratio test has some technical advantages over the Wald test, the Wald test is usually what people report—perhaps because it is a bit tedious to estimate the likelihood-ratio chi-squared in other statistical packages. If the Wald test and the likelihood-ratio tests differ, the likelihood-ratio test is preferred. These results are identical whether you are testing the regression coefficients with the `logit` command or the odds ratios with the `logistic` command. Because this `lrdrop1` command is so easy to use and because the likelihood-ratio test is invariant to nonlinear transformations, you may want to report the likelihood-ratio test routinely. When you do so, specify that it is the likelihood-ratio chi-squared test rather than the Wald test.

## 11.6.2   Testing sets of coefficients

There are three situations for which we would need to test a set of coefficients rather than a single coefficient. These parallel the tests we used for multiple regression in chapter 10.

1. We have a categorical variable that requires several dummies. Suppose that we want to know if marital status has an effect, and we have three dummies representing different marital statuses (divorced or separated, widowed, and never married, with married serving as the reference category). We want to know if marital status is significant, so we need to test the set of coefficients rather than just one coefficient.

2. We want to know if a set of variables is significant. The set might be whether peer and family variables are significant when age and gender are already included in a model to predict drinking behavior. We need to test the combination of coefficients for peer and family rather than testing these individually.

3. We have two variables measured on the same scale, and we want to know if their parameter estimates are equal. Suppose that we want to know if mother's education is more important than father's education as a predictor of whether an adolescent plans to go to college. Both predictors are measured on the same scale (years of education), so we can test whether the mother's education and the father's education have significantly different coefficients.

The first two situations are tested by the hypothesis that all coefficients in the set are zero. For the marital status set, the three simultaneous null hypotheses would be

$H_0$: Divorce effect $= 0$

$H_0$: Widow effect $= 0$

$H_0$: Never-married effect $= 0$

For the peer and family set, the set of two simultaneous null hypotheses would be

$H_0$: `pdrink97` effect $= 0$

$H_0$: `dinner97` effect $= 0$

We will illustrate how to test for the first two situations with the data we have been using. The full model is

```
. logistic drank30 age97 male pdrink97 dinner97
```

We want to know if `pdrink97` and `dinner97`, as a set, are significant.

There are several ways of testing this. We could run the model twice:

(a) `logistic drank30 age97 male`

(b) `logistic drank30 age97 male pdrink97 dinner97`

We would then see whether the difference of chi-squared values was significant. Because the second model added two variables, we would test the chi-squared with 2 degrees of freedom.

We can alternatively use the **test** command. We run this after we run the full logistic regression. In other words, we run the command `logistic drank30 age97 male pdrink97 dinner97`, and then we run the command `test pdrink97 dinner97`. Here are the results:

```
. test pdrink97 dinner97
 ( 1)  pdrink97 = 0
 ( 2)  dinner97 = 0
           chi2(  2) =    48.85
         Prob > chi2 =    0.0000
```

The variables `pdrink97` and `dinner97`, as a set, are statistically significant, chi-squared(2) = 48.85, $p < 0.001$. We can say that one or both of these variables are statistically significant. When the set of variables is significant, we should always look at the tests for the individual coefficients because this overall test only tells us that at least one of them is significant.

The third situation for testing a set of parameter estimates is to test the equality of parameter estimates for two variables (assuming that both variables are measured on the same scale). We might do a logistic regression and find that both mother's education and father's education are significant predictors of whether an adolescent intends to go to college. However, we may want to know if the mother's education is more or less important than the father's education. Say that we want to test whether the parameter estimates are equal. Our null hypothesis is

$H_0$: mother's education effect = father's education effect

I will not illustrate this test because it uses a different dataset, but the command is simple. Let's assume that mother's education is represented by the variable `maeduc` and father's education is represented by `faeduc`. The command is `test maeduc=faeduc`.

## 11.7   More on interpreting results from logistic regression

This section will stay with the same basic model except we will substitute the variable `black` for the variable `male`. We do this substitution because the effect of gender was not significant, and because we are interested in understanding possible differences between whites and blacks. To keep the example straightforward, we limit our analysis to whites and blacks and exclude other racial and ethnic groups. We first need to generate this new variable:

```
. generate black = race97 - 1
. replace black = . if race97 > 2
```

We also will label our race variable:

```
. label define black 0 "White" 1 "Black"
. label define drank30 0 "No" 1 "Yes"
. label values drank30 drank30
. label values black black
```

We discussed the special notation for factor variables in chapter 10 as used in modeling interaction and with a quadratic term. We can use this notation here as well. We can think of three of our independent variables, `age97`, `pdrink97`, and `dinner76`,

as having underlying continuums and hence as continuous variables even though some
are measured crudely. However, our variable `black` is strictly categorical because it
is coded 1 for blacks and 0 for whites. We can run the logistic regression using the
`i.` label for this categorical variable, `i.black`. This produces the same results for the
logistic regression as if we had simply used `black`, but the results will work properly if
we follow this command with other postestimation commands. Here are the command
and the results:

```
. logit drank30 age97 i.black pdrink97 dinner97

Iteration 0:   log likelihood = -935.86755
Iteration 1:   log likelihood = -901.48553
Iteration 2:   log likelihood = -901.37312
Iteration 3:   log likelihood = -901.37311

Logistic regression                              Number of obs    =        1413
                                                 LR chi2(4)       =       68.99
                                                 Prob > chi2      =      0.0000
Log likelihood = -901.37311                      Pseudo R2        =      0.0369
```

| drank30 | Coef. | Std. Err. | z | P>\|z\| | [95% Conf. Interval] | |
|---|---|---|---|---|---|---|
| age97 | .138153 | .0635579 | 2.17 | 0.030 | .0135818 | .2627241 |
| black | | | | | | |
| Black | -.3804608 | .1352133 | -2.81 | 0.005 | -.645474 | -.1154476 |
| pdrink97 | .2822417 | .048233 | 5.85 | 0.000 | .1877067 | .3767767 |
| dinner97 | -.069024 | .0246204 | -2.80 | 0.005 | -.1172791 | -.0207689 |
| _cons | -2.590308 | .8609411 | -3.01 | 0.003 | -4.277722 | -.9028946 |

We have already discussed how to interpret these results, but now we are going to
consider some new aids. The first `margins` command that we use below is useful for categorical variables. In our example, the categorical variable is `black` (see Williams [2012]
for an excellent discussion of the `margins` command or see http://www.ats.ucla.edu/
stat/stata/faq/margins_mlogcatcon.htm for applications of this command, including use
of interactions). The `margins` command will tell us the difference in the probability of
having drunk in the last 30 days if an individual is black compared with if an individual
is white. For this to be meaningful, we need to set the covariates at some meaningful
value. We will set them at the mean. So the question we want to answer is as follows: What is the difference between blacks and whites who are average on the other
covariates in the probability that they drank in the last 30 days? Here is our command:

```
. margins, dydx(black) atmeans
Conditional marginal effects                          Number of obs    =        1413
Model VCE      : OIM

Expression     : Pr(drank30), predict()
dy/dx w.r.t.  : 1.black
at             : age97          =    13.67445  (mean)
                 0.black        =     .7523001 (mean)
                 1.black        =     .2476999 (mean)
                 pdrink97       =    2.112527  (mean)
                 dinner97       =    4.760793  (mean)
```

|         |    dy/dx | Delta-method Std. Err. |     z | P>\|z\| | [95% Conf. Interval] |          |
|--------:|---------:|----------------------:|------:|------:|---------------------:|---------:|
| black   |          |                       |       |       |                      |          |
| Black   | -.0862436 | .0296054             | -2.91 | 0.004 | -.144269             | -.0282181 |

Note: dy/dx for factor levels is the discrete change from the base level.

We specified two options. First, we specified **dydx(black)** because **black** is the variable on which we are focused. Second, we specified **atmeans** because this option will set the covariates at their mean. From the results, we see that age was fixed at 13.674 and so on. The dy/dx value of $-0.086$ is the difference. Because blacks are coded 1 and whites are coded 0, a hypothetical black adolescent who is average in age, percent of peers who drink, and in how often his or her family eats together will have a probability of drinking in the last 30 days that is 0.086 lower than a hypothetical white adolescent with the same scores on the covariates. An easier way of saying this is that black adolescents who are average on covariates are 8.6% less likely to report having drunk in the last 30 days than white adolescents.

Rather than trying to interpret the dy/dx values for quantitative and continuous independent variables, we will use a different option with the **margins** command. Suppose we wanted to understand the relationship between the percentage of friends who drink and whether the adolescent drank in the last 30 days, but we want to control for the other covariates. As before, we will fix the other covariates at their means. This way we can estimate the probability a hypothetical adolescent drank in the last 30 days who is average on all the covariates, except they differ on the percentage of their friends who drank. The **pdrink97** variable was coded 1 for up to 20% of friends, 2 for 20 to 40% of friends, 3 for 40 to 60% of friends, 4 for 60 to 80% of friends, and 5 for 80 to 100% of friends.

In the following command, the **at()** option fixes the other covariates at their means for each value of **pdrink97** from 1 to 5. If we were doing this for a variable that ranged from 0 to 100, we might type **margins, at(var = (0(10)100))**, which would be equivalent to typing **margins, at(var = (0 10 20 30 40 50 60 70 80 90 100))**, meaning every 10 years.

```
. margins, at(pdrink97=(1 2 3 4 5)) atmeans
Adjusted predictions                                Number of obs    =      1413
Model VCE      : OIM

Expression     : Pr(drank30), predict()

1._at          : age97              =    13.67445 (mean)
                 0.black            =    .7523001 (mean)
                 1.black            =    .2476999 (mean)
                 pdrink97           =           1
                 dinner97           =    4.760793 (mean)

2._at          : age97              =    13.67445 (mean)
                 0.black            =    .7523001 (mean)
                 1.black            =    .2476999 (mean)
                 pdrink97           =           2
                 dinner97           =    4.760793 (mean)

3._at          : age97              =    13.67445 (mean)
                 0.black            =    .7523001 (mean)
                 1.black            =    .2476999 (mean)
                 pdrink97           =           3
                 dinner97           =    4.760793 (mean)

4._at          : age97              =    13.67445 (mean)
                 0.black            =    .7523001 (mean)
                 1.black            =    .2476999 (mean)
                 pdrink97           =           4
                 dinner97           =    4.760793 (mean)

5._at          : age97              =    13.67445 (mean)
                 0.black            =    .7523001 (mean)
                 1.black            =    .2476999 (mean)
                 pdrink97           =           5
                 dinner97           =    4.760793 (mean)
```

|      |          | Delta-method |       |       |                      |
|------|----------|--------------|-------|-------|----------------------|
|      | Margin   | Std. Err.    | z     | P>\|z\| | [95% Conf. Interval] |
| _at  |          |              |       |       |                      |
| 1    | .3011665 | .0168523     | 17.87 | 0.000 | .2681367    .3341963 |
| 2    | .3636614 | .0132156     | 27.52 | 0.000 | .3377592    .3895636 |
| 3    | .4311242 | .0169284     | 25.47 | 0.000 | .3979451    .4643033 |
| 4    | .501244  | .0261244     | 19.19 | 0.000 | .450041     .5524469 |
| 5    | .5713148 | .0361017     | 15.83 | 0.000 | .5005567    .6420729 |

From these results, we see that if our hypothetical average adolescent on the covariates had a code of 1 on `pdrink97`, then we estimate a probability of 0.301 that this adolescent drank in the last 30 days. This estimated probability goes up fairly sharply so that if our hypothetical average adolescent on the covariates was coded 5 on `prink97`, then he or she would have a probability of 0.571 of having drunk in the last 30 days. It would be nice to be able to see a figure showing these results, so after the `margins` command, we can type the `marginsplot`. This graph is shown as figure 11.7 with a few minor changes applied using the Graph Editor:

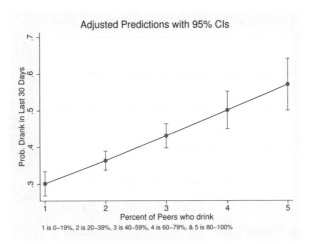

Figure 11.7. Estimated probability that an adolescent drank in last month adjusted for age, race, and frequency of family meals

## 11.8   Nested logistic regressions

In chapter 10, we discussed nested regression (sometimes called hierarchical regression), where blocks of variables are entered in a planned sequence. The `nestreg` command is extremely general, applicable across a variety of regression models, including logistic, negative binomial, Poisson, probit, ordered logistic, tobit, and others. It also works with the complex sample designs for many regression models. Although most of these models are beyond the scope of this book, I will illustrate how `nestreg` can be generalized with logistic regression.

We have been interested in whether the percentage of an adolescent's peers who drink and the number of days a week the adolescent eats with his or her family influence drinking behavior. In doing this, we controlled for gender and age. We might decide to enter the four predictors in three blocks. In the first block, we enter gender; in the second block, we enter age; and in the third block, we enter dinners with the family and peer drinking. We now apply the format used in chapter 10 for logistic regression to get the following results:

```
. nestreg: logistic drank30 (male) (age97) (dinner97 pdrink97)

Block  1: male

Logistic regression                          Number of obs   =       1654
                                             LR chi2(1)      =       4.37
                                             Prob > chi2     =     0.0365
Log likelihood = -1097.864                   Pseudo R2       =     0.0020
```

| drank30 | Odds Ratio | Std. Err. | z | P>\|z\| | [95% Conf. Interval] | |
|---|---|---|---|---|---|---|
| male | .8087911 | .0820956 | -2.09 | 0.037 | .6628816 | .9868173 |
| _cons | .6926503 | .0510996 | -4.98 | 0.000 | .599401 | .8004065 |

```
Block  2: age97

Logistic regression                          Number of obs   =       1654
                                             LR chi2(2)      =      27.87
                                             Prob > chi2     =     0.0000
Log likelihood = -1086.1173                  Pseudo R2       =     0.0127
```

| drank30 | Odds Ratio | Std. Err. | z | P>\|z\| | [95% Conf. Interval] | |
|---|---|---|---|---|---|---|
| male | .8304131 | .0849837 | -1.82 | 0.069 | .6794898 | 1.014858 |
| age97 | 1.304566 | .0723084 | 4.80 | 0.000 | 1.170271 | 1.454273 |
| _cons | .0178754 | .0137142 | -5.25 | 0.000 | .0039738 | .080409 |

```
Block  3: dinner97 pdrink97

Logistic regression                          Number of obs   =       1654
                                             LR chi2(4)      =      78.01
                                             Prob > chi2     =     0.0000
Log likelihood = -1061.0474                  Pseudo R2       =     0.0355
```

| drank30 | Odds Ratio | Std. Err. | z | P>\|z\| | [95% Conf. Interval] | |
|---|---|---|---|---|---|---|
| male | .9794922 | .1046935 | -0.19 | 0.846 | .7943646 | 1.207764 |
| age97 | 1.169241 | .0684191 | 2.67 | 0.008 | 1.042546 | 1.311332 |
| dinner97 | .942086 | .0208682 | -2.69 | 0.007 | .9020603 | .9838878 |
| pdrink97 | 1.329275 | .0598174 | 6.33 | 0.000 | 1.217056 | 1.451841 |
| _cons | .0524677 | .0415938 | -3.72 | 0.000 | .0110944 | .2481314 |

| Block | Wald chi2 | df | Pr > F |
|---|---|---|---|
| 1 | 4.37 | 1 | 0.0366 |
| 2 | 23.01 | 1 | 0.0000 |
| 3 | 48.85 | 2 | 0.0000 |

These results show that each block is statistically significant. Our main interest is in the third block, and we see that adding dinners with your family and peer drinking has $\chi^2(2) = 48.85$, $p < 0.001$. These results duplicate those obtained with the `lrtest` command. The results do not duplicate those for the first two blocks (age and gender) using the `lrtest`. Here we are testing if each additional block makes a statistically significant improvement above what was done by the preceding blocks. We would say that "Block 1", `male`, is significant. Gender was not significant in the overall model, controlling for the other three variables, but "Block 1" contains only the variable `male`. "Block 2", `age97`, increases the significance, but it does not control for "Block 3" variables because they have yet to be added to the model.

## 11.9   Power analysis when doing logistic regression

As with multiple regression, when doing logistic regression, it is helpful to perform a power analysis before beginning a study. Logistic regression, using maximum likelihood estimation tends to require a larger sample size than ordinary least-squares multiple regression. The program will run on a sample that is very small, but it is best to have at least 100 observations when doing a logistic regression for even a simple model. Philip Ender at UCLA adapted a SAS macro to create a command called `powerlog`. You can type the command `search powerlog` and follow the online instructions to install the command. This command is designed to work either with one continuous predictor or several predictors. We will illustrate it first for one predictor.

Suppose that a researcher wants to know if motivation (the predictor) can predict whether a person stays in a smoking cessation program or drops out. The question is whether a person with higher motivation has a higher rate of retention. Here is what we need to know to run the `powerlog` command:

1. What proportion of people would stay in the program if they had average motivation? Let's say that we found some results for similar programs, and about 70% of the participants stayed in the program and 30% dropped out. We would say the proportion for people with average motivation would be $p_1 = 0.70$.

2. What proportion of people would stay in the program if they were 1-standard-deviation change above average on motivation? We should think in terms of the smallest increase (or decrease) that would be considered important. Let's say that increasing the retention rate from 70% to 75% would be considered important. We would say the proportion of people who are retained for the entire program, if they were 1-standard-deviation change above average on motivation, would be $p_2 = 0.75$.

3. We need to set our alpha level at, say, 0.05.

This is all we need for the command.

```
. powerlog, p1(.70) p2(.75) alpha(.05)
Logistic regression power analysis
One-tailed test: alpha=.05  p1=.7  p2=.75  rsq=0  odds ratio=1.285714285714286

power           n
0.60          204
0.65          233
0.70          266
0.75          303
0.80          348
0.85          404
0.90          481
```

To have a power of 0.90, we would need a sample of 481 people. If we were satisfied with a power of 0.80, we would only need 348 people. Say that a person comes to you and says that she can collect retention data on a sample of 50 people and wants to do this analysis. You must advise her that she has too little power if she is interested in a difference that is this small. She says, What if I am interested in an increase of 0.10, so the retention rate would go from 70% to 80%. If you run the command using p2(80), you will see that she would have 0.80 power with just 89 cases, so you still need to advise her to get a larger sample.

When we are using this command, we set it up as a one-tailed test because we thought about increasing the retention rate. If you want a two-tailed test, where the proportion could go up or down, you could use $\alpha/2 = 0.025$ for your alpha value. You will also notice that we are thinking in terms of proportions rather than odds ratios. The command results report that the odds ratio is 1.286. Do you know how it got this value? For $p_1$, we had a proportion of 0.70, so the odds were $0.70/0.30 = 2.333$. For $p_2$, we had a proportion of 0.75, so the odds were $0.75/0.25 = 3.000$. Thus the odds ratio is $3.000/2.333 = 1.286$. This is a good reminder that when we say the odds are 28.6% greater, we are describing the odds and not the proportions, where the corresponding change from 0.70 to 0.75 represents only a 7.1% increase in the proportion.

Sometimes when you are doing a logistic regression, you have several predictors and want to know the power for adding one more predictor. Let's stick with our study of retention. Suppose that in addition to motivation being a predictor of whether a person stays in the smoking cessation program, we expect the following people to be more likely to stay for the entire program: people who are older, people who smoked fewer days in the last month, and people who are more educated. We want to add to this the number of other program participants who are their friends. If you have a group in which several of your friends are participating, you think this will substantially increase your retention rate. What do we need to know to do the power analysis?

- What proportion of people would stay in the program if they had average motivation, age, days smoked in last month, and education? Let's say that we found some results for similar programs, and about 70% of the participants stayed in the program and 30% dropped out. We would say the proportion for people with average scores would be $p_1 = 0.70$.

- What proportion of people would stay in the program if they were 1-standard-deviation change above average on the number of friends they had who were also in the program? Again we should think in terms of the smallest increase (or decrease) that would be considered important. Let's stay with the idea that increasing the retention rate from 70% to 75% would be considered important. We would say the proportion of people who are retained for the entire program, if they were 1-standard-deviation change above average on number of friends also participating, would be $p_2 = 0.75$.

- We need to set our alpha level at, say, 0.05.

- This time we need one additional estimate. When there are multiple variables on the right side of an equation, we need to worry about multicollinearity; thus we need an estimate of how serious the multicollinearity is. This estimate is the proportion of the variance in our new variable, which is the number of friends also participating. The proportion of the variance is explained by the variables already in the model (motivation, age, days smoked in last month, and education). We might find some data where we could do an $R^2$ for predicting number of friends from motivation, age, days smoked in last month, and education. Let's estimate the $R^2 = 0.30$.

Here are our command and results. We added the `help` option to get an explanation of the terms.

```
. powerlog, p1(.70) p2(.75) alpha(.05) rsq(.30) help
Logistic regression power analysis
One-tailed test: alpha=.05  p1=.7  p2=.75  rsq=.3  odds ratio=1.285714285714286

power            n
0.60            292
0.65            333
0.70            380
0.75            434
0.80            498
0.85            578
0.90            687

Explanation of terms
p1  -- the probability that the response variable equals 1
        when the predictor is at the mean
p2  -- the probability that the response variable equals 1
        when the predictor is one standard deviation above the mean
rsq -- the squared mulitple correlation between the predictor
        variable and all other variables in the model
```

As these results show, if we want a power of 0.90, we will need 687 participants in our study.

What do you do when some of your predictors are categorical, such as race/ethnicity or gender? One possibility is to use the `powerlog` command, but it will only give very approximate results because the command assumes you have continuous and normally distributed predictors.

It is not unusual for people to run a logistic regression command when they have less-than-optimal power. Suppose that you ran this analysis on a sample of just 200 people and all your variables were statistically significant. This can be interpreted two ways. First, you are a big risk taker because you had woefully inadequate power to run this analysis. You were most lucky to find a funding source that would support a study that had inadequate power. Second, you must be a lucky researcher because you did find significant results, and when you find significant results even when you do not have a lot of power, they are still significant.

## 11.10  Summary

Logistic regression is a powerful data-analysis procedure that on one hand is fairly complicated but on the other hand can be presented to a lay audience in a clear and compelling fashion. Policy makers are likely to understand what odds mean, and they want to know which factors raise the odds of a success and reduce the odds of a failure. Although the language surrounding logistic regression can be intimidating, the results are easy to understand and explain. To continue with the example from this chapter, policy makers are definitely interested in the factors that increase or decrease the odds that an adolescent will drink. With logistic regression, this is exactly the information you can provide.

Some inexperienced researchers, upon learning logistic regression, may use it for too many applications. Logistic regression makes sense whenever you have an outcome that is clearly binary. If you have an outcome that is continuous or a count of how often something happens, you may want to use other procedures. For example, if you have a scale measuring political conservatism, which is what you want to explain, it is probably not wise to dichotomize people into conservative versus liberal categories just so that you can do logistic regression. When you have a continuous variable and you dichotomize it, you lose a lot of variance. The difference between a person who is just a little right of center and a radical conservative is lost if they both go into the same category. In our example, we lost the difference between an adolescent who drank only occasionally and one who drank 30 days a month. This was probably okay because we wanted to find out what explained whether the child drank or did not drink. Therefore, the variance in how often an adolescent drinks was not part of our question.

This chapter covered the following topics that prepare you to do logistic regression:

- Examples of when the technique is appropriate

- Key concepts, including odds, odds ratios, and logit

- How a linear estimation of a logit value corresponds to a nonlinear relationship for the probability of a success

- Two Stata commands, `logit` and `logistic`, that are used for doing logistic regression

- The regression coefficients and the odds ratios

- How to interpret the results for categorical and continuous predictors

- How to associate a percentage change in the odds ratio with each variable and how to summarize this in a bar chart

- Alternatives for hypothesis testing, including two ways to test the significance of each variable's effect and how to test the significance of sets of predictors

Chapters 10 and 11 provided the core techniques used by many social scientists. Although multiple regression and logistic regression are both complex procedures, they can provide informative answers to many research and policy questions. I have covered many useful and powerful methods of data analysis using Stata, and you are prepared for the next steps. Many researchers can do everything they need to do with the techniques we have already covered, but we have tapped only the core capabilities of Stata. An exciting thing about working with Stata is that it is a constantly evolving statistical package, and it is one that few of us will outgrow.

## 11.11   Exercises

1. Use `severity.dta`, which has three variables. The variable `severity` is whether the person sees prison sentences as too severe (coded 1) or not too severe (coded 0). The variable `liberal` is how liberal the person is, varying from 1 for conservative to 5 for liberal. The variable `female` is coded 1 for women and 0 for men. Do a logistic regression analysis with `severity` as the dependent variable and `liberal` and `female` as the independent variables. Carefully interpret your results.

2. Use `gss2002_chapter11.dta`. What predicts who will support abortion for any reason? Recode `abort12` to be a dummy variable called `abort`. Use the following variables as predictors: `reliten`, `polviews`, `premarsx`, and `sei`. You may want to create new variables for these so that a higher score goes with the name; for example, `polviews` might be renamed `conservative` because a higher score means more conservative. Do a logistic regression followed by the `listcoef` command. Carefully interpret the results.

3. Using the results from exercise 2, create a bar graph showing the percentage change in odds of supporting abortion for any reason associated with each predictor. Justify using a 1-unit change or a 1-standard-deviation change in the score on each predictor.

4. Using the results from the logistic regression in exercise 2, compute likelihood-ratio chi-squared tests for each predictor, and compare them with the standard $z$ tests in the logistic regression results.

5. In exercise 1, you did a logistic regression using gender and how liberal a person was to predict whether they thought prison sentences were too harsh. Suppose that you think people with more education are likely to feel that sentences are too severe. Do a tabulation on severity to see what proportion overall think sentences are too severe. Then do a power analysis to see how big a sample you would need to have a power of 0.80 to detect whether a 1-standard-deviation increase in education would raise the proportion of people who think sentences are too severe by 0.10 when liberalism and gender are already in the model. You would be adding education as a third predictor. State the information you need and the values you assume. What assumption of the command you used did you violate?

# 12 Measurement, reliability, and validity

## 12.1   Overview of reliability and validity

The quality of our measurement is the foundation for all statistical analysis. We must always keep two facts in mind: First, our analysis can be no better than our measurement. If we have poor measurement of our variables, then the relationship we find will be inaccurate and will often underestimate the true strength of the relationship. Second, virtually every variable we measure has some error, and often this error is substantial.

Imagine doing a study of the relationship between postpartum depression and body fat among a group of women two months after childbirth. Suppose that we find a weak relationship, say, $r = 0.10$. Is this because postpartum depression is not strongly related to body fat percentage? Perhaps, but it could be due to poor measurement of depression and body fat. Our depression scale might have an alpha ($\alpha$) reliability of 0.90, and our measure of body fat percentage might have a test–retest reliability of 0.80. I will explain these measures of reliability later, but both of them indicate that at least some of the variance in our two measures is not related to what we call them but is simply measurement error. Reliability means that our measurement process will produce consistent results.

We also need to worry about validity. Many measures of body fat percentage are not only prone to error, that is, have low reliability, but also are biased. For example, a strong athlete may have a completely wrong estimated fat percentage. An NFL running back may be 5'11" and weigh 235 pounds. An estimator of his body fat—such as the BMI, which uses just height and weight to estimate body fat—might grossly overestimate his percentage of body fat. In his case, his weight is coming from musculature rather than from fat. Just adding his waist size of 32 inches would greatly improve the accuracy of the estimate. This is a problem of validity. The answer we get is off the mark. It may or may not be reliable, but it is not valid.

This chapter discusses what we can do to evaluate the reliability and the validity of our measures. Our discussion is largely limited to what is known as classical measurement theory.

## 12.2   Constructing a scale

The most typical situation is having a set of $k$ items that measure a concept. We might have 10 items measuring depression, 5 items measuring attitude toward abortion, or 15 items measuring perceived risk of sexually transmitted HIV. How does Stata generate a scale score? Some researchers like to generate the sum of the items. If each of 10 items is on a scale of 1 to 4 (strongly agree, agree, disagree, and strongly disagree), the sum (total score) would range from a minimum of 10 to a maximum of 40. Other researchers like to generate the mean of the items, and in this example, the mean would be between 1 and 4. One advantage of generating the mean is that it is on the same scale as the individual items and hence can be interpreted more easily. For example, a sum of 25 might not have an obvious meaning, but a mean of 2.5 would indicate that the participant was halfway between agree and disagree on the 4-point scale.

The 2006 General Social Survey (`gss2006_chapter12.dta`) asked people a series of seven items, `empathy1`–`empathy7`, designed to measure empathy. Each of these was scored from 1 for *Does not describe me very well* to 5 for *Describes me very well*. There was also a category of 8 for *Don't know*, which we treat as a missing value.

If you run a `tabulate` on the empathy items, you will see that some of the items, `empathy2`, `empathy4`, and `empathy5`, are worded so that a higher score goes with less empathy, and the others, `empathy1`, `empathy3`, `empathy6`, and `empathy7`, are worded so that a higher score goes with more empathy. When we cover factor analysis, we will see how such an arrangement introduces a methods factor. For now, we will recode the three negative items so that each item has a higher score indicating greater empathy. Always coding a scale so that a higher score represents more of whatever is being measured makes it easier for readers to understand what you are doing. We discussed how to recode variables by using the `recode` command. The command is

```
. recode empathy2 empathy4 empathy5 (1=5 "Does not describe very well")
> (2=4) (3=3) (4=2) (5=1 "Describes very well"), pre(rev) label(empathy)
```

Because we are recoding three variables with one command, this is a bit more complicated. We are applying value labels to two values, the new value 5 and the new value 1, and we did not bother to make labels for the other values. When we have value labels in the command, we need to give a nickname for this set of labels. This could be any value label name we have not used already in a dataset. I chose the `label(empathy)` option for the nickname. We never overwrite variables but instead generate new variables, and to name those variables, we chose to add a prefix of `rev` to each variable by using the `pre(rev)` option. Thus our seven variables are now named `empathy1`, `revempathy2`, `empathy3`, `revempathy4`, `revempathy5`, `empathy6`, and `empathy7`.

## 12.2.1   Generating a mean score for each person

Stata has a convenient command for generating the mean score for each person. Because each participant's data are in a row, we want the row mean for the selected items. The command name is `egen`, which stands for extended generation. We name the new variable that we want to generate `empathy`, and we make this equal to the row mean (with the `rowmean()` function) of the listed variables. The variables are listed inside parentheses and have no commas separating them. Here is the command:

```
. egen empathy = rowmean(empathy1 revempathy2 empathy3 revempathy4 revempathy5
> empathy6 empathy7)
```

**Requiring a 75% completion rate**

The `rowmean()` function calculates the mean of the items from the set that a participant answers. If Sara answered all but the last empathy item, she would get a mean for the six items she answered. If George answered only the first three items, he would get a mean for those three items. Some researchers will want to exclude people who did not answer at least 75% of the items (at least 5 of the 7 items). One way to do this is to count how many items each person answers and then exclude people who did not answer at least the minimum requirement. Here we would count the number of missing values by using `egen miss = rowmiss(empathy1 revempathy2 empathy3 revempathy4 revempathy5 empathy6 empathy7)`. This creates a new variable, `miss`, that is the count of the number of items that have a missing value. Because we require that 5 of the 7 questions be answered, we would compute the empathy score only for those who have a count of fewer than 3 (they answered 5, 6, or all 7 of the items). We can use this variable to restrict our sample; our new `egen` command will be: `egen empathya = rowmean(empathy1 revempathy2 empathy3 revempathy4 revempathy5 empathy6 empathy7) if miss < 3`.

## 12.3  Reliability

There are four types of reliability, and these are illustrated in Shultz, Whitney, and Zickar [2014].

Table 12.1. Four kinds of reliability and the appropriate statistical measure

| Type of reliability | Reliability focus | Name of procedure | Statistical measure |
|---|---|---|---|
| Stability | Changes in participants | Test–retest correlation<br>Test–retest agreement | correlation, $r$<br>intraclass correlation, $\rho_I$ |
| Equivalence | Sampling of content domain | Alternate forms | correlation, $r_{xx'}$ |
| Internal consistency | Sampling of content domain | Split-half<br>Alpha reliability | correlation, $r_{X_a X_b}$<br>alpha, $\alpha$ |
| Rater consistency | Interrater agreement | Agreement | kappa, $\kappa$ |

Stability means that if you measure a variable today using a particular scale and then measure it again tomorrow using the same scale, your results will be consistent. For depression, we would expect new mothers who are higher than average on depression today to still be higher than average tomorrow.

## A problem generating a total scale score

Many researchers generate a total score rather than a mean. I mentioned that this is often hard to interpret because it is not on the same scale as the original items. Another problem is that Stata, as well as other software such as IBM SPSS Statistics, assigns a value of zero to missing data when generating the total score. If you have items that range from 1 to 5, assigning a zero to missing values makes no sense. If a person answers four of the items and picks a 5 for each of them, the total score will be 20. If another person answers seven items and picks a 3 for each of them, the score will be 21. It does not make sense that a person who picked a neutral answer to every item has more empathy than the person who answered the most empathetic response to each of the items. We could restrict scores to those who answered all the items, but we would be throwing out a lot of information. In the box on page 364, Sara answered six of the seven items, and we would be acting like we had no information on her level of empathy. There is a simple solution if a person insists on having a total score: generate the row mean and then multiply it by the number of items. You could restrict your sample to those who answered at least 75% of the items.

Equivalence means that you have two measures of the same variable and they produce consistent results. For example, you might have mothers write an essay on motherhood, and then your coders would rate the essay from 1 to 10 based on indications in the essay that the mother is depressed. Then you might give the new mothers a questionnaire that includes a 20-item scale measuring depression. The alternate forms or alternate measures could be evaluated for reliability by correlating the score on one measure with the score on the other measure.

Internal consistency is widely used with scales that involve several items in the same format. Likert-type items that have responses varying from strongly agree to strongly disagree are one example. A reliability assessment might be done by generating the mean score on the first half of the items and the mean score on the second half and then correlating these two scores. A reliable test would be internally consistent if the score for the first half of the items was highly correlated with the score for the second half of the items. Below we will discuss the coefficient alpha, $\alpha$, which is a better measure of internal consistency.

Rater consistency is important when you have observers rating a video, observed behavior, essay, or something else where two or more people are rating the same information. Here reliability means that a pair of raters gives consistent results. If a person who wrote an essay is rated high on depression by one rater and low on depression by a second rater, then we have reason to say there is low reliability measured for rater consistency.

## 12.3.1   Stability and test–retest reliability

A researcher uses the identical measure twice and simply correlates the results. The data (`retest.dta`) consist of the score at time one, called `score1`, and the score at time two, called `score2`.

The test–retest correlation is computed by using the command

```
. pwcorr score1 score2, obs sig
                 score1    score2

        score1   1.0000

                     10

        score2   0.7702    1.0000
                 0.0091
                     10        10

```

The option `obs` reports the number of observations, and the option `sig` reports the significance level. From the results, we see that there is a high test–retest reliability, $r = 0.77, p < 0.01$. We would expect the correlation to be substantial, say, $r > 0.5$. If the correlation is not substantial, then the measure is not giving reliable results.

Test–retest reliability has some limitations as a measure of reliability. First, distinguishing between a lack of a reliable measure and real change can be difficult. If the length of time between the two measures is very long or if some event related to the concept has occurred between the measurements, there may be real change. In that case, the correlation would be low regardless of the reliability of the true measure. If you have a measure of health risks associated with smoking tobacco, a television special on the topic might air between the first and second measurements. The measure might be reliable, but the correlation could be low if those who viewed the television program changed their beliefs. If you try to minimize the risk of this happening by having the two measurements close together, say, one in the morning and the other in the afternoon, participants may recall their responses and try to be consistent.

A second limitation of using test–retest reliability to measure reliability is that correlation does not measure agreement in a strict sense. If the scores at time one had been 2, 3, 4, 5 and the corresponding scores at time two had been 4, 6, 8, 10, then the regression equation

$$\widehat{Y} = 0 + 2X$$

would have a correlation of $r = 1.0$ because each score at the second measurement is exactly twice the time one score. There is no strict agreement, only relative agreement, with the higher the score at time one being matched to the higher the score at time two.

## 12.3.2   Equivalence

An alternative approach to reliability uses a different form of the measure for the second measurement. This alternative form has different items and, hence, the issue of participants trying to remember how they answered items the first time is eliminated. The alternative forms can be administered at separate times or one right after the other. The main problem with this approach is that developing two versions of a measure that we can assume are truly equivalent is extremely difficult. With this approach, the correlation, $r$, between the scores on the two forms is all that is needed to measure reliability. A low correlation means either that the measure is not reliable or that the measures are not truly equivalent.

## 12.3.3   Split-half and alpha reliability—internal consistency

Instead of developing alternative forms of a measure, one approach is known as split-half reliability. This is easier to use because you only need to have one measurement time. Suppose that you have 10 items. You compute a score for the first five items and a separate score for the last five items. Then you correlate these two scores. There are various ways of splitting the items. A random selection would probably be best. One problem with this approach is that we are using only 5-item measures to estimate the reliability of a 10-item measure. A 10-item measure will be more reliable than a 5-item measure, so our correlation for the two 5-item measures will underestimate the correlation for the 10-item measure. To correct this, we can use the Spearman–Brown prophecy formula (developed by Spearman and Brown independently over a century ago).

The major limitation of split-half reliability is that we could get a different result with each split we make. Using the first half and the second half of the items would yield a different reliability estimate than would using the even versus the odd items or randomly selecting items. Also, if the ordering of items is important, this method could be misleading.

Given these concerns with the split-half correlation, the standard for measuring reliability within classical measurement-theory framework is to use a coefficient called Cronbach's alpha ($\alpha$). In general, an $\alpha > 0.80$ is considered good reliability, and many researchers feel an $\alpha > 0.70$ is adequate reliability.

There are two widely used interpretations of $\alpha$. The simplest is that it is the expected value of all corrected split-half correlations. This approach is a simple interpretation. The second interpretation of $\alpha$ is the proportion of the observed variance that represents true variance. If we believe our items are an unbiased sample of items from a domain of possible items and hence that $\alpha$ is valid, we can interpret $\alpha$ as the ratio of the true variance to the total variance (the total variance contains both the true variance and the error variance): $\alpha = \sigma_{\text{true}}^2 / (\sigma_{\text{true}}^2 + \sigma_{\text{error}}^2)$. This approach is elegant in that an $\alpha = 0.80$ means that 80% of the variance in the scale represents the true score on the variable, and 20% is random error. However, for this interpretation to be used, we need to assume that the scale is valid. We will discuss validity assessment techniques later in this chapter.

There are two formulas for computing $\alpha$. One of these uses the covariance matrix, and the other uses the correlation matrix. You can think of the correlation matrix as a special transformation of the covariance matrix, where all the variances are rescaled to be equal and fixed at a value of 1.0. Analyzing the covariance matrix is called the unstandardized approach, and analyzing the correlation matrix is called the standardized approach. If items have similar variances, the two approaches will yield similar results. The unstandardized alpha is more general because it allows for items to have different variances. This can happen where some items have most people giving the same response, say, most people strongly agree to the item. By contrast, other items may have considerable variance with some people strongly agreeing and some people strongly disagreeing. The unstandardized alpha is the default in Stata. This is appropriate when generating a total score or a mean of the items.

However, sometimes people are generating a scale score when items are measured on different scales. One item might have three response options (agree, undecided, and disagree) and another item might have four response options (strongly agree, agree, disagree, and strongly disagree). In this case, it is reasonable to first standardized the items before generating the total or mean score. With standardized variables pooled this way, the standardized alpha would be appropriate.

We now return to using the recoded `gss2006_chapter12.dta`. We will use the dialog box to generate the `alpha` command because there are several options that might be hard to remember if you write the command directly. Open the dialog box by selecting Statistics ▷ Multivariate analysis ▷ Cronbach's alpha. On the Main tab, simply enter the variables, `empathy1 revempathy2 empathy3 revempathy4 revempathy5 empathy6 empathy7`. Under the Options tab, there are several powerful options. The first one takes the sign of each item as is. This makes sense to select because we already reverse-coded the items that were worded negatively. If we do not check this option, Stata will decide if any items should have their signs reversed. The second option will delete cases with missing values on any item. We will not check this option for now. Next we have an option to list individual interitem correlations and covariances. We will not check this option now, but it is useful when combined with selecting a standardized solution to see if there are problematic items, that is, items that have low or negative correlations with the other items.

Next is an option that will generate the scale for us. I already showed you how to generate this score earlier by using the `egen rowmean` command. It is reasonable to check this if you want Stata to generate the mean score for you. But only do this after you are satisfied with your scale and its reliability. If you check this option, there is a box highlighted in which you enter the name of the computed scale score. We check the next option to display item-test and item-rest correlations. We do not check the box to include variable labels, though with big scales this can be useful. We do check the box to require a certain number of items and enter 5. This means that people who answer 5, 6, or 7 of the items will be included, and people who answer fewer than 5 items will be dropped. At the bottom of the dialog box is the option to select a standardized solution. This can be useful when combined with asking for correlations if your reliability is low to see where there is a problem. We will not check this for now. The command and results are as follows:

```
. alpha empathy1 revempathy2 empathy3 revempathy4 revempathy5 empathy6 empathy7,
> asis item min(5)
Test scale = mean(unstandardized items)
```

| Item | Obs | Sign | item-test correlation | item-rest correlation | average inter-item covariance | alpha |
|------|-----|------|----------------------|----------------------|------------------------------|-------|
| empathy1 | 1349 | + | 0.6701 | 0.5151 | .3722749 | 0.7034 |
| revempathy2 | 1346 | + | 0.5957 | 0.3933 | .3954405 | 0.7332 |
| empathy3 | 1348 | + | 0.5733 | 0.4027 | .4118086 | 0.7281 |
| revempathy4 | 1349 | + | 0.6303 | 0.4474 | .3827332 | 0.7191 |
| revempathy5 | 1342 | + | 0.6088 | 0.4277 | .3932707 | 0.7233 |
| empathy6 | 1347 | + | 0.6762 | 0.5358 | .3764256 | 0.7004 |
| empathy7 | 1349 | + | 0.6701 | 0.5204 | .3748609 | 0.7026 |
| Test scale | | | | | .3866902 | 0.7462 |

We have selected three options. The `asis` (as is) option means that we do not want Stata to change the signs of any of our variables. We want a display of the item analysis, and we require that there are a minimum of 5 answered items, `min(5)`. The bottom row of the output table, `Test scale`, reports the $\alpha$ for the scale (0.7462). Above this value is the $\alpha$ we would obtain if we dropped each item, one at a time. For example, if we dropped `empathy7`, our $\alpha$ would be 0.7026. Dropping this item would make our scale less reliable. Sometimes you will find an item that, if dropped, would improve $\alpha$. Carefully used, this information can maximize your $\alpha$. However, be cautious in doing this because you are capitalizing on chance. Imagine having 100 items and computing $\alpha$. Then drop one item to increase $\alpha$ and redo the analysis using 99 items. You could continue until you had a very high $\alpha$ with, say, 20 items, but you would have not tested to see if you are capitalizing on chance.

The column labeled `Obs` reports the number of observations. This varies slightly because we used pairwise deletion for people who answered at least 5 of the 7 questions. The column labeled `Sign` contains all plus signs, meaning that all the items are positively related to the scale. If any item had a minus sign in this column, you would need to reverse-code the item or drop it. The `item-test correlation` column reports the correlation of each item with the total score of the seven items. All of these are strong. Because the total score includes the item, the values in this column are artificially inflated. Therefore, Stata also reports the `item-rest correlation`. This is the correlation of each item with the total of the other items. For example, the correlation between `empathy1` and the total of the other six items is 0.5151. The correlations are useful, but the most valuable information is in the `alpha` column, where we can see whether each item is useful (dropping it would lower $\alpha$) or not useful (dropping it would increase $\alpha$).

**Alpha, average correlation, number of items**

The unstandardized alpha is a function of the number of items, the mean covariance of the items, and the mean variance of the items. The standardized alpha is a function of just the number of items and their mean correlation. We will have a large $\alpha$ whenever we have either a large average correlation among the items, $\bar{r}$, or a large number of items, $k$. This can lead to misleading results when you have just a few items that are reasonably consistent or when you have a large number of items that are not consistent.

For example, we might not consider a set of items with an average correlation of $\bar{r} = 0.1$ as having much in common. With a 5-item scale, this results in an alpha of 0.36; with a 20-item scale, the alpha becomes 0.69; and with a 50-item scale, the alpha is 0.85. This shows an important qualification on how we view alpha. A 50-item scale with a reliability of just 0.85 contains items that do not have much in common. Some major scales are like this where they gain consistency from the large number of items and not from the items sharing much common variance.

The opposite happens when we have scales with just a few items. Many large-scale surveys attempt to measure many concepts and have just a few items to measure each of the concepts. If we have an average correlation among items of 0.3, a 3-item scale yields a standardized alpha of just 0.56, whereas a 10-item scale yields an alpha of 0.81. Adding just a few items greatly enhances the reliability of a scale, and 10-item scales should be adequate even if the average correlation is only moderate. At the other extreme, some researchers feel the need to have many items. Increasing our scale from 10 items to 30 items, and measuring one-third as many concepts in the survey as a result, raises the alpha from 0.81 to 0.93. For many applications, this small improvement will not offset the resulting increase in the length of the questionnaire.

## 12.3.4   Kuder–Richardson reliability for dichotomous items

The equivalent of alpha for items that are dichotomous is the Kuder–Richardson measure of reliability. The dichotomous items need to be coded as a 1 for "Yes" or "Correct" and a 0 for "No" or "Incorrect". This measure is sometimes used with multiple-choice tests where students get each item right or wrong. It is also used when there are simple yes/no response options to the items. You can estimate the Kuder–Richardson reliability coefficient by simply running `alpha` on the set of items about using the web for art that were asked on the 2002 General Social Survey (`kuder-richardson.dta`):

```
. alpha newartmus1 newartmus2 newartview newartinfo newartmus3, asis item

Test scale = mean(unstandardized items)

                                                          average
                          item-test    item-rest       interitem
Item        Obs   Sign   correlation   correlation     covariance      alpha

newartmus1   90    +       0.5767        0.3432         .0578652        0.5626
newartmus2   90    +       0.6600        0.3966         .0485019        0.5327
newartview   90    +       0.5948        0.3123         .0563046        0.5784
newartinfo   90    +       0.6420        0.3831         .0506242        0.5404
newartmus3   90    +       0.6423        0.3749         .0506242        0.5448

Test scale                                             .052784         0.6066
```

The value of alpha reported by Stata is identical to the value given using the formula for the Kuder–Richardson reliability. The results show that the Kuder–Richardson reliability coefficient is 0.607.

## 12.3.5   Rater agreement—kappa ($\kappa$)

Reliable coding of observational data is often extremely difficult. If you have two people trained to code videotapes of interactions for evidence of conflict, it is hard to have both of them code everything the same way. If your two coders do not agree on how they are coding the observational data, then it is difficult to have much confidence in the analysis based on their coding.

When collecting observational data, we often have two or more raters code their observations, and we measure the reliability of their agreement. We might have two physicians rating the health of the same patients on a 5-point scale. We might have two clinicians identify the mental health condition of patients. A widely used measure of interrater agreement is the coefficient kappa ($\kappa$).

Suppose that we are observing people from the press who are asking questions of a political leader. We might have several categories such as

- Friendly questions that the political leader would like to have asked—Do you feel education is important?

- Difficult questions that the political leader would like to have asked—Could you elaborate on the benefits of your plan to reform the Social Security system?

- Unwanted questions that the political leader would not like to have asked—Why did your press secretary resign?

- Extremely unwanted and difficult questions that the political leader would not like to have asked—How can you justify the $125,000 donation you received from the Committee to Promote Higher Pay for Men?

Perhaps at the beginning of the politician's tenure, there is a honeymoon period in which most of the questions are ones the leader wants to answer because they are easy or because they allow the leader to clarify his or her plans. Perhaps later in their tenure, the questions become less desirable from the politician's perspective. You have two coders rate a press conference in which 20 questions were asked, and their hypothetical ratings are in `kappa1.dta`.

We can do a tabulation of these data:

```
. tabulate coder1 coder2
```

| | coder2 | | | | |
|---|---|---|---|---|---|
| coder1 | Easy | Hard want | Unwanted | Very unwa | Total |
| Easy | 5 | 1 | 0 | 0 | 6 |
| Hard wanted | 0 | 3 | 0 | 0 | 3 |
| Unwanted | 1 | 2 | 2 | 0 | 5 |
| Very unwanted | 0 | 1 | 1 | 4 | 6 |
| Total | 6 | 7 | 3 | 4 | 20 |

The coders seem to agree on the classification of the friendly questions and the unwanted questions. They seem to have a harder time distinguishing the difficult-but-wanted questions from the difficult-unwanted questions. Ideally, all the observations should fall on the diagonal where coder one and coder two have the same classification of the question.

To estimate $\kappa$, we enter the command `kap coder1 coder2`. Note that we typed `kap` and not `kappa`. The latter is only used for data entered in a special way. The output is as follows:

```
. kap coder1 coder2
```

| Agreement | Expected Agreement | Kappa | Std. Err. | Z | Prob>Z |
|---|---|---|---|---|---|
| 70.00% | 24.00% | 0.6053 | 0.1221 | 4.96 | 0.0000 |

We would describe this by writing that the judges agree exactly on the classification 70% of the time, that is, $(5+3+2+4)/20$. The coefficient of agreement is $\kappa = 0.61, z = 4.96, p < 0.001$, indicating a reasonable level of agreement.

Some people would use the percentage agreement instead of $\kappa$, but the percentage agreement exaggerates the amount of agreement because it ignores the agreement we would expect by chance. $\kappa$ only gives us credit for the extent the agreement exceeds what we would have expected to get by chance alone.

**What is a strong kappa?**

We would like alpha to be 0.80 or higher for a scale to be considered reliable. The expectations for kappa are different. Because kappa adjusts the percentage agreement for what would be expected by chance, kappa tends to be lower than alpha. Landis and Koch (1977) suggested these guidelines; see also Altman (1991):

| Alpha | Strength of agreement |
|---|---|
| Under 0.20 | Poor |
| 0.21 – 0.40 | Fair |
| 0.41 – 0.60 | Moderate |
| 0.61 – 0.80 | Good |
| 0.81 – 1.00 | Very good |

Thus our $\kappa = 0.605$ is between moderate and good.

If you conceptualize the classes as ordinal, it is possible to estimate a weighted kappa. This $\kappa_{\text{weighted}}$ gives partial credit for being close. If the two judges are close, the weighted kappa gives us partial credit. Stata allows us to use the default value for the weights or to enter our own criterion. This is explained in [R] **kappa** with a detailed example. Entering the command `help kap` provides a brief explanation. We will not cover weighted kappa here.

Some researchers can use three raters rather than two, and this offers much more information. With three raters, kappa not only reports the overall $\kappa$ but also can give us a $\kappa$ for each category. For example, we may find that the coders have a high level of reliability for coding items that are in the "Easy" or "Hard wanted" categories, but a very low reliability for the "Unwanted" or "Very unwanted" categories. Here we might find that combining the "Unwanted" and "Very unwanted" categories into one category is necessary to have interrater reliability. The extension of kappa to three raters is described in [R] **kappa**.

## 12.4 Validity

A good measure needs to be reliable, but it also needs to be valid. A valid measure is one that measures what it is supposed to be measuring. A measure of maternal depression may be reliable, meaning it gets consistent results, without being valid. In order to be valid, the measure should have a high score for mothers who actually are depressed and a low score for mothers who are not. A measure is not valid when it is measuring other phenomena. For example, asking a person if they have high anxiety is not an especially valid item to measure depression. A person may have high anxiety just before an examination, but this is different from being depressed. The gist of validity is simply that a measure needs to measure what you are trying to measure and not something else. Assessing validity involves comparing a measure with one or more standards. There are many possible standards, resulting in several types of validity checks.

### 12.4.1 Expert judgment

In making a scale, you first need to carefully define the content domain of what you are measuring. Once you have done this, you share your items and your definition of the domain with a small group of experts in the field and have them critique each item. This will provide a good measure of content validity if your definition of the content domain and the quality of your experts are both good.

Defining the content domain is often difficult. Suppose that you are measuring self-reported health. What does this include? Does it include mental health? Does it include living a healthy lifestyle? Is it limited to physical health? What about ability to perform activities of daily living? What about flexibility, strength, and endurance? Are you interested in a global rating of health or more specific aspects?

Once you have carefully defined the domain of your concept, your panel of experts can judge whether you have items that represent this domain. Judges may be academic scholars who have published in the area or others who have worked in the field. A psychiatric social worker, a physician, and an academic scholar on health might be good candidates for your panel of experts. In other cases, you might want to include people who are experiencing issues related to your content. A member of a youth gang would be an expert if you were measuring gang-related violence.

One measure of content validity is the content validity ratio (CVR) (Lawshe 1975; Shultz et al. 2014). Judges rate each item as *essential*, *useful*, or *not necessary*. For each item, you compute the CVR by using

$$\mathrm{CVR}_i = \frac{n_e - \frac{N}{2}}{\frac{N}{2}}$$

where $n_e$ is the total number of judges rating item $i$ as *essential*, and $N$ is the number of judges. Assume that you have a 10-item scale and four judges, with three of the four judges rating the first item as *essential*. The $\mathrm{CVR}_i$ would be $(3-2)/2 = 0.50$. You can keep the items that have a relatively high CVR and drop those that do not. However, a limitation of this is that it is no better than the quality of your definition of your domain and the quality of the judges.

*Face validity* is a type of content validity, but it does not require expert judgment. People like those who will be measured by your scale can often be most helpful for evaluating face validity. If there are subgroups that might have special understandings (gender, ethnicity, etc.), they should be asked to say whether the items make sense for measuring the variable you are trying to measure. Imagine that you are developing a scale to measure the self-regulation in a group of 8-year-olds. You could ask a small number of 8-year-olds to read each item aloud and explain what the item is trying to say or ask. Then ask each of these 8-year-olds to pick an answer and explain why they picked that answer. You will almost always find that a few of the items are interpreted differently by these 8-year-olds from the way you intended and that the reason they pick an answer is extraneous to what you are measuring.

## 12.4.2   Criterion-related validity

Criterion-related validity picks one or more criteria or standards for evaluating a scale. The standard may be a concurrent measure or a predictive measure. If you have a 10-item scale of self-reported health, you need to find a standard. This could be a more comprehensive 40-item measure that has been used in the literature but is too long for your purposes. You include both your 10-item measure and the 40-item measure in a pilot survey and correlate the two scale scores. You might also use as a standard a clinical evaluation of the person's health.

Predictive standards are criteria that your measure should predict, and these typically require a time interval. A measure of motivation to achieve given to high school juniors might use grade point averages during their senior year as a criterion. Where correlations are used, a value of 0.30 is considered moderate support, and a correlation of 0.50 is considered strong support.

If you are measuring intention of medical students to work in rural areas, you might wait until the group graduates and see how many of them actually do work in a rural area. This would involve predictive validity. Because working in a rural area is a binary 0/1 variable, you would use logistic regression with your scale score as the predictor. If your scale score is a substantively and statistically significant predictor of whether medical students work in a rural area, then this would be evidence of predictive criterion validity. If you are measuring the commitment to quit smoking, you might use the number of cessation sessions a person attends as your criterion, in which case you would compute a simple correlation.

Criterion-related validity can also be used to evaluate individual items. If you believe the overall scale is valid, you can correlate each item with the total score and drop those that have low correlations with the total score. Stata's command `alpha` with the option `item` provides this information, as was shown in the output on page 372 under the columns labeled `item-test correlation` and `item-rest correlation`.

You can also divide the sample into quartiles based on the total score. Then you would compare the top quartile to the bottom quartile on each item by using a *t* test. Try this as an exercise.

## 12.4.3 Construct validity

The term construct validity is reserved for analyses that involve more than one approach to assessing validity. Some of the analyses that fall under the rubric of construct validity include the following:

- Comparing known groups. A measure of depression might be evaluated by comparing the mean score for a group of people being treated for post-traumatic stress disorder with a group who are not being treated. With two groups, you would use a *t* test, and with more than two groups, you would use ANOVA.

- Internal consistency. Although I presented alpha reliability as a measure of reliability, it is also reasonable to think of it as tapping validity. If you have a series of items measuring commitment to stop smoking, they should all be internally consistent, as indicated by item-rest correlations and alpha, as well as by comparing the top and bottom quartiles. These techniques were covered in the previous section. If the scale is valid, then $\alpha = 0.80$ signifies that 80% of the variance represents true variance in the variable.

- Correlational analysis. We would expect criterion-related validity to be demonstrated. Better yet, with construct validity we would use multiple criteria. Our motivation scale given in the junior year of high school should predict grade point average in the senior year, college plans, etc. These are known as *convergent* criteria-related validity because we expect these correlations to all be positive. *Divergent* criteria-related validity would identify additional criteria for which we would expect a negative relationship. We would expect our motivational scale score to be negatively related to such criteria as frequency of arrest, use of hard drugs, likelihood of dropping out of high school, and unprotected sexual behavior. With construct validity, we are looking at a pattern of correlations. An extension of this is to use factor analysis (see the next section) and possibly confirmatory factor analysis.

- Qualitative analysis. We should interview people who complete our set of items and ask them to explain each answer they gave. Having a group of fifth-grade students answer questions about how much they like school by using a computer-assisted interview program that includes neat graphics may be measuring how much they like school, but it also may measure how much they like "foolin' around on the computer". Asked why they give a positive response to an item, we might learn that it is because they would rather play on the computer than sit in a class. A surprising number of scales have been developed by academics that have never been evaluated by the target population. We should always evaluate whether the people in our target population are interpreting the items and the options the way we expected.

## 12.5   Factor analysis

Ideally, a scale measures just one concept. We sometimes ignore this, and the results can be misleading. Imagine that Shanice has a GRE score on the quantitative reasoning part of the examination of 165 (very high),[1] but her score on the verbal reasoning part of the examination is just 140 (low). A second student, Emma, has just the opposite scores, that is, 140 on the quantitative reasoning part and 165 on the verbal reasoning part. Some graduate programs use a single number, the total of the quantitative and verbal sections, to evaluate applications. Using one number, the same 305 total for both Shanice and Emma, can be problematic if they both are applying for a statistics program or for a humanities program. This is because we are combining two distinct dimensions of academic ability—verbal and quantitative achievements. It would make more sense for graduate programs to consider the quantitative and verbal reasoning scores as two separate dimensions with some graduate programs emphasizing one or the other of these.

Some concepts are multidimensional. For example, health might have one dimension for physical health and a separate dimension for psychological health. We would expect these two dimensions of health to be correlated, but the correlation should be small enough that we recognize them as two distinct dimensions of health. In such a case, we might not want to average items. Instead, we might want to make two scales, one for physical health and one for psychological health. Some researchers would ignore our advice to have two scales and simply combine them. If you do this, it would make more sense to call this an index rather than to call it a scale. Ideally, a scale should represent one dimension.

How do we know a set of items represents one dimension? This is where we use factor analysis. If the items represent one dimension, then all of them should be moderately to strongly correlated with each other, and the pattern of correlations should be consistent. If we had a six-item scale, it might look something like table 12.2.

Table 12.2. Correlations you might expect for one factor

|        | Item 1 | Item 2 | Item 3 | Item 4 | Item 5 | Item 6 |
|--------|--------|--------|--------|--------|--------|--------|
| Item 1 | 1.00   |        |        |        |        |        |
| Item 2 | 0.60   | 1.00   |        |        |        |        |
| Item 3 | 0.45   | 0.40   | 1.00   |        |        |        |
| Item 4 | 0.50   | 0.55   | 0.45   | 1.00   |        |        |
| Item 5 | 0.70   | 0.55   | 0.45   | 0.40   | 1.00   |        |
| Item 6 | 0.35   | 0.40   | 0.45   | 0.40   | 0.35   | 1.00   |

---

1. Before August 2011, the GRE scores ranged from 200 to 800 for both the verbal reasoning and the quantitative reasoning sections. Since August 2011, the GRE scores have ranged from 130 to 170 for both sections.

What would happen if we had two dimensions with the first three items representing physical health and the last three items representing psychological health? This scenario is illustrated in table 12.3, where the first three items are highly correlated with each other (top left triangle), the last three items are highly correlated with each other (bottom right triangle), but the two sets of items are only moderately correlated with each other (bottom left rectangle).

Table 12.3. Correlations you might expect for two factors

|        | Item 1 | Item 2 | Item 3 | Item 4 | Item 5 | Item 6 |
|--------|--------|--------|--------|--------|--------|--------|
| Item 1 | **1.00** |        |        |        |        |        |
| Item 2 | **0.60** | **1.00** |        |        |        |        |
| Item 3 | **0.80** | **0.70** | **1.00** |        |        |        |
| Item 4 | 0.30   | 0.25   | 0.35   | **1.00** |        |        |
| Item 5 | 0.20   | 0.25   | 0.25   | **0.70** | **1.00** |        |
| Item 6 | 0.25   | 0.20   | 0.15   | **0.75** | **0.65** | **1.00** |

Factor analysis is a collection of techniques that does an exploratory analysis to see if there are clusters of items that go together. When constructing a scale, you would hope that all the items you include in your scale form one factor or principal component.

There are two types of analysis that are often lumped together under the label of factor analysis. The default is exploratory factor analysis, which Stata calls principal factor analysis. We will use PF to refer to this type of factor analysis. The second type is principal-component factor analysis. We will use PCF to refer to this approach. Each of these procedures has distinct objectives.

PF analysis attempts to identify a small number of latent variables or dimensions that explain the shared variance of a set of measures. If you believe that an attitude toward use of methamphetamines includes a cognitive component (beliefs), an affective component (emotional response, feelings), and a behavioral component (predispositions), you might do a PF analysis of a set of 30 items to see if the items fall into these three dimensions and label each dimension as a latent variable. Here the latent variables explain how you respond to the items. In PF analysis, the variance is partitioned into the shared variance and unique or error variance. The shared variance is how much of the variance in any one item can be explained by the rest of the items. This shared variance is estimated using a multiple $R^2$. In practical terms, factor analysis analyzes a correlation matrix in which the 1.00s on the diagonal are replaced by the square of the multiple correlation of each item with the rest of the items. PF analysis is not trying to explain all the variance, just the variance that is shared among the set of items.

PCF analysis has a different approach, and its strength is in the area of data reduction. It does not distinguish between the common or shared variance and the unique or error variance. In practical terms, PCF analysis analyzes a correlation matrix in which the 1.00s on the diagonal stay there. It is trying to explain all the variance. PCF analysis is used when developing a scale where one dimension is identified to represent the core of a set of items. If we have 20 items measuring your satisfaction with a drug, and we find that 15 of the items load highly on the first dimension in a PCF analysis, then we may want to use those 15 items in our scale and drop the other 5 items.

Factor analysis has a special vocabulary of its own. The following is a list of some of the most important terminology. You can read this now, but it may be most helpful to refer to it after you have read the section where each term is used.

- Extraction. There are different ways of extracting a solution. This extraction involves analyzing the correlation matrix (sometimes the covariance matrix) to obtain an initial factor solution. The initial factor solution may subsequently be rotated to aid interpretation.

- Eigenvalues. Both PCF analysis and PF analysis produce a value for each factor that is known as the eigenvalue for that factor. In the case of PCF analysis, because all the variance is being explained, the sum of the eigenvalues for all possible components is the number of items. If there are 10 items, the sum of the eigenvalues will be 10. Therefore, a component that has an eigenvalue of less than 1 is not helpful for data reduction because it accounts for less than one item. The factors will be ordered from the most important, which has the largest eigenvalue, to the least important, which has the smallest eigenvalue. In PF analysis, the sum of the eigenvalues will be less than the number of items, and the eigenvalues' interpretation is complex.

- Communality and uniqueness. Communality is how much of each item is explained by the set of factors, and uniqueness (shown in Stata) is how much is unexplained. PF analysis tries to explain the shared variance. PCF analysis tries to explain all the variance, which is why it is ideal for the uniqueness to approach zero. This ideal is rarely achieved in practice but is a caution that PCF may not be appropriate.

- Loadings. Both PCF analysis and PF analysis produce a matrix with the items in the rows and the factors in the columns. The elements of this matrix are called loadings, and they are used to assess how clusters of items are most related to one or another of the factors. If an item has a loading over 0.4 on a factor, it is considered a good indicator of that factor. Many researchers limit themselves to orthogonal rotation. This may be a problem if they conceptualize the factors as correlated. For example, we might have a physical health factor and a mental health factor. Conceptually, these would be correlated.

- Simple structure. This is a pattern of loadings where each item loads strongly on just one factor and a subset of items load strongly on each factor. When an item loads strongly on more than one factor, it is factorially confounded.

- Scree plot. This is a graph showing the eigenvalue for each factor. Its name comes from geology, where scree collects at the bottom of a cliff. It is used to decide how many factors are included by throwing out those that are considered scree. When doing a PCF analysis, we usually drop factors that have eigenvalues in the neighborhood of 1.0 or smaller.

- Rotation. Neither PF analysis nor PCF analysis has a unique solution without making a series of assumptions. It is often helpful to rotate the initial solution to obtain a solution that is more interpretable.

- Orthogonal versus oblique. Some rotations require that the factors be uncorrelated. These are orthogonal solutions. Other rotations allow the factors to be correlated and are called oblique solutions. There are many options for each type of rotation, and each type of rotation makes different assumptions to obtain identification.

- Factor score. This is a score on each factor. Rather than simply averaging or adding the items that load above some value, such as 0.4, on the factor to get a scale score, a factor score weights each item based on how related it is to the factor. Also the factor score is scaled to have a mean of 0.0 and a variance of 1.0.

Should you use PF analysis or PCF analysis? If you want to explain as much of the variance in a set of items as possible with one dimension, then PCF analysis is reasonable. Use PCF when you have a set of items that you believe all measure one concept. In this situation, you would be interested in the first principal factor. You would want to see if it explained a substantial part of the total variance for the entire set of items, and you would want most of the items to have a loading of 0.4 or above on this factor.

If, on the other hand, you want to identify two or more latent variables that represent interpretable dimensions of some concept, then PF analysis is probably best. For example, you might have 20 items asking about a person's level of alienation. A PF analysis might find that there are subsets of items that fall into different dimensions such as value isolation, powerlessness, and normlessness. With PF analysis, the focus is on finding two or more interpretable dimensions or factors rather than on maximizing the amount of variance that is explained.

In this chapter, we limit our coverage to PCF analysis because it relates to developing a measure of a concept. Using the Stata menu, select Statistics ▷ Multivariate analysis ▷ Factor and principal component analysis. Here you will find several options:

- Factor analysis

- Factor analysis of a correlation matrix

- Principal component analysis (PCA)

- PCA of a correlation or covariance matrix

- Postestimation

We will focus on the first of these, factor analysis. This option allows us to do either PF analysis or PCF analysis. (Be careful not to select Principal component analysis (PCA).) The terminology used in factor analysis is highly specialized, and the same terms can be used differently by different software packages.

The second menu option, Factor analysis of a correlation matrix, allows you to enter a correlation matrix and do the factor analysis of it. This is useful when reading an article that includes a correlation matrix because you do not need the raw data to do the factor analysis; you just need the correlation matrix.

The third and fourth options involve Stata's application of principal component analysis (PCA). We will not consider these options here. The last option, Postestimation, includes things we can do after we obtain the first solution to help us interpret the results, including rotating the solution, creating scree plots, and estimating factor scores.

## What's in a name?

Because factor analysis is an area of long-standing specialization, there are many differences of opinion as to what is the best approach to it. Many of these distinctions are not of great importance when we use factor analysis to develop or evaluate a scale. There are many technical terms that you rarely see in other areas of statistics. To make matters especially confusing, different software may use the same name for different procedures. We will only cover the procedures for factor analysis as Stata implements them. But do not be surprised if you hear the same names used differently with other packages.

In Stata, PCF analysis produces the same results as what IBM SPSS Statistics calls principal component analysis. However, what Stata calls principal component analysis (PCA—the third and fourth options in the menu) has important differences. We will use the name PCF analysis to refer to what IBM SPSS Statistics calls principal component analysis.

The default in Stata is to do a PF analysis; in IBM SPSS Statistics, the default is to do a PCF analysis.

## 12.6   PCF analysis

Here is an example of PCF analysis using 14 items from the 2006 General Social Survey
(gss2006_chapter12_selected.dta). There are four options for how Stata extracts
factors from your data. Other packages offer more options not available in Stata, but
the options in Stata should meet most needs for people developing or evaluating a scale.

We will examine the 14 items from the 2006 General Social Survey that measure
desire to invest in different national programs. Our goal is to select items that we can
use in a scale. Here is a compact codebook of the items:

```
. codebook natspac natenvir natheal natcity natcrime natdrug nateduc natrace
> natarms natfare natroad natsoc natchld natsci, compact

Variable     Obs Unique  Mean   Min  Max  Label

natspac     1407    3  2.237385   1    3   SPACE EXPLORATION PROGRAM
natenvir    1446    3  1.375519   1    3   IMPROVING   PROTECTING ENVIRONMENT
natheal     1451    3  1.297726   1    3   IMPROVING   PROTECTING NATIONS HEALTH
natcity     1349    3  1.634544   1    3   SOLVING PROBLEMS OF BIG CITIES
natcrime    1448    3  1.446133   1    3   HALTING RISING CRIME RATE
natdrug     1428    3   1.45028   1    3   DEALING WITH DRUG ADDICTION
nateduc     1462    3  1.317373   1    3   IMPROVING NATIONS EDUCATION SYSTEM
natrace     1348    3  1.780415   1    3   IMPROVING THE CONDITIONS OF BLACKS
natarms     1442    3  2.144938   1    3   MILITARY, ARMAMENTS, AND DEFENSE
natfare     1434    3  2.110879   1    3   WELFARE
natroad     2899    3  1.756468   1    3   HIGHWAYS AND BRIDGES
natsoc      2864    3  1.407472   1    3   SOCIAL SECURITY
natchld     2758    3  1.517041   1    3   ASSISTANCE FOR CHILDCARE
natsci      2786    3  1.688442   1    3   SUPPORTING SCIENTIFIC RESEARCH
```

Running `tabulate` with the `codebook` command or the `fre` command, if you have
that command installed, we see that these items have a higher score, implying less
support for spending. We can reverse these codes, reverse the corresponding labels, and
add a prefix of "r" to each variable to remind us that we have reversed them:

```
. recode natspac natenvir natheal natcity natcrime natdrug nateduc natrace
> natarms natfare natroad natsoc natchld natsci
> (1=3 "Too little") (2=2 "About right") (3=1 "Too much"), prefix(r) label(revnat)
(756 differences between natspac and rnatspac)
(1081 differences between natenvir and rnatenvir)
(1169 differences between natheal and rnatheal)
(839 differences between natcity and rnatcity)
(992 differences between natcrime and rnatcrime)
(1027 differences between natdrug and rnatdrug)
(1152 differences between nateduc and rnateduc)
(746 differences between natrace and rnatrace)
(967 differences between natarms and rnatarms)
(919 differences between natfare and rnatfare)
(1392 differences between natroad and rnatroad)
(1973 differences between natsoc and rnatsoc)
(1756 differences between natchld and rnatchld)
(1570 differences between natsci and rnatsci)
```

Now a higher score implies greater support. A researcher may think there is one dimension representing how liberal or conservative a person is regarding government spending. This researcher may feel that a liberal would score high on all of these items (big spenders) and a conservative would score low, meaning that they want to minimize government spending initiatives. Another person may think there are two separate dimensions, one focused on investing in social welfare issues—for example, education, health, welfare, environment—and one focused on investing in the national infrastructure. A third researcher may believe it is more complicated and want to see what dimensions are needed to explain this set of 14 items.

To run a PCF analysis, we select Statistics ▷ Multivariate analysis ▷ Factor and principal component analysis ▷ Factor analysis. The first tab is labeled Model, and here we list our 14 recoded variables. The second tab is labeled Model 2. Here we select *Principal-component factor*. This generates the following results:

```
. factor rnatspac rnatenvir rnatheal rnatcity rnatcrime rnatdrug rnateduc
> rnatrace rnatarms rnatfare rnatroad rnatsoc rnatchld rnatsci, pcf
(obs=1082)

Factor analysis/correlation                   Number of obs    =      1082
    Method: principal-component factors       Retained factors =         4
    Rotation: (unrotated)                     Number of params =        50
```

| Factor | Eigenvalue | Difference | Proportion | Cumulative |
|--------|-----------|-----------|-----------|-----------|
| Factor1 | 3.00301 | 1.62066 | 0.2145 | 0.2145 |
| Factor2 | 1.38235 | 0.12372 | 0.0987 | 0.3132 |
| Factor3 | 1.25863 | 0.22702 | 0.0899 | 0.4031 |
| Factor4 | 1.03161 | 0.06974 | 0.0737 | 0.4768 |
| Factor5 | 0.96187 | 0.05047 | 0.0687 | 0.5455 |
| Factor6 | 0.91140 | 0.09801 | 0.0651 | 0.6106 |
| Factor7 | 0.81339 | 0.02930 | 0.0581 | 0.6687 |
| Factor8 | 0.78409 | 0.05945 | 0.0560 | 0.7247 |
| Factor9 | 0.72464 | 0.01659 | 0.0518 | 0.7765 |
| Factor10 | 0.70804 | 0.04088 | 0.0506 | 0.8271 |
| Factor11 | 0.66717 | 0.06485 | 0.0477 | 0.8747 |
| Factor12 | 0.60231 | 0.01414 | 0.0430 | 0.9178 |
| Factor13 | 0.58817 | 0.02484 | 0.0420 | 0.9598 |
| Factor14 | 0.56333 | . | 0.0402 | 1.0000 |

```
LR test: independent vs. saturated:  chi2(91) = 1701.47 Prob>chi2 = 0.0000
```

Factor loadings (pattern matrix) and unique variances

| Variable | Factor1 | Factor2 | Factor3 | Factor4 | Uniqueness |
|---|---|---|---|---|---|
| rnatspac | -0.0405 | 0.6150 | -0.4468 | -0.0108 | 0.4204 |
| rnatenvir | 0.5562 | 0.0698 | -0.1925 | 0.1629 | 0.6221 |
| rnatheal | 0.6298 | 0.0420 | 0.0923 | 0.2638 | 0.5235 |
| rnatcity | 0.5264 | -0.0580 | -0.0677 | -0.4598 | 0.5036 |
| rnatcrime | 0.4533 | 0.1160 | 0.5154 | -0.4234 | 0.3361 |
| rnatdrug | 0.5381 | 0.1048 | 0.3346 | -0.2872 | 0.5051 |
| rnateduc | 0.5997 | 0.0035 | 0.0425 | 0.3838 | 0.4912 |
| rnatrace | 0.5921 | -0.2286 | -0.2660 | -0.1885 | 0.4909 |
| rnatarms | -0.1087 | 0.4926 | 0.4682 | 0.1719 | 0.4968 |
| rnatfare | 0.4725 | -0.1726 | -0.3547 | -0.0629 | 0.6172 |
| rnatroad | 0.1421 | 0.5022 | 0.0704 | -0.1269 | 0.7065 |
| rnatsoc | 0.4196 | -0.1209 | 0.3374 | 0.4575 | 0.4862 |
| rnatchld | 0.5369 | -0.1049 | -0.1130 | 0.1343 | 0.6699 |
| rnatsci | 0.3057 | 0.6059 | -0.2865 | 0.0493 | 0.4549 |

If these 14 items all fall along one dimension, we would expect the first factor to have a dominant effect. We can assess this by using the Eigenvalues. Remembering that in PCF analysis the total of the eigenvalues for all factors is the number of items, 14 in this case, we can see that the first factor has an eigenvalue of 3.00, and this is 21% of the total possible of 14. By comparison, the second factor has an eigenvalue of just 1.38. It looks like there is a strong first factor.

If you look at the matrix showing the "Factor loadings", you can go down the column labeled Factor1 to see the loadings for each item on the first factor (component). Notice how many of the items have a loading greater than 0.40. The exceptions are for the space programs, armaments, roads, and science. The rest of the items have strong loadings and include the environment, health, crime, drugs, education, race, welfare, social programs, and children. This seems to be a general liberal spending program, and we could make a scale using these 10 items, dropping the other 4 items.

There is a problem with our results that many researchers ignore. Because PCF analysis is trying to explain all the variance in the items, the uniqueness for each item should approach zero. For example, rnatchld has a loading of 0.537 on the first factor and this is considered satisfactory. However, the uniqueness for this item is 0.670. In other words, the set of factors does not explain about two-thirds of the variance in rnatchld. This calls to question the use of PCF analysis and suggests that another type of factor analysis would be more appropriate. Although we note this problem with PCF, we will still illustrate PCF because of its wide use in scale construction. You will need to explore the *Stata Multivariate Statistics Reference Manual* to learn about these more advanced options.

Generally, we should consider any factor that has an eigenvalue of more than 1.0. A visual way to examine the eigenvalues is with a scree plot. This is a postestimation command that we run just after doing a factor analysis. Select Statistics ▷ Multivariate analysis ▷ Factor and principal component analysis ▷ Postestimation ▷ Scree plot of eigenvalues. This dialog box gives us a variety of options, but it takes less work to enter the command directly than it does to find the dialog box. The command is `screeplot`, and it produces the graph in figure 12.1.

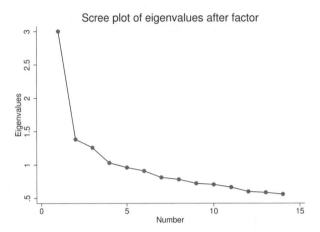

Figure 12.1. Scree plot: National priorities

This scree plot starts leveling off (going across more quickly than it is going down) with the fourth factor. This would suggest that we need three factors, and we could drop the rest of them. Deciding on the number of factors when doing a PCF analysis is not an easy task. We had four factors that had an eigenvalue of more than 1, although the fourth factor was just barely over 1. The scree plot suggests that we might just focus on the first three factors, and the scree plot makes it clear that the first factor is dominant because there is such a big drop-off in the size of the eigenvalue between the first and second factors.

When there is more than one factor, it is usually useful to rotate the initial solution to see if a different solution is easier to interpret. We have two ways of doing this. One is an orthogonal rotation where we force the factors to be uncorrelated with each other. The other is an oblique rotation in which we allow the factors to be correlated. There are multiple options for both types of rotation (15 for orthogonal rotation and 8 more for oblique rotation). I will illustrate just one approach for each type.

## 12.6.1  Orthogonal rotation: Varimax

Stata has kept the factor analysis solution in memory, so to do an orthogonal rotation, all we need to do is enter the command `rotate` (this is the default rotation). If you forget this command, you can find it under the Statistics ▷ Multivariate analysis ▷ Factor and principal component analysis ▷ Postestimation ▷ Rotate loadings menu. Here is what we obtain:

```
. rotate
Factor analysis/correlation                        Number of obs     =     1082
    Method: principal-component factors            Retained factors =        4
    Rotation: orthogonal varimax (Kaiser off)      Number of params =       50
```

| Factor | Variance | Difference | Proportion | Cumulative |
|---|---|---|---|---|
| Factor1 | 1.96564 | 0.29211 | 0.1404 | 0.1404 |
| Factor2 | 1.67353 | 0.04093 | 0.1195 | 0.2599 |
| Factor3 | 1.63260 | 0.22876 | 0.1166 | 0.3766 |
| Factor4 | 1.40384 | . | 0.1003 | 0.4768 |

```
LR test: independent vs. saturated:  chi2(91) = 1701.47 Prob>chi2 = 0.0000
Rotated factor loadings (pattern matrix) and unique variances
```

| Variable | Factor1 | Factor2 | Factor3 | Factor4 | Uniqueness |
|---|---|---|---|---|---|
| rnatspac | -0.1295 | -0.0143 | -0.1449 | 0.7360 | 0.4204 |
| rnatenvir | 0.4623 | 0.3085 | 0.0890 | 0.2471 | 0.6221 |
| rnatheal | 0.6328 | 0.1374 | 0.2186 | 0.0975 | 0.5235 |
| rnatcity | 0.0262 | 0.4831 | 0.5052 | 0.0839 | 0.5036 |
| rnatcrime | 0.0941 | -0.0276 | 0.8066 | -0.0599 | 0.3361 |
| rnatdrug | 0.2187 | 0.0910 | 0.6618 | 0.0296 | 0.5051 |
| rnateduc | 0.6899 | 0.1329 | 0.0932 | 0.0808 | 0.4912 |
| rnatrace | 0.2352 | 0.6335 | 0.2259 | 0.0389 | 0.4909 |
| rnatarms | 0.1103 | -0.6491 | 0.1967 | 0.1763 | 0.4968 |
| rnatfare | 0.2255 | 0.5642 | 0.0480 | 0.1061 | 0.6172 |
| rnatroad | 0.0039 | -0.1708 | 0.2889 | 0.4252 | 0.7065 |
| rnatsoc | 0.6718 | -0.1136 | 0.0967 | -0.2007 | 0.4862 |
| rnatchld | 0.4475 | 0.3393 | 0.1075 | 0.0559 | 0.6699 |
| rnatsci | 0.1777 | 0.0334 | 0.0832 | 0.7110 | 0.4549 |

```
Factor rotation matrix
```

|  | Factor1 | Factor2 | Factor3 | Factor4 |
|---|---|---|---|---|
| Factor1 | 0.6858 | 0.4868 | 0.5116 | 0.1758 |
| Factor2 | -0.0311 | -0.4717 | 0.1951 | 0.8594 |
| Factor3 | 0.1676 | -0.6530 | 0.5615 | -0.4798 |
| Factor4 | 0.7075 | -0.3379 | -0.6204 | -0.0190 |

The second table shows the factor loadings. With a varimax rotation, we can think of the loadings as being the estimated correlation between each item and each factor. For example, the last item, **rnatsci** has a high correlation with the fourth factor, 0.711. We hope each item has a loading of 0.40 or higher on one factor and a smaller loading on the other factors. We use the items that have relatively large loadings on each factor to decide on what the factor is to be called. Let's work backward, starting with the fourth and weakest factor. It has just two items with high loadings, **rnatspac** and **rnatsci**. This factor might be labeled liberal spending support for science as a national priority. **Factor3** has three items with loadings higher than 0.4, namely, **rnatcity**, **rnatcrime**, and **rnatdrug**. This might be called a priority on urban problems, although we need to acknowledge that crime and drugs are problems outside of urban places as well. We could call this liberal spending support for urban problems.

Factor2 has four items with loadings higher than 0.4. Support for minorities (rnatrace) and support for welfare programs (rnatfare) both have strong positive loadings. By contrast, support for the military (rnatarms) has just as strong a loading, but it is negative. This factor seems to be pitting together support for the disadvantaged against support for the military. This might make sense considering the disproportional burden minorities pay for military engagements and how military expenses can drain money from social welfare. The problem with this factor and the third factor is that rnatcity loads strongly on both factors. It is compounded. Although Stata distinguishes between these two factors, the public sees both of these as special concerns in urban areas. We could call this factor liberal spending support for the disadvantaged.

Factor1 is the strongest, and it has five items with loadings greater than 0.4. These involve the environment, health, education, Social Security, and child care. One might think of this as representing a liberal–conservative factor, where liberals want to put a higher spending priority on each of these than do conservatives.

## 12.6.2    Oblique rotation: Promax

Sometimes it is unreasonable to think that the factors are orthogonal. We could argue that the first factor, which we have labeled liberalism, would be correlated with the second factor, which we have labeled support for the disadvantaged. It is possible through an option for the rotate command to do an oblique rotation in which we allow the factors to be correlated. Here we illustrate one of the seven options for oblique rotation. We will do a promax rotation by using the command rotate, promax. When we allow the factors to be correlated, it is useful to see how correlated they are. When we run any command in Stata, the program saves some statistics that it does not report in the Results window. These saved statistics can be obtained with the estat command. To obtain the correlations between the common factors in an oblique rotation, the command is estat common. Here are the results of running these two commands in sequence:

```
. rotate, promax
Factor analysis/correlation                    Number of obs    =      1082
    Method: principal-component factors        Retained factors =         4
    Rotation: oblique promax (Kaiser off)      Number of params =        50
```

| Factor  | Variance | Proportion | Rotated factors are correlated |
|---------|----------|------------|--------------------------------|
| Factor1 | 2.46691  | 0.1762     |                                |
| Factor2 | 1.99629  | 0.1426     |                                |
| Factor3 | 1.97508  | 0.1411     |                                |
| Factor4 | 1.53936  | 0.1100     |                                |

```
LR test: independent vs. saturated:  chi2(91) = 1701.47 Prob>chi2 = 0.0000
```

Rotated factor loadings (pattern matrix) and unique variances

| Variable | Factor1 | Factor2 | Factor3 | Factor4 | Uniqueness |
|---|---|---|---|---|---|
| rnatspac | −0.1749 | −0.0294 | −0.2002 | 0.7787 | 0.4204 |
| rnatenvir | 0.4216 | 0.2378 | −0.0437 | 0.2156 | 0.6221 |
| rnatheal | 0.6293 | 0.0361 | 0.0830 | 0.0460 | 0.5235 |
| rnatcity | −0.1558 | 0.4893 | 0.5021 | 0.0281 | 0.5036 |
| rnatcrime | −0.0338 | −0.0575 | 0.8526 | −0.1366 | 0.3361 |
| rnatdrug | 0.1026 | 0.0462 | 0.6555 | −0.0410 | 0.5051 |
| rnateduc | 0.7179 | 0.0258 | −0.0616 | 0.0370 | 0.4912 |
| rnatrace | 0.1043 | 0.6229 | 0.1546 | −0.0083 | 0.4909 |
| rnatarms | 0.1864 | −0.7131 | 0.2047 | 0.1762 | 0.4968 |
| rnatfare | 0.1329 | 0.5519 | −0.0371 | 0.0790 | 0.6172 |
| rnatroad | −0.0535 | −0.2080 | 0.2760 | 0.4176 | 0.7065 |
| rnatsoc | 0.7637 | −0.2163 | −0.0129 | −0.2463 | 0.4862 |
| rnatchld | 0.4121 | 0.2818 | −0.0035 | 0.0167 | 0.6699 |
| rnatsci | 0.1175 | −0.0307 | −0.0209 | 0.7162 | 0.4549 |

Factor rotation matrix

| | Factor1 | Factor2 | Factor3 | Factor4 |
|---|---|---|---|---|
| Factor1 | 0.8518 | 0.6478 | 0.6627 | 0.3341 |
| Factor2 | −0.0024 | −0.4155 | 0.2418 | 0.8357 |
| Factor3 | 0.1455 | −0.5818 | 0.5015 | −0.4341 |
| Factor4 | 0.5032 | −0.2632 | −0.5008 | −0.0396 |

. estat common

Correlation matrix of the promax(3) rotated common factors

| Factors | Factor1 | Factor2 | Factor3 | Factor4 |
|---|---|---|---|---|
| Factor1 | 1 | | | |
| Factor2 | .3358 | 1 | | |
| Factor3 | .3849 | .1688 | 1 | |
| Factor4 | .1995 | .1322 | .2256 | 1 |

We can see that the factors are all positively correlated. The factor loadings can no longer be interpreted as simple correlations like before, but they tend to be "cleaner" than with the oblique rotation. By this I mean that the big loadings tend to be even bigger, and the small loadings tend to be even smaller. This sometimes simplifies the interpretation.

## 12.7  But we wanted one scale, not four scales

The rotations helped us gain insight into how people cluster different items and helped us recognize that there are some differences between people who would support each of these dimensions. However, we are trying to develop one scale, and we need to select items that represent one dimension. For this, we go back to our original PCF analysis results.

We clearly had one dominant factor as the first principal component, and most of the items loaded strongly on that first component. Indeed, 10 of the 14 items had loadings over 0.4. The items that failed to meet this standard are support for the space program, support for spending on science, support for roads, and support for the military. Space and science are special areas, as are roads and the military. Both liberals and conservatives may support or not support more spending for roads, space, and science. Some liberals may not support spending on the military in general, but many liberals and conservatives support spending to support the soldiers. By contrast, the 10 items with strong loadings on the first factor clearly differentiate liberals and conservatives. We could think of the 10 items as representing a general liberal-to-conservative dimension for federal spending policies.

We need to rerun our PCF analysis, including just the 10 items. Try this as an exercise. When we do this, the eigenvalue for the first factor is 2.941, explaining almost 30% of the variance in the set of 10 items. There are a second and third factor, but the eigenvalue for each of them is just barely over 1. All the items load over 0.4 on the first factor.

Not all researchers use factor analysis when they are constructing a scale. However, if you use this approach and obtain a set of 10 items that have loadings over 0.4 on a first factor, and the first factor has a much larger eigenvalue than the other factors, your results of doing a reliability analysis are almost certainly going to be positive. For these 10 items, the alpha is 0.72 and dropping any of the items would reduce alpha.

## 12.7.1   Scoring our variable

We have used factor analysis to identify a set of items that load highly on one dimension. We can use the 10 items to construct our scale in one of two ways. One of these is exactly what we did before by using the `egen` command to generate the row mean of the 10 items. We could do that and then do an alpha for the 10 items to estimate the reliability of our scale.

There is a more complex alternative, which is to estimate what is called a factor score. This is a standardized value that has a mean of 0 and a standard deviation of 1. If our variable is normally distributed, about two-thirds of the observations will have a score between $-1.0$ and $+1.0$, and about 95% of the observations will have a score between $-2.0$ and $+2.0$.

An advantage of a factor score over a mean or total score is that the factor score weights each of the items differently, based on how central it is to the first factor. For example, education has a loading of 0.6, and health has a loading of 0.63. These should be weighted more in the factor score than Social Security, which has a loading of 0.43. By contrast, when we generate a mean or total for the set of items, each item counts as if it were equally central to the concept.

It is easy to estimate a factor score. Using the default options for the command just after doing the PCF analysis, we enter the command `predict libfscore, norotate`, where `libfscore` is the name we are giving the factor score.

Which method is better? There is some controversy about how to best estimate a factor score, so I have simply used the default option, which is a regression-based approach. Although there is some disagreement about the best method, there is rarely a substantive difference in the result.

We should not forget that there is an advantage of a factor score over a mean or total score in that the factor score weights less-central items less and more-central items more, which makes sense. However, this distinction rarely makes a lot of practical difference. The factor score may make a difference if there are some items with very large loadings, say, 0.9, and others with very small loadings, say, 0.2. But we would probably drop the weakest items. When the loadings do not vary a great deal, computing a factor score or a mean/total score will produce comparable results. Here are the commands you would use to compute a factor score and a mean score:

```
. factor rnatenvir rnatheal rnatcity rnatcrime rnatdrug rnateduc rnatrace
> rnatfare rnatsoc rnatchld, pcf

. predict libfscore, norotate

. egen libmean = rowmean(rnatenvir rnatheal rnatcity rnatcrime rnatdrug
> rnateduc rnatrace rnatfare rnatsoc rnatchld)
```

## 12.8   Summary

This chapter introduced some of the classical procedures for developing a scale. This is an area of social research that often is given too little attention. Without measures that are both reliable and valid, it is impossible for social science to develop. The intent of this chapter was to introduce the most basic ideas for developing a scale and how to execute basic Stata commands that help demonstrate the reliability and validity of a scale. We covered

- How to use the `egen` command to generate a scale score.

- Test–retest reliability.

- How to estimate alpha and how to use it to evaluate the internal consistency standard for a reliable scale. We saw that alpha depends on the average correlation between the items and the number of items in the scale; too few items makes it hard to obtain an adequate alpha, and too many items makes a questionnaire needlessly long. We also covered the Kuder–Richardson reliability coefficient for dichotomous items.

- Kappa as a measure of interrater reliability. We briefly discussed weighted kappa and how kappa can be used when there are three raters.

- Various types of validity, including expert judges, face validity, criterion-related validity, predictive validity, and construct validity.

- Principal factor (PF) analysis and principal-component factor (PCF) analysis, with an example of how to use PCF analysis.

- Key terminology for understanding factor analysis, along with orthogonal and oblique rotations of the PCF analysis solution.

- How to construct a factor score and how this compares with generating a mean score.

## 12.9   Exercises

1. Use gss2002 and 2006_chapter12.dta. Compute a split-half correlation for the 10 items that load most strongly on the first factor in the PCF analysis (section 12.6). Correlate the mean of the first five items and the mean of the second five items. Why does this correlation underestimate the reliability of the 10-item scale?

2. Using the data from above, compute the alpha reliability for this scale. Interpret alpha to decide if this scale is adequate. Interpret the item-test correlations, item-rest correlations, and what happens to alpha if any of the items is dropped. Why is alpha a better measure of reliability for this 10-item scale than the split-half correlation?

3. Using the data from above, do an alpha analysis including all 14 items. Use the values of alpha to drop the worst item and repeat alpha for the remaining 13 items. Repeat this process, dropping an item each time, until there is no item to drop that would increase alpha. Compare this set of items with what the factor analysis indicated was the best set of items.

4. Use gss2002 and 2006_chapter12.dta. Do a PCF analysis (section 12.6) of the 10 items that load on the first factor for the spending items. Interpret the results.

5. Using the results from exercise 4, create the factor score for the 10 items and use egen to create a mean score for the 10 items. Compare the means and standard deviations of both scores, and explain the differences. What is the correlation between the two scale scores? What does it mean that it is so high? When would it not be so high?

6. Use gss2002 and 2006_chapter12.dta. Do a PCF analysis of the 14 items, and construct a factor score. Do a PCF analysis for the 10 items as done above, and construct a factor score. Correlate these two scores, and explain why they are so highly correlated even though they are based on different items.

# 13 Working with missing values—multiple imputation

## 13.1 The nature of the problem

Most data that social scientists analyze have missing values. These can occur for many reasons.

- A participant in a randomized trial may decide to withdraw from the study halfway through data collection.

- In a panel study, some participant may have no data at one wave. For example, they were ill when measurement was done halfway through the project. They would have missing values on every item at that wave of the study.

- There are sensitive items that people are reluctant to report, such as criminal behavior. A business may conduct an anonymous survey of employees that includes some questions about employee theft. Although the survey is anonymous, some employees may be reluctant to answer these questions, resent being asked these questions, or even be suspicious of the stated anonymity of their responses.

- About 20–30% of people, when asked their income in national surveys, simply do not give an answer.

- Interviewers sometimes accidentally skip a section of a questionnaire.

- A survey participant simply gets tired of answering questions and turns in a partially complete questionnaire.

- Participants in an experimental condition do poorly on the baseline and decide the intervention is going to be too hard so they drop out.

In the first 12 chapters of this book, we used the default approach to missing values, which drops any participant who does not have complete information on every item used in the analysis. This approach goes by several names, including full case analysis, casewise deletion, or listwise deletion. Perhaps the term listwise deletion is most common in the social sciences. If you use 50 items in an analysis, anybody who did not give a complete answer to all 50 items (the list) is deleted. It is not unusual for this approach to result in 20–50% of the participants being dropped from a complex analysis. This loss of information has two serious implications:

- There will be a substantial loss of power because of the reduced sample size. When doing a power analysis using listwise deletion, you need to take this loss of power into account.

- Listwise deletion can introduce substantial bias. For example, if people who are poorly motivated at the start of an intervention drop out of a study, it will look like the intervention was more successful than it really was.

One alternative to listwise deletion involves substituting the mean on a variable for anybody who does not have a response. This is based on the idea that if you do not know anything about an observation and need to guess a value, the mean is the best guess. This has two serious limitations. People who are average on a variable are often more likely to give an answer than are people who have an extreme value. Income provides a good example of this. People who have extremely low income are hesitant to reveal this. If the average income in a community were $52,500, this would be a terrible estimate for many people who have refused to report their income because they felt that their income was too low. The second problem with mean substitution is that when you give several people the same score on a variable, these people have zero variance on the variable. Sticking with income, if we have 30% of our people not reporting their income and give each of them the same estimated value of $52,500, we will dramatically attenuate the variance on income. Because this makes income a constant for 30% of our sample, there is no explanatory power for 30% of our sample—they all are treated as if they had the same incomes. This artificially reduced variance will seriously bias our parameter estimates. You can imagine the effect this will have on $R^2$. We will not be able to explain any variance in any outcome for the 30% of the people who have identical estimated incomes. This will attenuate the variance explained by income but can have a complex effect on other variables that have fewer missing values.

## 13.2 Multiple imputation and its assumptions about the mechanism for missingness

Beginning with Stata 11, there is a comprehensive suite of commands for doing multiple imputation of missing values. These are in addition to the excellent user-written commands `ice` (Royston 2004, 2009) and `mim` (Carlin, Galati, and Royston 2008; Royston, Carlin, and White 2009). All of these procedures make an important assumption about the mechanism for missingness but are fairly robust against violations of these assumptions.

When a value is missing, we need to make some assumption about the underlying process that generated this missingness (Rubin 1987). Ideally, the missing value could be missing completely at random (MCAR). Suppose that you have a research team developing a survey that will be given to children in the third grade. Each member of the team is likely to want his or her own set of variables measured. When you get a good measure of each of these variables, your draft questionnaire explodes to 150 items. You are not likely to give a 150-item questionnaire to a third grader; they would skip over many of the items or give answers without reading each of the items carefully. Thoughtful completion of a 150-item questionnaire is just too much to expect of a typical third grader. You decide the most items you can ask a third grader to expect thoughtful responses is 50. What can you do? You could randomly select 50 items from your list of 150 and given them to the first child. Then randomly select a new set of 50 items and give them to the second child. You could use the `sample` command in Stata to randomly generate a different set of 50 items for each participant. Each child would have missing values on 100 of the 150 items, 67% missing values, but these missing values would be MCAR. The process that produced the missingness in this example is unrelated to whether a child was asked any particular item or to how the child might answer any item. Because the process that generated the missingness is completely random, there is no bias introduced in your study.

MCAR is probably unreasonable in most research settings because people with missing values are likely to be different from people without missing values. A second mechanism for missingness occurs whenever the reason for missingness can be explained by variables that are in your analysis. For example, we know that men are less likely to answer items than are women. So long as we have gender in our model, we can explain that part of the differential missingness. The missing data would not be MCAR, but a variable in our analysis would explain the missingness. This missing-value type is called missing at random (MAR).

MAR is not to be confused with MCAR. Many researchers are confused by the name MAR, thinking that it literally means that the missing values are like a random sample of all values, which is not the case. The variables that explain missingness are either included in the model or added as auxiliary variables when imputing missing values. Fortunately, many of our studies include the key variables that predict who will and who will not be as likely to answer individual items. We know that in addition to gender, race/ethnicity and education are related to missingness, and many studies include measures of these variables.

The MAR assumption is more reasonable than you might think. Suppose that you have an intervention that is given at three different levels of intensity: low, medium, and high. Let's imagine an exercise program for people who are obese. Some people who were randomly assigned to the low-intensity condition may decide that the intervention is not likely to be effective because they recognize that not much is happening. They drop out. Conversely, the high intensity of exercise may overwhelm some of the people who were randomly assigned to the high-intensity intervention and they also drop out. Certainly, the dropouts are not MCAR. However, they are MAR because the reason they have missing data, namely, the intensity of the intervention, is a variable that is included in the model.

The key to understanding multiple imputation is that the imputed missing values will not contain any unique information once the variables in the model and the auxiliary variables are allowed to explain the patterns of missing values and predict the score of the missing values. The imputed values for variables with missing values are simply consistent with the observed data. This allows us to use all available information in our analysis.

Unfortunately, there is no test of the validity of the MAR assumption. We may include gender, race/ethnicity, and education as more or less universal mechanism variables. If we know a pretest score and believe people who score poorly on that measure will drop out, we want to include the pretest score even if it is not relevant to our model. The reason we do not have an explicit test of the MAR assumption is that we never know if some additional unobserved variable also provides a mechanism for missingness—because we would not have the variable in our analysis. We can never be certain that we have specified the exact right imputation model.

How can we locate important auxiliary variables? One way is to generate an indicator variable coded as 1 if the value is missing and 0 if the value is present, regardless of what the person's score on the variable might be. This is a new indicator variable, and it may or may not be related to the observed score. It is simply an indicator of missingness. We will have one such indicator variable for each variable that has any missing values. Next we can run a series of logistic regressions, where each of these indicator variables is regressed on a list of potential auxiliary variables. We probably should err on the side of inclusion and might have 10 to 20 candidate auxiliary variables. Any of these candidate auxiliary variables that is significantly related to any of the indicator variables should be included as an auxiliary variable. If a variable has very few missing values, the logistic regressions will encounter estimation problems and will drop some variables and cases.

You may want to ignore these results because there are so few missing values on those variables anyway. Doing this makes the MAR assumption reasonable. Extensive studies including simulations have shown that including a reasonable set of auxiliary variables leads to excellent results.

## 13.3   What variables do we include when doing imputations?

Some applied researchers are reluctant to include the main outcome variable from the analysis model when doing multiple imputation. They think including the outcome variable is somehow cheating; they are wrong. When doing multiple imputation, it is important to properly specify the imputation model. If we leave out any relevant variable, we are effectively assuming that the deleted variable is unrelated to the other variables. Because the outcome variable should be related to each of the predictors in the model, leaving the outcome variable out of the imputation will bias correlations to zero.

In addition to the variables in our model and the auxiliary variables, we should include any available variables that would help us predict the score on the variable that has missing values. Of course, this can be a problem if the additional variable has many missing values itself. Here is a checklist:

1. Include all variables in the analysis model, including the dependent variable,

2. Include auxiliary variables that predict patterns of missingness, and

3. Include additional variables that predict a person's score on a variable that has missing values.

Excluding variables from the imputation process is a misspecification of the imputation model. This probably does not matter much when there are few missing values; say, no variable has more than a couple percent of missing values. But it can lead to substantial biases when many missing values exist. This book has not fully addressed the more advanced topic of working with complex sample designs, but in those cases, it would also be important to include sampling characteristics such as weights and strata (StataCorp 2013a).

All this discussion of what variables to include to properly specify the imputation model needs to be balanced with common sense. First, you probably only want to impute values for variables where a score would be meaningful. It would not make sense to ask men the age they were when they had their first period. Theoretically, you could impute this without bias, but you might want to leave it missing. People who were not asked a question on purpose might be left missing. Many questionnaires use complex skip patterns where you get a different set of questions based on your answers to initial questions. If there were a legitimate skip of an item because it would make no sense to ask it, you probably do not want to impute the missing value.

The advice is to include a lot of variables to ensure proper specification, but this can make matters worse if the included variables are themselves adding a lot of missing values. If you were using education as a variable in your model, but not income, you might say income would be a good additional variable because it would help predict education. However, income might have so many missing values that it will not be helpful. Once you have included a reasonably large number of variables, adding additional variables may not be helpful because of multicollinearity. If you were to put 100 variables in your imputation model, the contribution of many of these variables may be virtually redundant, adding complexity without materially improving the proper specification of your imputation.

## 13.4   Multiple imputation

Beginning with Stata 11, there is a powerful way of working with missing values that involves multiple imputation. The command `mi` will be illustrated in this chapter. This command involves three straightforward steps:

1. Create $m$ complete datasets by imputing the missing values. Each dataset will have no missing values, but the values imputed for missing values will vary across the $m$ datasets.

2. Do your analysis in each of the $m$ complete datasets.

3. Pool your $m$ solutions to get one solution.

   a. The parameter estimates—for example, regression coefficients—will be the mean of their corresponding values in the $m$ datasets.

   b. The standard errors used for testing significance will combine the standard errors from the $m$ solutions plus the variance of the parameter estimates across the $m$ solutions. If each solution is yielding a very different estimate, this uncertainty is added to the standard errors. Also the degrees of freedom is adjusted based on the number of imputations and proportion of data that have missing values.

Is this cheating? This really is not cheating! The uncertainty caused by missing values comes into the process at multiple points. In generating $m$ complete datasets, each of which will have somewhat different imputed values, the variability of the resulting solutions is added to the estimated standard errors. If the imputations are doing a good job, not much variance will be added. The results report the average relative increase in the variance (RVI) of the estimates (squared standard errors) because of missing values. Because this cannot be done with a single imputation, studies that rely on single imputation will potentially underestimate standard errors and inflate $t$-values. If the RVI is close to zero, then the imputation will increase power, but power may not increase over listwise deletion if the RVI is substantial. If the imputations are problematic, then considerable variance is added. The imputed values do not contain any additional information than that contained in all the variables in your model and those used as auxiliary variables. The imputed values allow Stata to use all available data. Imagine having 10 waves of data in a study of the transition to adulthood. John participated in all but the third wave because he was sick the day that measurement was taken. It makes no sense to ignore all the information John provides on the other nine waves; that would be a waste of valuable information. Using multiple imputations will usually increase power by using all available data on all participants.

## 13.5   A detailed example

Stata has a comprehensive set of commands for working with missing values. We will only consider the most general case. The *Stata Multiple-Imputation Reference Manual* (StataCorp 2013a) provides details on alternatives. The most widely used approach is using multivariate normal regression (MVN). Stata's `mi impute mvn` command implements an iterative Markov chain Monte Carlo method (Schafer 1997) that is known as data augmentation. The MVN method is designed for continuous variables. Where there are several categorical or count variables, Stata's `mi impute chained` command implements the chained equation approach for imputing missing values (van Buuren, Boshuizen, and Knook 1999) and provides a useful alternative. If you are using a version earlier than Stata 12, there is a pair of user-written commands available, `ice` (Royston 2004, 2009) and `mim` (Carlin, Galati, and Royston 2008; Royston, Carlin, and White 2009), that have many of the same capabilities as `mi impute chained` and `mi estimate`.

The chained equations approach has less theoretical justification than the MVN approach, but it has the advantage of using an appropriate estimator for each imputed variable based on the measurement level and distribution characteristics of the variable, for example, logistic regression, multinomial logistic regression, Poisson regression, negative binomial regression, and others. In practical experience, the final model estimation usually produces similar results whether the `mi impute mvn` command or the `mi impute chained` command is used. This section focuses on the MVN method and is intended to be introductory. Users of multiple imputation are encouraged to consult the *Stata Multiple-Imputation Reference Manual*, which provides a comprehensive treatment on over 350 pages.

We will use a modified version of a dataset, `regsmpl.dta`, that is available from the Stata menu, File ▷ Example Datasets... ▷ Stata 13 manual datasets. The modified version we will use, `chapter13_missing.dta`, can be thought of as a nonrandom subsample of `regsmpl.dta`.

We are interested in the natural log of wages (`ln_wagem`). We have as predictors the person's highest grade completed (`gradem`), age (`agem`), total work experience (`ttl_expm`), time in years at his or her present job (`tenurem`), whether they are from a non-SMSA area (`not_smsa`), whether they are from the South (`south`), and whether they are black (`blackm`). The letter `m` has been appended to each variable where there are missing values to distinguish them from the original dataset.

## 13.5.1  Preliminary analysis

To get an idea of the extent of the problem with `chapter13_missing.dta`, we use the `misstable` command: first, to get a summary of missing values, and second, to get a sense of the pattern of missingness.

```
. misstable summarize ln_wagem gradem agem ttl_expm tenurem not_smsa south
> blackm
```

|          |       |       |        | Obs<.            |         |          |
|---------:|------:|------:|-------:|-----------------:|--------:|---------:|
| Variable | Obs=. | Obs>. | Obs<.  | Unique<br>values | Min     | Max      |
| ln_wagem |   300 |       |  1,393 |              393 | .0682788| 4.242752 |
|   gradem |   148 |       |  1,545 |               13 |       0 |       18 |
|     agem |   165 |       |  1,528 |               12 |      18 |       30 |
| ttl_expm |   206 |       |  1,487 |             >500 |.0833333 | 15.53846 |
|  tenurem |   200 |       |  1,493 |               74 |       0 |     15.5 |
|   blackm |    99 |       |  1,594 |                2 |       0 |        1 |

This is similar to the regular `summarize` command except that it is for missing values. A missing value will have a code of `.`, `.a`, `.b`, etc. Observations with these codes are tabulated in the columns labeled `Obs=.` and `Obs>.`. Remember that a missing value is recorded in a Stata dataset as an extremely high value. Within `mi`, a missing-value code, `.` (dot), has a special meaning. It denotes the missing values eligible for imputation. If you have a set of missing values that should not be imputed, you should record them as extended missing values, that is, as `.a`, `.b`, etc. The column labeled `Obs<.` contains the number of observations that are not missing. For example, our dependent variable has 300 missing values we want to impute, no missing values we do not want to impute (`.a`, `.b`, etc.), and scores on 1,393 observations. Notice that we listed all our variables in the command, but neither `not_smsa` nor `south` appear in the table. This is because they have no missing values.

Some datasets use the extended missing-value codes `.a`, `.b`, `.c`, etc. One of these extended missing-value codes might be assigned for people who answer "Don't know". If you want to impute missing values for any of the extended missing-value codes, you will want to recode it to a system missing value of dot (`.`) before you perform the multiple imputation. For example, if you used `.a` for "Don't know" responses on age and want to include these responses when you do the multiple imputation, you should type `recode agem (.a = .)` prior to performing multiple imputation.

Next we type the `misstable patterns` command. This command can produce a lot of output when there are many different patterns. Here are the patterns that each variable has for at least 2% of the total $N$.

```
. misstable patterns ln_wagem gradem agem ttl_expm tenurem not_smsa south
> blackm
```

```
        Missing-value patterns
           (1 means complete)

               |     Pattern
   Percent     | 1  2  3  4    5  6
   -----------------------------------
     51%       | 1  1  1  1    1  1

      8        | 1  1  1  1    1  0
      7        | 1  1  0  1    1  1
      6        | 1  1  1  0    1  1
      5        | 1  1  1  1    0  1
      4        | 1  0  1  1    1  1
      4        | 0  1  1  1    1  1
      3        | 1  1  1  1    0  0
      2        | 1  0  1  1    1  0
```
      *(output omitted)*
```
   -----------------------------------
     100%      |
```

```
Variables are   (1) blackm   (2) gradem   (3) agem   (4) tenurem   (5) ttl_expm
                (6) ln_wagem
```

As noted in the results, a 1 indicates that the value is present. Thus 51% of the sample have a value on all the variables. If we did a listwise deletion, we would have lost 49% of our cases! Below this row are different patterns. The most common pattern is to have answered all but the sixth variable, which we see from the bottom of the table is ln_wagem. Eight percent of the participants are in this pattern. This table is not especially useful here because there are no dominant patterns of missingness.

Next we will generate a dummy variable for each of our variables that has missing values. We want the dummy variable to be coded 0 if the variable is observed and 1 if the variable has a missing value. We will use the misstable summarize command again, but this time we include the generate(miss_) option. This option generates the dummy variables and adds the prefix miss_ to each variable name. For example, agem is a variable we want to impute, and miss_agem will be a dummy variable coded 1 if the person has a missing value on agem and 0 if the person has an observed value. We use the quietly prefix because we already have seen the standard misstable summarize output. We also use describe miss_* to review the changes. Here are the commands and results:

```
. quietly misstable summarize ln_wagem gradem agem ttl_expm tenurem not_smsa
> south blackm, generate(miss_)

. describe miss_*

                 storage  display    value
variable name    type     format     label      variable label
----------------------------------------------------------------------------
miss_ln_wagem    byte     %8.0g                  (ln_wagem>=.)
miss_gradem      byte     %8.0g                  (gradem>=.)
miss_agem        byte     %8.0g                  (agem>=.)
miss_ttl_expm    byte     %8.0g                  (ttl_expm>=.)
miss_tenurem     byte     %8.0g                  (tenurem>=.)
miss_blackm      byte     %8.0g                  (blackm>=.)
```

The `generate()` option is nice because it not only generates the dummy variables, but also it adds a variable label to each of the generated variables to remind us what has been done.

We can treat each of these dummy variables as a dependent variable and do a logistic regression to locate possible auxiliary variables. We do not have any candidates for auxiliary variables in our dataset. Can you think of some candidates we should have included? For instance, gender would be an auxiliary variable. Here we show the six logistic regression commands, recognizing that, ideally, we should have included additional variables that are not part of our model but may help explain missingness. If your data happen to have variables with very few missing values, the `logit` command will have estimation problems and will drop one or more variables and cases. You might simply ignore those results because there are few missing values on those variables. You are looking for the results of the logistic regressions for variables that have several missing values, say, 5% or more.

```
. logit miss_ln_wagem gradem agem ttl_expm tenurem not_smsa south blackm
> if ln_wages <= .
. logit miss_gradem ln_wagem agem ttl_expm tenurem not_smsa south blackm
> if gradem <= .
. logit miss_agem ln_wagem gradem ttl_expm tenurem not_smsa south blackm
> if agem <= .
. logit miss_ttl_expm ln_wagem gradem agem tenurem not_smsa south blackm
> if ttl_expm <= .
. logit miss_tenurem ln_wagem gradem agem ttl_expm not_smsa south blackm
> if tenurem <= .
. logit miss_blackm ln_wagem gradem agem ttl_expm tenurem not_smsa south
> if blackm <= .
```

We are not interested in predicting the indicator variables if the person has an extended missing value such as `.a`, `.b`, etc. We ensure the avoidance of predicting these variables by using the `if` condition to drop any case that has an extended missing value such as `.a`. This works because these extended missing values are recorded in the Stata dataset as a larger value than the value assigned to the `.` (dot). Any variable that is statistically significant in these logistic regressions should be included in the imputation step. In our example, these are already included because they are all variables used in the analysis stage.

## 13.5.2   Setup and multiple-imputation stage

Next we tell Stata how to arrange the imputed datasets; that is, we choose a multiple-imputation storage style. We could store the datasets in a wide format, using the wide style, where we have new variable names for each imputed dataset and they go to the right. Or we could store the datasets in a long format, where the datasets are stacked. Using the long style, we have our initial dataset, labeled 0. Below that, we have a dataset labeled 1; below that is a dataset labeled 2; and so on. We will use this long style in this example.

You could imagine how the stacked datasets could get very, very long. If we imputed 100 datasets on a base sample of 10,000 observations, we would need 1,000,000 records. Stata has a way of minimizing the storage issues with this approach by including only the records that have imputed values in each of the subsequent sets. In our example, we would have all our records in the first set, labeled 0. In the second and subsequent sets, we would only have cases for the 49% of the participants who had missing values. This would cut the number of records in our file in half. The style containing the complete sets is `flong` (full and long), which is the style we will use; the style that only has the cases with missing values is `mlong` (marginal and long). If you run out of memory space, you could change from the `flong` to the `mlong` style. Stata has other storage styles described in the *Stata Multiple-Imputation Reference Manual* (StataCorp 2013a).

To prepare our data for multiple imputation, we need to run three commands:

```
. mi set flong
. mi register imputed ln_wagem gradem agem ttl_expm tenurem blackm
. mi register regular not_smsa south
```

The `mi set flong` command tells Stata how to arrange our multiple datasets. The `mi register imputed` command registers all the variables that have missing values and need to be imputed. The `mi register regular` command registers all the variables that have no missing values or for which we do not want to impute values.

We are now ready to create the multiple completed datasets. Here we use the `mi impute` command to obtain imputations and the `rseed()` option to set a seed at some arbitrary value so that the results can be duplicated. You can enter any value in `rseed()`; I use 2121 in this example.

```
. mi impute mvn ln_wagem gradem agem ttl_expm tenurem blackm, add(20) rseed(2121)

Performing EM optimization:
  observed log likelihood = -5347.0692 at iteration 12

Performing MCMC data augmentation ...

Multivariate imputation              Imputations =        20
Multivariate normal regression             added =        20
Imputed: m=1 through m=20                 updated =         0

Prior: uniform                        Iterations =      2000
                                         burn-in =       100
                                         between =       100
```

| | Observations per *m* | | | |
|---|---|---|---|---|
| Variable | Complete | Incomplete | Imputed | Total |
| ln_wagem | 1393 | 300 | 300 | 1693 |
| gradem | 1545 | 148 | 148 | 1693 |
| agem | 1528 | 165 | 165 | 1693 |
| ttl_expm | 1487 | 206 | 206 | 1693 |
| tenurem | 1493 | 200 | 200 | 1693 |
| blackm | 1594 | 99 | 99 | 1693 |

```
(complete + incomplete = total; imputed is the minimum across m
 of the number of filled-in observations.)
```

The `mi impute mvn` command specifies that we are using the multivariate normal model to impute missing values. It would be good to check that the variables are reasonably normal. With a large sample, violations of this assumed normality may not be too serious. We specified the `add(20)` option to generate 20 datasets in addition to the initial dataset that has missing values. Each of the 20 datasets will be complete; each will have no missing values. The imputed values will vary from one of these datasets to the next.

Why did we impute 20 complete datasets? Much of the literature, especially the early applications, uses relatively few imputed datasets, often picking $m = 5$. Early software was much harder to use than Stata is, so this might have been one motivation for a small $m$. The theory behind multiple imputation as presented by Rubin (1987) was based on an infinite number of imputed, complete datasets. However, high relative efficiency for even small $m$ has been documented when the assumptions are appropriate. Rubin (1987) and van Buuren, Boshuizen, and Knook (1999) report that $m = 5$ can produce 95% relative efficiency when half the values are missing, given that the MAR assumption is justified. However, others have shown cases where a much larger value of $m$ is needed (Kenward and Carpenter 2007). The *Stata Multiple-Imputation Reference Manual* (StataCorp 2013a) recommends an $m$ of at least 20. Because of the ease with which Stata can implement multiple imputations, you should adhere to this recommendation and may experiment with even larger values of $m$. If you obtain comparable results with 10, 20, 40, and 60 imputed datasets, this is reassuring. The multiple imputation and the analysis take a bit longer with more imputations, but with modern computers, there is no reason to limit the number to 5. The current example took 50 seconds for the imputation stage and 3 seconds for the analysis stage (described later) using $m = 20$, compared with just over 3 minutes and 23 seconds when estimated using $m = 100$. This was using a not-so-new iMac with five other programs running in the background. There is rapidly decreasing marginal utility to making $m$ even larger. However, the cost of computation time is fairly small for making $m$ large.

Along with the creation of our 20 new complete datasets, the `mi set` command created three new variables at the bottom of the list of variables in our Variables window. _mi_id is a new ID variable. It ranges from 1 for our first case to 1,693 for our last case. If you do a tabulation of it, you will see that each ID value has a frequency of 21; that is, 1 for each of our 20 imputed datasets and 1 for our original dataset. Our new data file now has 35,553 observations. The _mi_miss variable is simply an indicator variable coded 0 if there are no missing values and 1 if there are any in the original data. It contains a missing value for all records in imputed data. As we saw before, 51.09% of the observations had no missing values. The _mi_m variable identifies datasets and ranges from 0 to 20. All 1,693 cases in the first imputed dataset have _mi_m $= 1$; all 1,693 cases in the second imputed dataset have _mi_m $= 2$; and so on. Original records have _mi_m $= 0$.

The `mi impute mvn` command uses data augmentation, an iterative imputation method, to fill in missing values. Thus it is important to check the convergence of data augmentation after imputation; however, a discussion of this step is beyond the scope of this book. For details, you can refer to *Convergence of the MCMC method* in [MI] **mi impute mvn**.

## 13.5.3 The analysis stage

Now we are ready to run our 20 regressions and pool the results. We use the prefix command **mi estimate:** in front of the regression command. Stata then knows to do the 20 regressions and pool the results. We have added the option **dftable** to the **mi estimate** prefix command to see coefficient-specific degrees of freedom. You might want to run this without the **dftable** option to obtain confidence intervals. Here are the command and results:

```
. mi estimate, dftable: regress ln_wagem gradem agem ttl_expm tenurem not_smsa
> south blackm
```

```
Multiple-imputation estimates             Imputations      =         20
Linear regression                         Number of obs    =       1693
                                          Average RVI      =     0.4925
                                          Largest FMI      =     0.4706
                                          Complete DF      =       1685
DF adjustment:    Small sample            DF:       min    =      82.21
                                                    avg    =     188.33
                                                    max    =     340.51
Model F test:        Equal FMI            F(   7,  672.6)  =      66.25
Within VCE type:          OLS             Prob > F         =     0.0000
```

| ln_wagem | Coef. | Std. Err. | t | P>\|t\| | DF | % Increase Std. Err. |
|---------:|------:|----------:|---:|-------:|------:|-----:|
| gradem | .0764098 | .0066494 | 11.49 | 0.000 | 97.1 | 31.27 |
| agem | .0201299 | .0041309 | 4.87 | 0.000 | 118.3 | 26.73 |
| ttl_expm | .0155184 | .0104858 | 1.48 | 0.143 | 82.2 | 35.90 |
| tenurem | .062319 | .0091294 | 6.83 | 0.000 | 129.3 | 24.97 |
| not_smsa | -.1124143 | .0239484 | -4.69 | 0.000 | 285.4 | 13.81 |
| south | -.0831899 | .0222689 | -3.74 | 0.000 | 323.9 | 12.55 |
| blackm | -.0247194 | .0243708 | -1.01 | 0.311 | 340.5 | 12.08 |
| _cons | .133795 | .1084921 | 1.23 | 0.220 | 130.0 | 24.87 |

These results look similar to what we get with the **regress** command on one dataset, but some information is additional and some information is missing. On the right, we have the number of imputations, 20; the number of observations after imputations, 1,693; and what the degrees of freedom would have been if there were no missing values, 1,685. We have a different number of degrees of freedom for different variables, ranging from 82.2 to 340.5. These numbers are estimated using complex formulas described in the *Stata Multiple-Imputation Reference Manual* (StataCorp 2013a). The calculated degrees of freedom depends on the number of imputations and, loosely speaking, the rates of missing values for different variables, because both of these factors influence our inference. If you repeat this analysis using more imputations, you will have more information for estimating the parameters and will gain additional degrees of freedom. For example, repeating this analysis with 100 imputations rather than 20 increases the average degrees of freedom from the 188.33 reported here to 597.03 degrees of freedom.

The `dftable` option gives us the specific degrees of freedom for each of the predictors. Although having more imputations will increase the degrees of freedom, this does not necessarily result in a higher level of statistical significance.

There is an overall $F$ test of the model, $F(7, 672.6) = 66.25$, $p < 0.001$, and you will notice an adjustment for the degrees of freedom. In a separate analysis using 100 imputations, the $F(7, 1374.2) = 69.58$, $p < 0.001$. In both cases, the model is highly significant. We could report the significance for each of the variables. For instance, we could say that each additional year of tenure on a job increases the natural log of income by 0.06, $t(129.3) = 6.83$, $p < 0.001$. In writing a report, you should explain that the degrees of freedom has been adjusted based on the number of imputations and fractions of information about parameter estimates lost because of missing values.

The `average RVI = 0.4925` line is an estimate of the average relative inflation in variance of the estimates, that is, in the squared standard error, that is caused by the missing values. Ideally, this would be close to zero. The obtained value of 0.4925 is relatively high because we had missing values for almost half the participants. `mi estimate` explicitly recognizes this inflation, which increases the standard errors, thus reducing the $t$-values and increasing the probabilities. This increase in the standard errors is conservative and an example of how this approach is not cheating or taking advantage of made-up new data.

### 13.5.4  For those who want an $R^2$ and standardized $\beta$s

Stata's `mi estimate` command does not pool the $R^2$ that is obtained for each of the 20 solutions. It also does not pool the standardized beta weights, $\beta$s. Rubin's rule (1987) for pooling may not apply to pooling the $R^2$s and the $\beta$s—both parameter estimates are not normally distributed because of the upper limit of 1.0 (although there are rare cases where $\beta$s can exceed 1). However, the values of these coefficients in social-science research rarely approach these limits, and a simple average may be satisfactory. Because the sampling distribution of the $R^2$ and $\beta$s may have a substantial negative skew, a better solution is to first transform the values by using an inverse hyperbolic-tangent transformation. Then the final estimate is transformed back by using the hyperbolic tangent. This approach was originally recommended in Fisher's tests of correlations and is sometimes labeled a $z$ transformation. This approach is recommended by Harel (2009).

Fortunately, there is a user-written command, `mibeta`, that was written by Yulia Marchenko. You can find and install this command by typing `search mibeta`. Let's say that we want to estimate how much variance in `ln_wage` our set of predictors explains. We need a pooled estimate of $R^2$. We might also be interested in the standardized $\beta$ for education. We can use the `mibeta` command to obtain these estimates.

```
. mibeta ln_wagem gradem agem ttl_expm tenurem not_smsa south blackm, fisherz
> miopts(vartable)
```

| Multiple-imputation estimates | Imputations | = | 20 |
|---|---|---|---|

Linear regression

Variance information

|  | Within | Imputation variance Between | Total | RVI | FMI | Relative efficiency |
|---|---|---|---|---|---|---|
| gradem | .000026 | .000018 | .000044 | .723264 | .430175 | .978944 |
| agem | .000011 | 6.1e-06 | .000017 | .606162 | .386527 | .98104 |
| ttl_expm | .00006 | .000048 | .00011 | .846905 | .470153 | .977032 |
| tenurem | .000053 | .000029 | .000083 | .561662 | .368201 | .981923 |
| not_smsa | .000443 | .000125 | .000574 | .295244 | .232133 | .988527 |
| south | .000391 | .000099 | .000496 | .266842 | .214296 | .989399 |
| blackm | .000473 | .000115 | .000594 | .256283 | .207465 | .989733 |
| _cons | .007549 | .00402 | .011771 | .559136 | .367129 | .981974 |

Note: FMIs are based on Rubin's large-sample degrees of freedom.

Multiple-imputation estimates                      Imputations      =          20
Linear regression                                  Number of obs    =        1693
                                                   Average RVI      =      0.4925
                                                   Largest FMI      =      0.4702
                                                   Complete DF      =        1685
DF adjustment:    Small sample                     DF:       min    =       82.21
                                                             avg    =      188.33
                                                             max    =      340.51
Model F test:        Equal FMI                     F(   7,  672.6) =       66.25
Within VCE type:        OLS                         Prob > F        =      0.0000

| ln_wagem | Coef. | Std. Err. | t | P>\|t\| | [95% Conf. Interval] | |
|---|---|---|---|---|---|---|
| gradem | .0764098 | .0066494 | 11.49 | 0.000 | .0632128 | .0896067 |
| agem | .0201299 | .0041309 | 4.87 | 0.000 | .0119499 | .02831 |
| ttl_expm | .0155184 | .0104858 | 1.48 | 0.143 | -.0053404 | .0363772 |
| tenurem | .062319 | .0091294 | 6.83 | 0.000 | .0442566 | .0803814 |
| not_smsa | -.1124143 | .0239484 | -4.69 | 0.000 | -.1595523 | -.0652764 |
| south | -.0831899 | .0222689 | -3.74 | 0.000 | -.127 | -.0393799 |
| blackm | -.0247194 | .0243708 | -1.01 | 0.311 | -.0726557 | .0232169 |
| _cons | .133795 | .1084921 | 1.23 | 0.220 | -.0808444 | .3484344 |

Standardized coefficients and R-squared
Summary statistics over 20 imputations

|  | mean* | min | p25 | median | p75 | max |
|---|---|---|---|---|---|---|
| gradem | .3199883 | .276 | .3119789 | .3182819 | .3356211 | .347 |
| agem | .1394257 | .106 | .1275456 | .138892 | .1534131 | .167 |
| ttl_expm | .0556211 | .0167 | .0358535 | .0590751 | .0712295 | .104 |
| tenurem | .2253669 | .189 | .2165822 | .2267747 | .2371736 | .265 |
| not_smsa | -.1132504 | -.133 | -.1215403 | -.1120459 | -.1060444 | -.093 |
| south | -.0927989 | -.118 | -.1008746 | -.0927794 | -.0842485 | -.0755 |
| blackm | -.0248217 | -.0475 | -.0267832 | -.0223056 | -.0204527 | -.00461 |
| R-square | .2937282 | .273 | .287215 | .2932148 | .3003806 | .316 |
| Adj R-square | .2907927 | .27 | .2842539 | .2902786 | .2974742 | .313 |

* based on Fisher's z transformation

Notice that we have included two options. The `fisherz` option implements the $z$ transformation to help normalize the sampling distribution. The `miopts(vartable)` option implements an option from the `mi estimate` command's set of options. This option provides very helpful additional information that we discuss below. The mean $R^2 = 0.294$ and the mean standardized weight for `gradem` is 0.320. We should not treat these values as formal parameter estimates, but they give a sense of how much of the variance we can explain and the approximate standardized strength for the effect of education. To see if means are reasonable estimates, we can look at other summaries of the distribution of $R^2$ and the standardized $\beta$ weights over imputed data as reported by `mibeta`.

From the results table labeled "Standardized coefficients and R-squared", we have some useful information to help us assess the quality of our estimates. We have seen that the $\beta$ for `gradem` is 0.320. To the right of that in the table, we see that for over 20 regressions, the minimum value is 0.276, the 25th percentile is 0.312, the median is 0.318 (close to the mean, suggesting that the sampling distribution is symmetrical), the 75th percentile is 0.336, and the maximum value is 0.347. We can conclude that the estimate is quite stable across the 20 imputations. Even more useful is the top table in the results, which is labeled "Variance information". This information is available because we used the `miopts(vartable)` option. Perhaps the most useful information in this table is in the last column on the right, "Relative efficiency". For `gradem`, the relative efficiency is 0.979. This is how efficient our estimate is when using 20 imputed datasets, compared with what we would have obtained with use of an infinite number of imputed datasets. When this value is very close to 1.000, we can be confident that $m$ is large enough. There is no convention as to how close this needs to be to 1.000, but a value of at least 0.98 is reasonable. In general, when the variance in an estimate from one dataset to another (between the datasets) is fairly large, increasing the value of $m$ will be very beneficial. When there is little variance between datasets, there is little to gain by using a large value of $m$.

## 13.5.5   When impossible values are imputed: Binary variables, squares, and interactions

The MVN assumption assumes a multivariate normal distribution. As a consequence, we may have imputed values that are outside the possible range of a variable. Remember that a normal distribution can range from minus infinity to plus infinity, and this is not the case with any of our variables. Here is a summary of the 33,860 observations in our 20 completed datasets. Notice the restriction of `if _mi_m > 0` so that we do not include the original dataset in our summary results. The dataset included the original data where the value of `_mi_m` was 0 and we are only interested in the 20 created datasets; that is, `_mi_m` equals 1 to 20.

```
. summarize ln_wagem gradem agem ttl_expm tenurem not_smsa south blackm
> if _mi_m > 0

    Variable |        Obs        Mean    Std. Dev.         Min         Max
-------------+--------------------------------------------------------------
    ln_wagem |      33860    1.612628    .4384923   -.2661006    4.242752
      gradem |      33860    12.15729    1.835523           0    18.23831
        agem |      33860    22.81233     3.03532    14.77328    33.67864
    ttl_expm |      33860    3.336014    1.569637   -1.765149    15.53846
     tenurem |      33860    1.736031    1.585292    -3.53045        15.5
-------------+--------------------------------------------------------------
    not_smsa |      33860    .2658004     .441765           0           1
       south |      33860    .3951565    .4888915           0           1
      blackm |      33860    .2621881    .4402057   -1.512018    1.606588
```

The scores for education, `gradem`, range from 0 to 18.238. In our original data, the range was 0 to 18. Because a normal distribution does not have a fixed upper or lower limit, we can understand how we might have an imputed value of slightly over the maximum. In our original data, `agem` ranged from 18 to 30, and here it ranged from 14.773 to 33.679. This change is a bit more worrisome because the study was intended to be limited to people ages 18 to 30. The minimum imputed value for `tenurem` is $-3.530$, which is confusing. A person cannot be on a job for less than 0 years. If we do a tabulation on `tenurem`, we see that these impossible negative values are rare, so the imputed values may not be too problematic. We might be more concerned about the `blackm` variable, an indicator variable coded as 1 if the person is black and 0 otherwise. A minimum value of $-1.512$ and a maximum value of 1.607 for `blackm` make no intuitive sense.

We know the cause of these "impossible" values. It is that the MVN solution does not limit the imputed values to be in a specified range as with `tenurem` or to only take on certain values as with `blackm`. There has been considerable controversy over how to handle the "impossible" values. One approach that was widely used in early applications of multiple imputation involved fixing the problem by restricting the range of continuous variables or by rounding to the nearest integer for binary variables. We could take all the negative values for `tenurem` and change them to 0. We could change the `agem` scores below 18 to 18 and those above 30 to 30. We could assign anybody who has a score on `blackm` that is greater than 0.5 a value of 1 and anybody who has a score of 0.5 or lower a value of 0. Intuitively, this is an appealing solution.

More recently, the choice of how to handle "impossible" values is to leave them alone. Although this may not seem intuitively reasonable, it may be the best solution. The imputation model and the analysis model need to be compatible or else the results will be biased. If the analysis stage, where we did the regressions, used different datasets (because of the rounding or truncating) than the one generated by the imputation model, then the results would be biased (Horton, Lipsitz, and Parzen 2003). Today, the recommended approach is to leave the imputed values as they are. Statisticians will continue to work on this problem, but usually the conclusions you draw will not change whether you make the adjustment or not.

Not illustrated here is what you do when you have an interaction term $(X_1 X_2)$ or squared variables $(X^2)$. The recommended approach for these situations is to create the interactions or squared variables in the original dataset, before doing the multiple-imputation stage (von Hippel 2009). Because of the way the interactions and squared terms are imputed as distinct variables, the values of the imputed $X^2$ may not be equal to the value of the imputed $X$ times itself. You might have something like an imputed $X = 10.3$ and an imputed $X^2 = 98.6$. Similarly, the values of the imputed $X_1 X_2$ may not equal the value of the imputed $X_1$ times the imputed $X_2$. It would be tempting to leave the interaction terms and squared terms out of the original dataset, do the imputation, and then compute them such that if an imputed $X = 10.3$, the $X^2$ will equal 106.09. $X^2$ will then be defined as a "passive" variable in Stata. This is not recommended; von Hippel has shown that doing this will introduce a bias. The best general advice is to impute everything and do nothing to change any imputed values for the analysis stage.

## 13.6   Summary

We have only touched the surface of Stata's multiple-imputation capabilities. The *Stata Multiple-Imputation Reference Manual* (StataCorp 2013a) is well written and gives numerous examples for more specialized applications. I included this brief introduction to multiple imputation because it would serve most needs of researchers.

Professional journals are increasingly demanding that some sort of imputation be done so that all available data can be used in the analysis. Several simulations suggest that most of the modern approaches yield remarkably similar findings. We have discussed some issues, such as rounding imputed values for binary variables. There is an enormous amount of scientific investigation by statisticians into what is the optimal approach. It can be shown that there are situations where different strategies lead to very different parameter estimates and tests of significance. However, in most applications, one type of multiple imputation produces results that are similar to other types. We may be worrying about more subtle distinctions than is necessary. The multivariate normal approach outlined in this chapter should work for a wide range of situations; see, for example, Allison (2001).

I only illustrated the use of the `regress` command with the `mi estimate` prefix command. This one example can be used in a wide range of research applications. However, the `mi estimate` prefix command works with many Stata commands. Among those not covered in this book is `logit` used for logistic regression. The *Stata Multiple-Imputation Reference Manual* (StataCorp 2013a) lists 43 estimation commands in Stata that work with `mi estimate` prefix. These cover count models, ordinal models, categorical models, quantile regression models, specialized regression commands, descriptive statistics for means, proportions and ratios, many survey regression commands, and many panel-data regression commands.

Here is a checklist of what we need to do for multiple imputation:

1. Include in the imputation model all variables that will be used in the analysis model, including the dependent variable.

2. Include auxiliary variables that predict patterns of missingness.

3. Include additional variables that predict a person's score on a variable that has missing values.

4. Drop observations that have variables you do not want to impute because the item was deliberately skipped.

5. Avoid adding variables that have a lot of missing values themselves because including them may increase the uncertainty of your imputations and greatly reduce your degrees of freedom.

6. Explore patterns of missing values to see if there are certain variables or combinations of variables that explain the lion's share of missing values.

7. Although you should include all important variables in the imputation stage, including too many can create estimation problems due to multicollinearity.

8. Generate all interactions and squared variables in the original dataset before the imputation stage.

9. Impute a reasonable number of completed datasets, at least $m = 20$, and experiment with different numbers of imputed datasets.

10. Do not change the imputed values as you move to the analysis stage, even if they seem to be "impossible" values.

11. Use the `mibeta` command if you want pooled estimates of $R^2$ and the $\beta$s and also to obtain the relative efficiency of the value you have used for $m$.

Those who have been plagued by missing-data problems may think these procedures can change lead to gold. They cannot. These procedures should be used with caution, and the MAR assumption should be given serious consideration. Properly specified imputation models, however, can do better than any of the alternative ways of working with missing values.

# 13.7   Exercises

1. Describe a study in which the data are MCAR.

2. Describe a regression model using survey data in which the data are not MCAR but are instead MAR.

3. How can you have an important auxiliary variable that is not related to a person's score? Why is such a variable important to include when doing the imputations?

4. If you use $m = 10$, what statistical results would indicate that you need a much larger $m$?

5. A person says that when you get significant results by using multiple imputation, it is just because you made up data. Give an intuitive explanation of why this argument is wrong.

6. Why do you lose degrees of freedom when you do multiple imputations? When would you lose the most degrees of freedom?

7. Use the dataset we used in chapter 10, `ops2004.dta`. Using this dataset, we saw a multiple regression predicting a person's concern for the environment by using their education, income, involvement in the community, health problems, and how much they feel the environment impacts people's health. Answer each of the following:

   a. How much missing data are there? Which variables contain the biggest number of missing values?

   b. Past research has shown that males are less likely to complete all questions. Evaluate whether gender needs to be included as an auxiliary variable. Why is it a good idea to include gender as an auxiliary variable based on this analysis?

   c. Impute 20 datasets and include the auxiliary variable. How can you do this so that you get the same result if you run your do-file another time?

   d. Use the `mi estimate` prefix command to do the regression and compare this with the results reported in chapter 10.

   e. Why does income have far fewer degrees of freedom than do the other variables?

8. Repeat exercise 7, but use `mibeta` to obtain estimates for $R^2$ and for the $\beta$s.

# 14 The sem and gsem commands

The sem and gsem commands extend everything we have covered with regression and factor analysis. These commands are more fully discussed in terms of structural equation modeling (SEM) for continuous outcome variables and growth curve analysis in Acock (2013). In this chapter, we will show how to use the sem and gsem commands to fit basic regression models and logistic regression models, and we will extend this to an introduction to path analysis including mediation models with direct and indirect effects. We will not cover latent variables, confirmatory factor analysis, or growth curves; see Acock (2013) for more information on these topics.

## 14.1 Ordinary least-squares regression models using sem

The application of sem to models that we have been fitting using the regress command offers two distinct advantages:

1. Stata's SEM Builder provides a production-quality graphic interface that provides an enhanced way to report your regression results using a figure rather than a table. Many readers find that a figure is a great way to understand what is happening in a regression model and a valuable tool for remembering their work.

2. In the presence of missing data when the missing at random (MAR) assumption and the assumption of multivariate normality are both reasonable, the `sem` command provides a much simpler way of handling missing values than the use of multiple imputation that we covered in chapter 13. Sometimes called full-information maximum likelihood (FIML), this method uses all available data and is equivalent to multiple imputation when the number of imputed datasets approaches infinity (Graham et al. 2007). Like multiple imputation, this approach works best with large samples. As with multiple imputation, the FIML estimation makes the less restrictive assumption about the missing values than listwise (casewise) deletion, which is the default with the `regress` command. Where the `regress` command's default of listwise deletion assumes missing completely at random (MCAR), the FIML approach assumes only MAR. This important distinction was discussed in chapter 13.

To show how to use `sem` to represent a basic regression model, we will use data from the Flourishing Families Study (Day and Acock 2013). I have modified the data to simplify the example so generalizations of our findings are problematic; thus our results will not replicate the results obtained by analyzing the unmodified Flourishing Families Study Data. To be in the Flourishing Families Study, the women had to have a child who was approaching adolescence, so this does not represent a nationally representative sample of adult women. We are predicting a mother's body mass index (BMI). Many variables are related to BMI among women. For this example, we have selected five predictors (independent variables). These include the mother's age, the number of children in the household, the natural log of the family income, the mother's education, and how often the family members eat at fast food restaurants. We hypothesize that the BMI is positively associated with age, the number of children, and how often the family eats at fast food restaurants. Conversely, we hypothesize that BMI is negatively associated with the log of household income and the mother's education. If we wanted to use the `regress` command, we would type

```
. use http://www.stata-press.com/data/agis4/flourishing_bmi.dta
. regress bmi age children income educ quickfood, beta
```

| Source   | SS         | df  | MS         |
|----------|-----------|-----|-----------|
| Model    | 2921.16092 | 5   | 584.232183 |
| Residual | 15264.4284 | 442 | 34.5349059 |
| Total    | 18185.5893 | 447 | 40.683645  |

Number of obs = 448
F( 5, 442) = 16.92
Prob > F = 0.0000
R-squared = 0.1606
Adj R-squared = 0.1511
Root MSE = 5.8766

| bmi      | Coef.     | Std. Err. | t     | P>\|t\| | Beta      |
|----------|-----------|-----------|-------|-------|-----------|
| age      | .0849882  | .0471673  | 1.80  | 0.072 | .085525   |
| children | .6167338  | .2881417  | 2.14  | 0.033 | .0983799  |
| incomeln | -1.750445 | .4360436  | -4.01 | 0.000 | -.2080029 |
| educ     | -.6900809 | .2208598  | -3.12 | 0.002 | -.1627779 |
| quickfood| .9260216  | .2496412  | 3.71  | 0.000 | .1716324  |
| _cons    | 37.48497  | 3.890134  | 9.64  | 0.000 | .         |

We obtain an $R^2 = 0.161$, $F(5,442) = 16.92$, and $p < 0.001$. The $p$-values are for a two-tailed test. Because our hypotheses are all one-tailed tests and because the sign of the estimated coefficient is in the same direction as our hypotheses, we cut the $p$-values in half. Thus all the variables have a $p < 0.05$ using a one-tailed test. When using the more conservative two-tailed test, all the variables except for age have a $p < 0.05$, and age is marginally significant, $p < 0.10$. The standardized beta's ($\beta$'s) are fairly small, but all of them are in the predicted direction.

## 14.1.1   Using the SEM Builder to fit a basic regression model

If you type the command `sembuilder` in the Command window, the graphic interface for building `sem` and `gsem` models will open. Figure 14.1 shows part of the SEM Builder as it appears on a Mac.

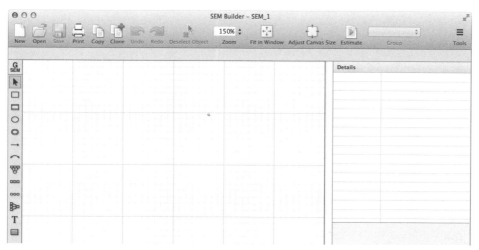

Figure 14.1. SEM Builder on a Mac

The interface is essentially the same when using Windows or Unix. The Mac requires you to click on the Tools icon in the upper-right corner to open the menus, whereas Windows and Unix have the names of the menus distributed across the top of the screen. The menus include Object, Estimation, Settings, and View. Before you start using the Builder, make sure you have the dataset open, which we did by typing

```
. use http://www.stata-press.com/data/agis4/flourishing_bmi.dta
```

The drawing tools appear on the left side of the Builder. Let's start drawing our model. We will do this the hard way at first to get you familiar with the interface. We have one dependent variable, `bmi`, and five predictors. Click on the rectangle tool, □; it is third down on the drawing toolbar on the left side. Next click on the interface where you want to locate each of the six variables. (Hint: look at figure 14.7 to see the finished drawing and how we should arrange the rectangles.) Generally, we place independent variables on the left side of the figure and the dependent variable on the right side of the figure. If your five rectangles for the independent variables are not aligned, you can click on the selection tool, ↖, and drag over the five predictors. Then select Object ▷ Align ▷ Vertical Left (remember on a Mac you first need to click on the Tool icon) from the Builder menu.

I used the blue grid lines to help me set up the figure. This grid should appear by default, but if it does not, you can select View ▷ Show Grid to change the setting. The grid will not appear when you print or copy your drawing.

Next click on the tool with the straight line with an arrow on the right side, →. It is the seventh tool down on the drawing toolbar. You use this drawing tool to draw the paths. If you click on the edge of a rectangle, it turns green. Drag to the dependent variable, and when it turns green, you let go. Notice that the Builder knows to add the error term for your dependent variable. Figure 14.2 shows the diagram thus far:

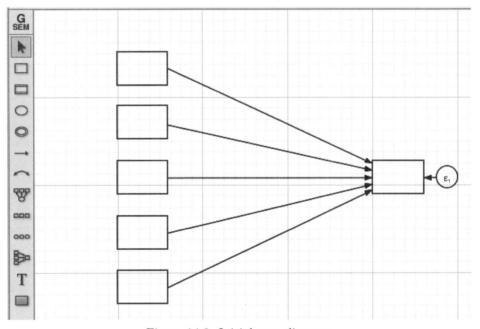

Figure 14.2. Initial SEM diagram

Click on the selection tool and then click on one of the rectangles. Just above the drawing area on the left side of the screen is a box labeled *Variable*. You can type the name of the variable, or you can use the drop-down menu and select the variable from the list. Click on the next rectangle and enter that variable name. Do this so that each rectangle has a variable name inserted. You must use the exact variable names.

One thing is missing. In regression, the independent variables are often correlated. For example, older mothers might have more children, or mothers with more education might eat fast food less often. We should allow for all the independent variables to be correlated. Stata will do this automatically and drawing these does not change the assumptions of the model, but for now let's see how we can insert these correlations. Click on the tool with the curved line with an arrow at both ends, ⌒. When there are several independent variables, entering all of these correlations becomes tedious. I start at the bottom and enter all adjacent variable correlations. That is, `quickfood` up to `educ`, `educ` up to `incomeln`, etc. When I connect `quickfood` to `educ`, going up, three tiny circles appear. One circle appears at each arrow point, and one circle appears to the left of the curved line. The circles at the arrows can be used to move the place where the arrows connect to the rectangles. The circle to the left of the line can be used to move the arc so it is steeper or flatter. If you started with `educ` and drew the arc down to `quickfood`, you could use the tiny circle to move the arc to the other side. Next I start at `quickfood` and link it to `incomeln`, then `educ` to `children`, and then `incomeln` to `age`. I then connect `quickfood` to `children` and `educ` to `children`. Finally, I connect `quickfood` with `age`. Figure 14.3 shows what we have done:

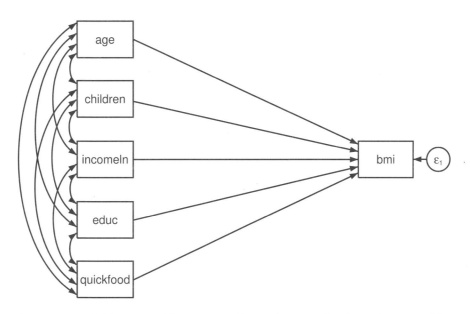

Figure 14.3. Adding variable names and correlations of independent variables

The first few times you add the correlations among the independent variables you may not have it appear as neat as in figure 14.3. You can use the tiny circles to move things around. In this example, there are four arrows attached to each independent variable and you can space them evenly.

We are ready to fit our model. Select Estimation ▷ Estimate... (on a Mac, select Tools ▷ Estimation ▷ Estimate...) from the Builder menu. The first tab in the dialog box that opens, Model, has three estimation options. The *Maximum likelihood* method is similar to the `regress` command in that it uses listwise deletion; it is different in that it uses maximum likelihood estimation rather than ordinary least-squares (OLS) estimation. The *Maximum likelihood with missing values* method indicates to select a FIML solution so that all available information is used. The *Asymptotic distribution free* method does not assume normality. We will check the radio button for *Maximum likelihood with missing values* so that we use all available data.

On the Group tab, we will select *Standard analysis (no groups)*, because we only have one group. If we wanted to fit this model on mothers and on fathers, we could select *Group analysis* and then use `gender` as the *Group variable*. The if/in tab allows you to restrict the sample, such as restricting it to mothers who are under 50 years old or to mothers who live in a certain geographic area. The Weights tab allows you to weight the data by using a weighting variable, if this were appropriate.

The SE/Robust tab provides several options for estimating the standard errors. Neither the *Robust* standard-error type nor the *Bootstrap* standard-error type will help if your parameter estimates are biased, but these types can help estimate unbiased standard errors and thus tests of significance without assuming normality of error terms. It should be stressed that neither *Robust* nor *Bootstrap* will give you unbiased parameter estimates if the assumptions of MAR and normality are violated. If we selected *Bootstrap*, we could select the number of replications, say, 200. This would fit the model on 200 subsamples of our data and then pool those results to obtain unbiased standard errors. This can take considerable time to run. By selecting *Bootstrap*, we also can enter a cluster variable. If we had a sample of 20 children from each of 15 schools, we could enter the school variable name as the cluster variable. If we had clustered data and wanted a robust solution, we could have selected *Clustered robust* as the standard-error type. For our example, we will use the *Default standard errors* type.

On the Reporting tab, check the box for *Display standardized coefficients and values*, which will report beta weights. Figure 14.4 shows our results.

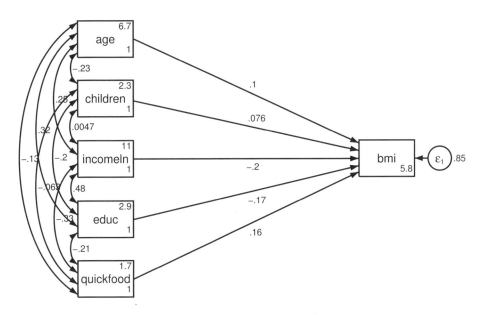

Figure 14.4. Result without any reformatting

These results are a bit overwhelming, and we should eliminate some of the reported coefficients. Select Settings ▷ Variables ▷ All observed.... If we had variable names that were too long to fit in the rectangles, we could change the default size of 0.60 inches by 0.38 inches to be a little wider, say, 0.80 inches by 0.38 inches. On the Results tab under the *Exogenous variables* section, change *Variance* to *None* and *Mean* to *None*. Under the *Endogenous variables* section, change *Intercept* to *None*. Click on OK. The term exogenous refers to independent variables or predictors that are not explained by our model. Their explanation is external to the model. The term endogenous refers to the dependent or outcome variable that is explained by our model.

It would be nice to have all the beta weights have the same number of decimal places. Let's fix these at 3 digits to the right of the decimal place. Select Settings ▷ Connections ▷ Paths.... On the Results tab, click on the Result 1... button in the *Appearance of results* section. In the resulting dialog box, change the *Format* from %7.2g to %5.3f and then click on OK twice.

It is nice to indicate the significance level of each path. We are no longer using OLS regression where *t* tests were used. With maximum likelihood estimation, we rely on *z* tests. The values will be different from the *t*-tests values because they are *z* tests, but also because we obtain different point estimates and standard errors when using all available information (or FIML) in this example. The results would be more similar to the *t* tests had we used the default maximum likelihood estimation, which does listwise deletion. Using the Select tool, ⬉ , click on the first path, which goes from age to bmi.

On the right side of the Builder, the details appear when you click on a parameter estimate in the model. Click on the 0.103 for the path from `age` to `bmi`. We are interested in the "Std. Coef" results: $z = 2.210$ and $p = 0.027$. This standardized coefficient is significantly different from zero based on a two-tailed test and using a 0.05 level. The `regress` command only reports one significance level, and this is valid only for the unstandardized coefficient, although many researchers also use it for the standardized solution as well. The `sem` command offers separate probabilities for the unstandardized and standardized solutions. The `sem` results happen to be stronger than the `regress` results for `age`, but this will not always be the case. For example, the effect of `children` is now weaker and less significant (although both variables are significant using a one-tailed test here and with the OLS regression result).

We can signify the level of significance using one-tailed tests by selecting the Text tool, $^\text{T}$ , from the drawing tools. Click just to the right of the 0.103, and in the resulting textbox, type a single asterisk, *. Click on OK, and then repeat this for each path, entering the appropriate one-tailed level of significance—* for 0.05, ** for 0.01, or *** for 0.001. If the asterisks are not exactly where you want them, you can click on the Select tool, and then click on the asterisk and move it to exactly where you want it. Figure 14.5 shows what we have so far.

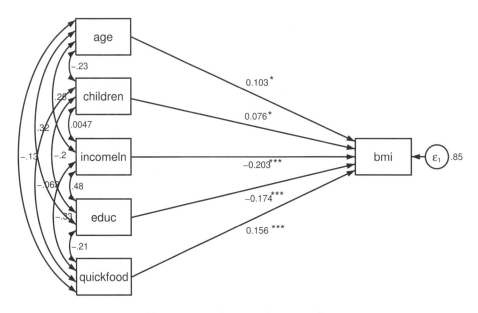

Figure 14.5. Intermediate results

Notice the correlations among our independent variables. With five independent variables, this is messy, and if you had more independent variables, the correlations would be unreadable. We should inspect these correlations ourselves. We can see that more-educated mothers eat at fast food restaurants less frequently than less-educated mothers do. However, including these correlations in the figure is too much detail for many readers. To omit these correlations, simply click on each of them and then press the Delete key. We will soon see that it was not necessary to draw them in the first place.

After deleting the correlations, the figure looks pretty good. We might want to move the independent variables closer to the dependent variable. You might want to do this before adding the asterisks. Click on the Select tool and drag over the set of independent variables. Then click on any one of them and move them all to the right. When we do this, the asterisks are in the wrong places, so we need to click on each of them and move them where they belong.

The figure is looking great at this point, but it is nice to add some additional information to a regression model. We need to show the sample size and $R^2$. When we fit our model, the Builder generated the appropriate Stata command and reported the results in Stata's Results window. Look at the Results window, and you see that the number of observations is 480. To obtain the $R^2$ by using the Builder, select Estimation ▷ Goodness of fit ▷ Equation-level goodness of fit, click on *Equation-level goodness-of-fit statistics (eqgof)*, and click on OK. Returning to Stata's Result window, we see the $R^2$ for `bmi` is 0.152.

Click on the Text tool and then click under the `bmi` variable. In the resulting textbox, enter the information shown in figure 14.6. The textbox automatically wraps text without making a new line, unless you press the Enter key. Therefore, after `z tests`, I pressed the Enter key and inserted five spaces to indent the next line. I did the same after `simplify` in the last line.

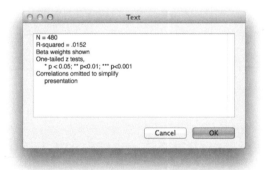

Figure 14.6. The SEM *Text* dialog box in Stata for Mac

When you click on OK, the textbox appears in your figure. Click on the Select tool and then click on the textbox to move it or to change the font. In the box labeled Size, located just above the drawing area on the left side of the screen, change the font size to 8.

Now we are ready to copy the final figure into our report. Click on the selection arrow and drag it over the entire figure, that is, highlight all the rectangles, paths, and text. Next click on the rectangle labeled `quickfood` and drag it toward the lower left corner, leaving a little space to the left and below the drawing. We have the grid lines turned on and each big square represents a square inch. Notice that our figure is about 4 inches wide and about 3.5 inches high—yours may vary. We need to reshape our drawing surface to be $4.5 \times 3.5$ inches. Select View ▷ Adjust Canvas Size..., type 4.5 and 3.5 in the textboxes, and click on OK. From the top toolbar, select the Copy icon, 🗎. In your Word document, select Edit ▷ Paste. Figure 14.7 shows the final result as it would appear in a Word (or other word processing) document.

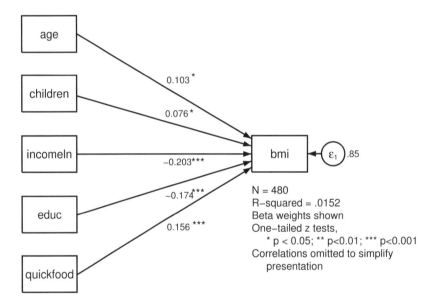

Figure 14.7. Final result

## 14.2   A quick way to draw a regression model and a fresh start

Now that you are more familiar with the SEM Builder, we will look at a great simplification. This works for any estimation method, but we will illustrate it using the `method(ml)` option (the default), which uses listwise deletion. Because we are going to use the maximum-likelihood listwise deletion estimator, the results will be closer to those results of the `regress` command. However, the point of this section is to provide a quicker way to draw a regression model. The method of drawing works regardless of the estimator you are using.

Because we have already drawn a model and changed several settings, such as the number of digits displayed, let's start fresh by restoring the default settings for the Builder. To do this, select Settings ▷ Setting Defaults. When you do this, a message notifies you that the SEM Builder is not empty; click on Yes. We should then close the SEM Builder to complete the resetting process and clear the interface. We can now type the `sembuilder` command again.

Select the third tool up from the bottom, 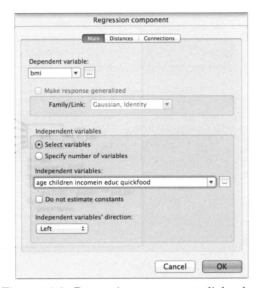, of the drawing toolbar. This tool is designed explicitly for drawing regression models. With this tool highlighted, click on the right side of the SEM Builder about halfway from the top and bottom. Doing this opens the dialog box shown in figure 14.8.

Figure 14.8. Regression component dialog box

Notice that we typed `bmi` as the dependent variable, entered the independent variables, and selected `Left` for the direction of the independent variables. When entering the independent variables, be sure that you use the exact names. Now you can click on OK and have the regression drawn without showing the correlations of the independent variables; see figure 14.9.

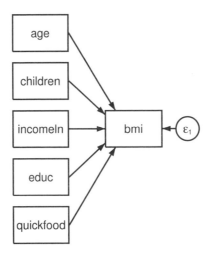

Figure 14.9. Quick drawing of regression model

To fit this model, we select Estimation ▷ Estimate.... In the resulting dialog box, check the radio button for *Maximum likelihood* on the Model tab, and then click on OK. Figure 14.10 shows the results.

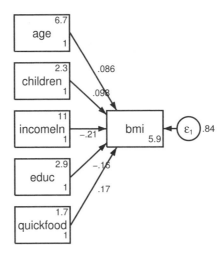

Figure 14.10. Maximum likelihood estimation of model using listwise deletion

Although the diagram does not show the correlations among the independent variables, these correlations are taken into account. If we had selected *Maximum likelihood with missing values*, we would have received a message box saying "Connections implied by your model are not in your diagram: Shall I add them?" We can click on "No" to keep our figure simple. This opens another message box that asks if we want to see the results anyway, and here we click on "Yes". The correlations will be taken into account but not shown in the figure. You can modify the figure the same way as was done in section 14.1.1.

After you have done this a few times, making a figure like this is a relatively quick process. Do you see the value of the figure? It is easier for most readers who are not statisticians themselves to interpret a figure than a regression summary table. We see that education and income are both contributing to lower obesity. Obesity is a significant problem for mothers with limited education and low income. There are many reasons for this, such as the elevated cost of a high quality diet. Eating at fast food restaurants also leads to great risk of obesity. The more often family members eat at fast food restaurants, the higher the mother's estimated BMI. As a writer, it is often easier to tell your story by using a figure like this than it is by using a regression summary table.

### 14.2.1  Using sem without the SEM Builder

A popular cliché is that a picture is worth a thousand words, but in programming, a few words can be a lot quicker than drawing a picture of your model. We can fit models like this without using the SEM Builder. If you examine the Results window within Stata, you will see that the Builder created an **sem** command. These are sometimes hard to read, and there are usually simpler ways to write the **sem** command. To fit a regression model, we use the **sem** command as the first word. The dependent variable, **bmi**, is the second word. This is followed by an arrow and the list of independent variables. In our model, we will use the FIML estimator, so after the comma, we write **method(mlmv)**. We would omit this option if we wanted to do listwise deletion with maximum likelihood estimation. Here is the command:

```
. sem bmi <- age children incomeln educ quickfood, method(mlmv)
```

To obtain the $R^2$, we enter a postestimation command:

```
. estat eqgof
```

That is it. Try it and see the result.

## 14.3  The gsem command for logistic regression

To show how to use **gsem** to fit a logistic regression model, we will dichotomize BMI. Rather than trying to predict a BMI for each mother, we can try to predict whether the mother is obese. Here we define obesity as having a BMI score of 30 or more. There are 111 mothers (24.3%) who have a BMI of 30 or more and 346 (75.7%) who have a BMI of less than 30. We can construct a variable called **obese**, which is coded 1 if the mother is obese and 0 if she is not obese.

```
. recode bmi (0/29.999=0) (30/60=1), gen(obese)
```

Alternatively, we could type

```
. generate obese = bmi>=30 & bmi < .
```

## 14.3.1  Fitting the model using the logit command

In chapter 11, we covered logistic regression, which is appropriate for a binary outcome. We can fit the logistic regression model with the `logistic` command. To help simplify our interpretation, we will obtain odds ratios for a one-unit difference in each variable and for a one standard-deviation difference by using the `listcoef` postestimation command. If you do not already have the user-written command `listcoef` installed, type `search spost9` in the Command window. Then click on the spost9_ado package and install this package. This command was previously discussed in section 11.5. Here are the results:

```
. logit obese age children incomeln educ quickfood
Iteration 0:    log likelihood = -247.43569
Iteration 1:    log likelihood = -221.17297
Iteration 2:    log likelihood = -220.50785
Iteration 3:    log likelihood = -220.50712
Iteration 4:    log likelihood = -220.50712

Logistic regression                            Number of obs   =        448
                                               LR chi2(5)      =      53.86
                                               Prob > chi2     =     0.0000
Log likelihood = -220.50712                    Pseudo R2       =     0.1088
```

| obese | Coef. | Std. Err. | z | P>\|z\| | [95% Conf. Interval] | |
|---|---|---|---|---|---|---|
| age | .0288883 | .0190012 | 1.52 | 0.128 | -.0083534 | .06613 |
| children | .2782205 | .1189227 | 2.34 | 0.019 | .0451362 | .5113047 |
| incomeln | -.6097175 | .1797268 | -3.39 | 0.001 | -.9619756 | -.2574595 |
| educ | -.2011274 | .0917525 | -2.19 | 0.028 | -.3809591 | -.0212957 |
| quickfood | .2603238 | .108617 | 2.40 | 0.017 | .0474383 | .4732093 |
| _cons | 2.301053 | 1.574824 | 1.46 | 0.144 | -.7855441 | 5.387651 |

```
. listcoef
logit (N=448): Factor Change in Odds

  Odds of: 1 vs 0
```

| obese | b | z | P>\|z\| | e^b | e^bStdX | SDofX |
|---|---|---|---|---|---|---|
| age | 0.02889 | 1.520 | 0.128 | 1.0293 | 1.2037 | 6.4187 |
| children | 0.27822 | 2.340 | 0.019 | 1.3208 | 1.3272 | 1.0175 |
| incomeln | -0.60972 | -3.392 | 0.001 | 0.5435 | 0.6299 | 0.7579 |
| educ | -0.20113 | -2.192 | 0.028 | 0.8178 | 0.7389 | 1.5045 |
| quickfood | 0.26032 | 2.397 | 0.017 | 1.2974 | 1.3604 | 1.1822 |

Examining the results of the `listcoef` command, we can see that having more children and eating more meals at fast food restaurants is associated with greater odds of obesity. The odds ratio appears in the column labeled `e^b` because the odds ratio is the exponentiated value of the regression coefficient. Having more education and more income is associated with lower odds of obesity. The mother's age and controlling for the other predictors do not have a significant effect. You may want to review chapter 11 to provide a more detailed interpretation.

The `logit` command is a special application of the generalized linear model. We can obtain the same results by using the `glm` command. The `glm` command requires us to specify the family of our model, `family(binomial)`, and the link function, `link(logit)`. To obtain the odds ratio, we can replay these results by using `glm, eform`. We show this only because the `gsem` command is using the generalized linear model, and we will need to enter both the family and the link function. Here are the `glm` commands and partial results:

```
. glm obese age children incomeln educ quickfood, family(binomial) link(logit)

Iteration 0:   log likelihood = -220.76408
Iteration 1:   log likelihood = -220.50737
Iteration 2:   log likelihood = -220.50712
Iteration 3:   log likelihood = -220.50712

Generalized linear models                         No. of obs      =         448
Optimization     : ML                             Residual df     =         442
                                                  Scale parameter =           1
Deviance         =   441.0142362                  (1/df) Deviance =    .9977698
Pearson          =   429.8899828                  (1/df) Pearson  =    .9726018

Variance function: V(u) = u*(1-u/1)               [Binomial]
Link function    : g(u) = ln(u/(1-u))             [Logit]

                                                  AIC             =    1.011192
Log likelihood   = -220.5071181                   BIC             =   -2257.304
```

| obese | Coef. | OIM Std. Err. | z | P>\|z\| | [95% Conf. Interval] | |
|---|---|---|---|---|---|---|
| age | .0288883 | .0190012 | 1.52 | 0.128 | -.0083534 | .06613 |
| children | .2782205 | .1189227 | 2.34 | 0.019 | .0451362 | .5113047 |
| incomeln | -.6097175 | .1797268 | -3.39 | 0.001 | -.9619756 | -.2574595 |
| educ | -.2011274 | .0917525 | -2.19 | 0.028 | -.3809591 | -.0212957 |
| quickfood | .2603238 | .108617 | 2.40 | 0.017 | .0474383 | .4732093 |
| _cons | 2.301053 | 1.574824 | 1.46 | 0.144 | -.7855441 | 5.387651 |

```
. glm, eform
Generalized linear models                    No. of obs      =       448
Optimization    : ML                         Residual df     =       442
                                             Scale parameter =         1
Deviance       =   441.0142362               (1/df) Deviance =  .9977698
Pearson        =   429.8899828               (1/df) Pearson  =  .9726018

Variance function: V(u) = u*(1-u/1)          [Binomial]
Link function    : g(u) = ln(u/(1-u))        [Logit]

                                             AIC             =  1.011192
Log likelihood  = -220.5071181               BIC             = -2257.304
```

|          |            | OIM       |       |      |             |            |
|---------:|-----------:|----------:|------:|-----:|------------:|-----------:|
| obese    | Odds Ratio | Std. Err. | z     | P>\|z\| | [95% Conf. | Interval]  |
| age      | 1.02931    | .0195581  | 1.52  | 0.128 | .9916814   | 1.068366   |
| children | 1.320777   | .1570704  | 2.34  | 0.019 | 1.04617    | 1.667465   |
| incomeln | .5435044   | .0976823  | -3.39 | 0.001 | .3821372   | .7730129   |
| educ     | .8178083   | .075036   | -2.19 | 0.028 | .6832059   | .9789295   |
| quickfood| 1.29735    | .1409143  | 2.40  | 0.017 | 1.048582   | 1.605137   |
| _cons    | 9.984695   | 15.72413  | 1.46  | 0.144 | .4558716   | 218.6891   |

## 14.3.2  Fitting the model using the gsem command

We draw a figure in the same way we did before, except this time we need to click on the top tool, $\overset{G}{\underset{\text{SEM}}{\blacksquare}}$, from the drawing toolbar. This is the generalized SEM tool, GSEM, which uses the generalized linear model. This tool allows Stata to work with outcomes that are not continuous. In this example, our outcome is binary but it could be a count of a rare event when using a Poisson model, or it could be a nominal outcome that has more than two possibilities, such as religion (Muslim, Christian, Jewish, Hindu, Other), when using a multinomial model. Although we only illustrate the results of a binary outcome, there are many other options. Table 14.1 shows several of these options. Beginning with Stata 13.1, the gsem command can also be used with censored variables.

Table 14.1. Selected families available with gsem

| Dependent or outcome variable | Traditional command | Selected families | Link function |
|---|---|---|---|
| Continuous | regress | gaussian | identity |
| Binary | logit | binomial or bernoulli | logit |
| Binary | probit | binomial or bernoulli | probit |
| Multinomial (nominal) | mlogit | multinomial | logit |
| Ordinal | ologit | ordinal | logit |
| Ordinal | oprobit | ordinal | probit |
| Count | poisson | poisson | log |
| Count | nbreg | nbreg | log |
| Continuous | | gamma | log |

With the GSEM tool selected, click on the add regression tool, ◄. We specify the family and the link function for our estimation by checking the box for *Make response generalized* and selecting Binomial, Logit in the *Family/Link* box. By doing this, we are identifying the variable obese as being in the binomial family and in the logit link. Figure 14.11 is how we represent the model.

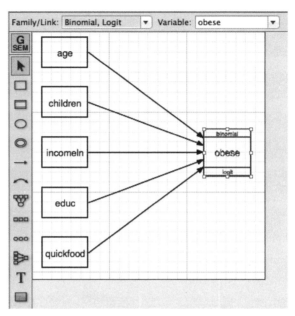

Figure 14.11. A logistic regression model with the outcome, `obese`, clicked to highlight it

Using the Select tool, you can click on the outcome variable, `obese`, as shown in figure 14.11. The top left of the interface then displays a new toolbar that shows *Family/Link*, which currently shows `Binomial, Logit`; you can click on the dropdown arrow to see the other possibilities.

With the GSEM logistic regression model, we cannot use the `method(mlmv)`. The default estimator, `method(ml)`, with listwise deletion is our only choice. We can click on Estimation ▷ Estimate... and then click on OK. This produces the results in figure 14.12 (I changed the format of the paths to `%5.3f`):

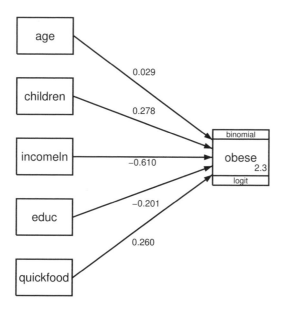

Figure 14.12. Initial results

We can now use the Add Text tool to add asterisks to indicate the level of significance and odds ratios. We obtain the odds ratios by selecting Estimation ▷ Other ▷
Report exponentiated coefficients, selecting *Table of exponentiated coefficients (eform)*
from the *Reports and statistics* box, and then clicking on OK. Doing this runs the estat
eform postestimation command. The odds ratios are not placed on the drawing but
are reported in the Results window. Notice that because a single unit change in age,
incomeln, educ, and quickfood is difficult to interpret, the odds ratios for those variables might be replaced by using the odds ratio for a 1-standard-deviation change in
the predictor rather than a one-unit change. You could obtain these results by running
the logit command followed by the listcoef command. To compute the results from
the gsem command, you need to calculate these results by hand using Stata's display
command.

To calculate the odds ratios from the gsem results, we need the standard deviations
for each independent variable. When running a summarize command to obtain the
standard deviations, we need to make sure that we are using listwise deletion. We use
the if !missing (not missing) restriction to do this and include all variables in our
logistic model. It is necessary to include the comma delimiter between each variable in
this list:

```
. summarize age incomeln educ quickfood
> if !missing(age,incomeln,educ,quickfood,obese)

    Variable |        Obs        Mean    Std. Dev.         Min         Max
-------------+--------------------------------------------------------------
         age |        448    43.10491    6.418665          27          74
     incomeln |        448    8.403083    .7579329    5.298317    11.15625
        educ |        448    4.430804    1.504546           1           7
    quickfood |        448    2.064174    1.182192          .5        11.5
```

For `age` with a 1-standard-deviation change, 6.419, we obtain the odds ratio for a 1-standard-deviation change by using $e^{(B \times \text{SD})}$ or by typing `display exp(.029*6.419)`. We do this respectively for `age`, `incomeln`, `educ`, and `quickfood` by typing the following `display` commands:

```
. display exp(.029*6.419)
1.2046041
. display exp(-.610*.758)
.62978298
. display exp(-.201*1.505)
.73896479
. display exp(.260*1.182)
1.359776
```

These results differ slightly from those of the `listcoef` command, which uses more significant figures. We will use the `listcoef` result because of its greater precision.

Thus a 1-standard-deviation difference in `incomeln` is associated with a 47% reduction in the odds of being obese $[(1 - 0.630) \times 100)]$. In the case of `chldren`, a one-unit difference, that is, one additional child, is meaningful and is associated with a 32.1% increase in the odds of obesity $[(1.321 - 1) \times 100]$.

After inserting the textboxes, I felt the size of the font was too large. I changed the font size to 7 points in the size control box that appears in the Contextual Toolbar at the upper left corner of the Builder when the textbox is selected. If you click on a completed textbox, it opens the dialog box shown in figure 14.13. Here you can modify the text and make other changes to the appearance.

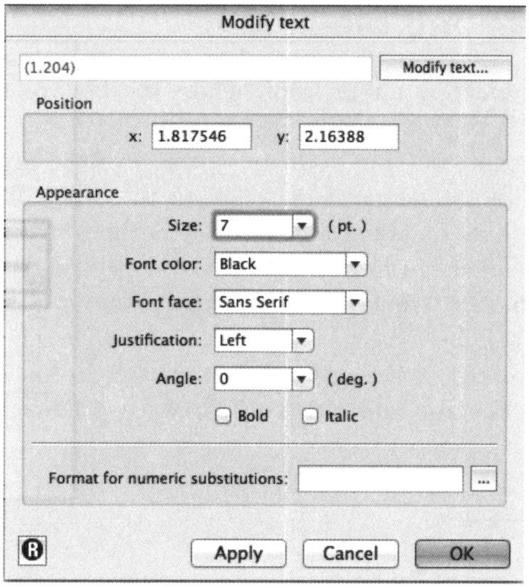

Figure 14.13. Dialog box for changing information in a textbox

Our final generalized SEM diagram is shown in figure 14.14. Asterisks show the significance levels. Odds ratios are given in parentheses.

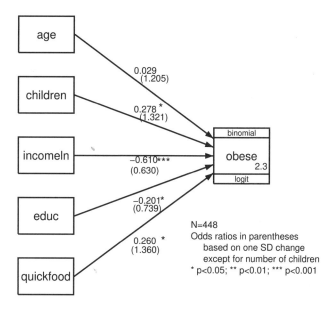

Figure 14.14. Final results for logistic regression

## 14.4   Path analysis and mediation

Path analysis is an extension of regression where we believe there is a causal order to the variables. In our regression model, we had all the predictors on the left and allowed them to be correlated. Can you see how one of the predictors might actually be an endogenous variable, meaning it depends on the other predictors? Consider the quickfood variable. It is reasonable to argue that mothers who have higher income and more education are less likely to have more meals at fast food restaurants. Because of their education, these mothers may be more aware of the health issues associated with eating more meals at fast food restaurants. Mothers from families with more income may be more able to afford options other than eating in fast food restaurants. For simplicity, we will drop the age and chldren variables. Figure 14.15 is our model that does not include the quickfood variable:

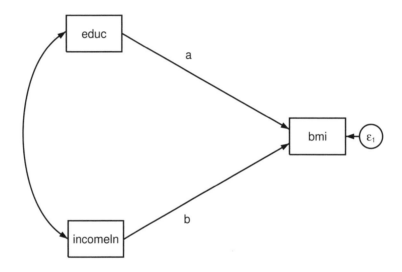

Figure 14.15. BMI predicted without using the quickfood variable

When we add the quickfood variable, we show it is an intervening mechanism (mediator) between educ and incomeln on the one hand and bmi on the other hand. The underlying argument is that education and income lead to eating fast food less often, and eating fast food less often in turn leads to a lower BMI. This appears in figure 14.16.

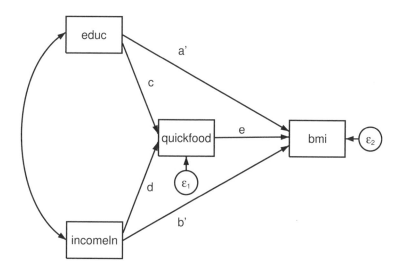

Figure 14.16. A path model with the `quickfood` variable mediating part of the effect of `educ` and `incomeln` on `bmi`

To evaluate whether `quickfood` mediates the influence of `educ` and `incomeln` on `bmi`, we normally fit both of these models. First, make sure you have the dataset opened:

```
. use http://www.stata-press.com/agis4/nourishing_bmi
```

Figure 14.15 had only two paths with the coefficients a and b. If these coefficients are not significant, then we would usually not go farther to test for mediation because there was not a significant effect to begin. There are special cases where we have multiple mediators that have the opposite effects. For example, divorce may indirectly harm the well-being of children mediated by a loss of income, but divorce may indirectly benefit the well-being of children by reducing parental conflict.

If a in figure 14.15 is significant, but a' in figure 14.16 is not significant while c × e is significant (indirect effect of `educ` on `bmi` that is mediated by `quickfood`), we say that `quickfood` mediates the effect of education on BMI (full mediation). If a in figure 14.15 is significant, and a' in figure 14.16 is smaller but also significant while c × e is significant, we say that `quickfood` mediates part of the relationship (partial mediation) between education and BMI. Whether there is full mediation or partial mediation, the mediation model has extended our understanding of the relationship between education and BMI. A similar analysis would be done by using `incomeln`.

When you read an article that says income predicts a person's BMI, you should think of possible mediators like we did here when we considered how often the person eats at fast food restaurants. Often there are two or more possible mediators. For example, food that is better for your health may cost more. People with higher income can afford a better diet. You might include dietary variables such as consumption of food that is high in fat. If we had two possible mediators, namely, eating food high in fat and eating at fast food restaurants, both of these or just one of these might be the mediator.

Path models are often called causal models. This is problematic because we are only examining associations and partial associations. A causal argument could be strengthened if we had longitudinal data. For example, we might measure education and income one year, measure the use of fast food restaurants the second year, and then measure BMI the third year. Instead, we measured all three variables simultaneously. When we measure all the variables concurrently, we cannot rule out the possibility that the true model goes in the opposite direction. A mother who has a very high BMI may have had this obesity problem all of her life, and that may have influenced her education level and her income. There is considerable evidence, for example, of discrimination against overweight women in the workforce. In any event, path models never prove anything with certainty, and this is especially true when the variables are measured concurrently. This is an important limitation that we need to keep in mind so that we do not overstate our findings. Used with caution, however, path models can enrich our understanding of influence processes. Path models offer a more elaborate explanation of causal processes and may be consistent or inconsistent with the observed data.

To fit a causal model that involves mediation, it is common to fit the model without the mediator or mediators first. For example, if a in figure 14.15 were not significant, then many researchers would not go on because they would conclude there was not a significant effect to be mediated. As noted earlier, there are special cases where this is not the case. The standardized results for the model that were shown in figure 14.15 are seen in figure 14.17:

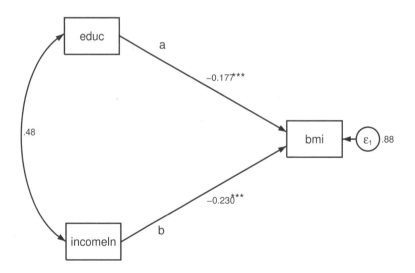

Figure 14.17. Direct effects without the mediator

We see that both education and income have significant negative effects on BMI. We meet our first goal of having **a** and **b** both statistically significant. Next we fit the model that includes the mediation term. In figure 14.18, I had to click on $\epsilon_1$, drag it to the right, and then click on the path from it to `quickfood` and move the line to improve the alignment. I also moved the letters to improve their alignment with the standardized regression coefficients, $\beta$'s.

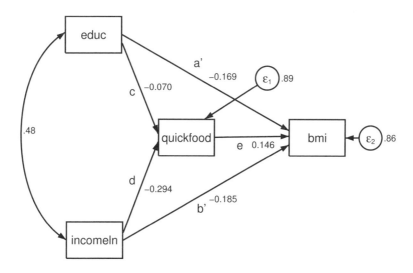

Figure 14.18. Final mediation model

We can obtain the indirect effects by calculating the product of the path coefficients. That is, the indirect effect of education on BMI that is mediated by `quickfood` is $c \times e = 0.070 \times 0.146 = -0.010$, and the indirect effect of income on BMI that is mediated by `quickfood` is $d \times e = -0.294 \times 0.146 = -0.043$. However, this does not tell us whether these indirect effects are statistically significant. We can obtain the indirect effects and their significance by clicking on Estimation ▷ Testing and CIs ▷ Direct and indirect effects. In the resulting dialog box, we select *Decomposition of effects into total, direct, and indirect (teffects)*. We click on *Report standardized effects* and then click on OK. Looking at the Results window, we see a section of `Direct effects`, a second section of `Indirect effects`, and a final section of `Total effects`.

We see that the indirect effect of `educ` through `quickfood` on `bmi` is $-0.010$ and this is not significant (as a caution, we should note that the test of significance is actually for the unstandardized solution). The indirect effect of `incomeln` through `quickfood` on `bmi` is $-0.043$ and this is significant, $p = 0.005$. We summarize these results in table 14.2. We include the total effect in the table that is simply the sum of the direct and indirect effects.

Table 14.2. Direct and indirect effects of mother's education and family income on her BMI

| Relationship | Direct effect | Indirect effect | Total effect |
|---|---|---|---|
| Education on BMI | $-0.169^{***}$ | $-0.010^{ns}$ | $-0.179^{***}$ |
| Income on BMI | $-0.185^{***}$ | $-0.043^{**}$ | $-0.228^{***}$ |

With these results, we can go back to our question of whether eating in fast food restaurants mediates the effect of education and income on a mother's BMI. Using the rules for whether mediation is partial, full, or insignificant, we will first look at the effects of education on BMI without controlling for `quickfood`. Education has a highly significant direct effect on a mother's BMI without controlling for `quickfood`. The more education she has, the lower her BMI ($\beta = -0.169^{***}$). However, there is no significant indirect effect mediated of education that is mediated by `quickfood` ($\beta = -0.010^{ns}$).

Now consider the effects of income on the mother's BMI. The larger the family income, the lower the mother's BMI, ($\beta = -0.185^{***}$) when we do not control for `quickfood`. There is also a small but statistically significant indirect effect mediated by eating at fast food restaurants ($\beta = -0.043^{*}$). When we compare these results to figure 14.15 where there was no mediation, the direct effect was larger ($\beta = -0.230^{***}$) than the $\beta = -0.185$ we have for figure 14.16. Thus we conclude that there is partial mediation whereby part of the direct effect of income on BMI is mediated by eating at fast food restaurants. Although significant, we should acknowledge that the indirect effect is quite small.

## 14.5    Conclusions and what is next for the sem command

This chapter has just given you a taste of what the `sem` command can do. If you are fitting more complex path models, you should also read the companion book; see Acock (2013). Nonetheless, what we have covered here is useful for presenting results for regression and logistic regression models that will have greater potential for impact on your reader than would presenting tabular results. Also there are many path models that can enrich a simple regression model by inserting potential mediators. Whenever you add a significant mediator to your model, you are helping explain the influence process. Consider the effects of divorce on child well-being. Generally, we have seen a negative effect where divorce is associated with somewhat worse child well-being. We can add to our understanding by thinking of possible mediators. What happens after divorce that might lead to problems for the children? Perhaps some things happen that will be good for the children.

- Possible adverse mediators

  1. Loss of household income post divorce
  2. Difference in peer group educational aspirations

- Possible positive mediators

  1. Reduced parental conflict
  2. Enhancement in mother–child relationship

How could we put all of these together? Figure 14.19 shows what the model predicts. Divorce is shown to have a direct effect on child well-being, but to also have two negative indirect effects mediated by a loss in income and a difference in the peer group educational aspirations, plus two positive indirect effects that include reduced parental conflict and enhancement of the relationship between the mother and child.

When we have endogenous mediators that do not have a causal relationship specified between them, it is important to consider whether the errors should be correlated. Otherwise, our model will not fit the data very well. For example, a loss of income and a difference in peers may be fully explained by divorce (errors would be zero in this case), but more likely there are other variables that account for these variables and the correlation of the error terms provides for this possibility.

What's next? The sem command has so many possibilities that it is hard to provide more than a rudimentary inventory. Acock (2013) discusses many of these extensions. Consider the model in figure 14.19; it is possible for the paths for daughters to be quite different from what they are for sons. Perhaps the mother–child relationship might become stronger for the daughters but not for the sons. Perhaps reduced parental conflict is more important for daughters than it is for sons. The sem command allows us to simultaneously fit the model for daughters and sons and test for all possible differences.

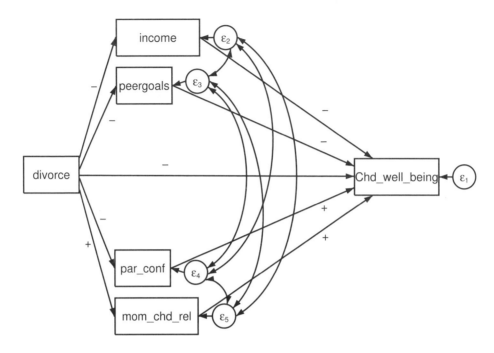

Figure 14.19. More complex path model

We have limited our discussion to observed variables. The **sem** command allows us to work with latent variables, as well as use confirmatory factor analysis models where we specify a specific form for a factor solution. We might be interested in internal and external political efficacy. We would specify observed indicators of each and confirm or disconfirm this model. Extending this, we can combine confirmatory factor analysis and our path models (the structural model) where instead of linking observed variables, we are linking latent variables.

## 14.6  Exercises

1. Using gss2002_chapter10.dta, fit the model using all the variables except for polviews.

   a. Do this using the **regress** command.

   b. Do this using the **sem** command and the SEM Builder.

   c. List the commands as you would enter them in a do-file.

2. Use a subset of data from the National Longitudinal Survey (NLS) of women who were aged 14–28 in 1968. You can obtain these data by typing

   . use http://www.stata-press.com/data/r13/regsmpl.dta

Predict wages by using a person's age, their highest grade completed, whether they live in a non-SMSA (rural) area, whether they are from the south, their total years of work experience, and their years at this job.

    a. Fit the regression model using the `regress` command (standardized) and interpret the results.

    b. Fit the regression model using the `sem` command and the SEM Builder, showing the results in a diagram. Interpret these results.

    c. Fit the regression model again using the `method(mlmv)` option of `sem`. Explain what this option does.

3. Redo exercise 2 from chapter 11 using the SEM Builder.

    a. Show your results in the figure and include both the significance level and the odds ratios.

    b. Interpret the results.

4. Open http://www.ats.ucla.edu/stat/stata/faq/pathreg.htm. This page was developed by the Institute for Digital Research and Education at UCLA. Fit the model using the SEM Builder.

    a. Does math fully mediate, partially mediate, or not significantly mediate the effect reading and writing have on science?

    b. Present a table summarizing your findings.

# A What's next?

## A.1    Introduction to the appendix

The goal I had in writing this book was to help you learn how to use Stata, including creating a dataset, managing the dataset, changing variables, creating graphs and tables, and doing basic data analysis. There is much more to learn about each of these topics, as well as about entirely new topics. At this point, you are ready to pursue these more advanced resources.

The purpose of this concluding chapter is to give you some guidance on what material is most useful as you develop greater expertise using Stata. What will give you a quick start? What requires a lot of statistical background? What are the most accessible resources? A word of caution as we start: new supporting resources appear regularly, so the suggestions here are current only at the time this book is published.

## A.2    Resources

Many resources can help you expand your knowledge of Stata. More importantly, many of these resources are absolutely free and can be obtained over the Internet, including some online "movies" about specialized Stata techniques. Other resources include books about Stata and online courses.

## A.2.1   Web resources

The premier web resource about Stata is at UCLA. They have an extraordinary webpage at http://www.ats.ucla.edu/stat/stata/. It has movies and extensions beyond what I have included in this book. For example, I only briefly mentioned how to work with complex samples that involve clusters, stratification, and weighting. The UCLA webpage has a link to a pair of movies about how to work with complex surveys. Some of the content is based on earlier versions of Stata, but even that content can be helpful. Stata maintains remarkable consistency between versions, with new versions simply adding capabilities. The UCLA webpage also has many links to statistics and data-analysis courses where you can obtain lecture information. It has many Stata commands that you can install on your machine to simplify your work. It even has Stata do-files that match examples in a few of the standard statistics textbooks. Much of what this webpage does not have itself appears in the links it provides to other sources.

Stata maintains a webpage on resources for learning Stata at http://www.stata.com/links/resources1.html, and this is an excellent place to start your resource search. This webpage shows you links to a wide variety of sources that can help you learn Stata and extend your knowledge of Stata. Stata has a searchable frequently-asked-questions (FAQs) webpage at http://www.stata.com/support/faqs/. This page includes many examples of how to do different procedures. Although some of these examples are extremely technical, many of them are accessible to a person who has completed this book. The way these pages work through examples and help you interpret results is way ahead of what we usually see in FAQs support.

We have installed a few commands from a webpage at Boston College. You can check out the full list of available commands at http://ideas.repec.org/s/boc/bocode.html. This site is often referred to by Stata users as the "SSC" (Statistical Software Components). This is the largest collection of user-written Stata commands and is maintained by Christopher F. Baum. For example, go down the list of commands for 2005, and find a command called `optifact`. This webpage has a brief description of the command, and when you click on the link, it takes you to a more detailed description. From there, you can click on the name of the author, Paul Millar, to get a list of papers he has written about this procedure. All of these commands are made available at no cost, and collectively, they represent a considerable extension of the basic capabilities of Stata itself. To install a package from the SSC, simply type `ssc install` *filename*, for example, `ssc install optifact`. The highly developed, user-driven extensibility of Stata is a feature that sets it above competing statistical analysis software.

The downloads of commands from the Statistical Software Components website have increased dramatically over the last 10 years. Figure A.1 shows the number of downloads and the number of abstracts that have been viewed. In 2000, there were fewer than 5,000 files downloaded, and in 2013, there were 100,000 files downloaded. The Statistical Software Components archive is an extraordinary resource for all Stata users.

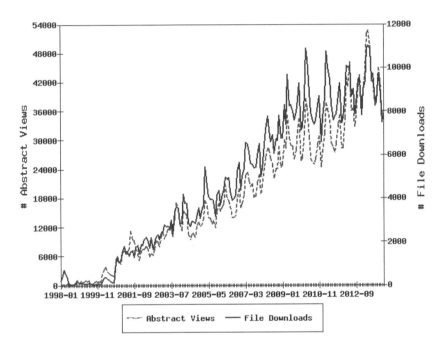

Figure A.1. Growth of downloads of files from Statistical Software Components (source: http://logec.repec.org/scripts/seriesstat.pf?item=repec:boc:bocode)

Stata also has a Facebook page and a Twitter resource. You can find these at http://www.facebook.com/StataCorp and http://twitter.com/stata, respectively. In addition, there is a very valuable Stata Blog at http://blog.stata.com. The Stata Blog has entries that show you how to get the most out of Stata and features new and improved capabilities. Many of these entries have been written by the top authorities at Stata, including William Gould, President of Stata, and Vince Wiggins, Vice President for Scientific Development. You can receive automatic updates of this blog by subscribing to the RSS feed.

There is a Stata listserver, Statalist, to which people submit questions and anybody who wants to can provide an answer. It is hosted at the Harvard School of Public Health and maintained by Marcello Pagano. You can find details on how to subscribe at http://www.stata.com/statalist/. You probably want to subscribe as a digest, which gives you one or two messages per day, each of which contains many questions and answers. If you do not pick the digest option, you might get 20–50 or more messages per day. Much of the content on Statalist is for professional programmers and statisticians. However, the subscribers are often willing to answer questions from beginners, and many of these answers would cost you hundreds of dollars if you were paying for a consultant. The list will also keep you aware of new commands that you can install.

Stata is a completely web-aware package, and web-based resources are constantly expanding. If you enter the command `search` followed by a keyword, Stata will search the web for relevant information. You might try it with a keyword, such as `search missing`, if you were concerned about having a lot of missing values in your dataset. Stata will give you a list of relevant resources followed by a list of commands you can install that will help you work with missing values.

We have discussed only a few of the available resources on the web. By the time you have checked these out, you will see more resources for your special interests and needs. Fortunately, many of these webpages have links to other webpages.

## A.2.2  Books about Stata

Stata has a collection of reference manuals containing over 11,000 pages of information that includes details about each command, detailed examples, calculation formulas, and references to other literature. These reference manuals are available as PDFs that can be accessed from within Stata. Printed manuals are available either separately or as a set from the Stata Bookstore website at http://www.stata.com/bookstore/documentation.html.

Many universities participate in a special program that Stata offers, the GradPlan, which lets you buy the printed manuals at a reduced price. Even at the full price, however, they are inexpensive compared with other books about statistics and software. The manuals are among the best provided by a software company. In my experience, the books that Stata Press publishes are less expensive if you buy them directly from StataCorp than they are if you buy them online from other sources. Before buying them elsewhere, check the Stata Press pricing.

All reference manuals are available as PDF files and installed when you install Stata. These manuals are an excellent resource, and many of the examples should be understandable now that you have finished this book. The examples for each command typically begin with a fairly simple problem, and subsequent examples deal with more complex applications of the command.

The *Stata Journal* is a peer-reviewed, indexed journal on Stata, which is published quarterly. The editors are H. Joseph Newton and Nicholas J. Cox. It includes articles on commands people have written that you might want to install; tutorials on how to do different tasks using Stata; articles on statistical analysis and interpretation, data management, and graphics; and book reviews. You can find out more about it at http://www.stata-journal.com/.

Several excellent books about Stata are available. In 2012, Stata Press published the third edition of the English-language version of Ulrich Kohler and Frauke Kreuter's German book *Data Analysis Using Stata*; see http://www.stata-press.com/books/daus3.html. This book builds on what I have done in *A Gentle Introduction to Stata* and covers some more-advanced data-analysis techniques. It goes much further than I did on programming Stata. Lawrence Hamilton published a book on Stata 12 in 2013, *Statistics with Stata*. The publisher is Brooks/Cole, and the book is available from the Stata Bookstore; see http://www.stata.com/bookstore/sws.html. This is a useful reference book, and you can quickly find commands that do specific tasks. It has a simple organization and many interesting empirical examples.

You have seen how graph commands are the longest and most complicated. The *Stata Graphics Reference Manual* is competently written, but many find it hard to follow. Remember, if a picture is worth a thousand words, it takes a lot of words to tell a computer how to make a good picture of your data. Michael Mitchell, who was previously associated with the UCLA Stata Portal, published a book in 2004 with Stata Press called *A Visual Guide to Stata Graphics* (now in its third edition); see Mitchell (2012) and http://www.stata.com/bookstore/vgsg3.html. This book shows a few graphs on each page, and you scan the book until you see something similar to what you want to do. Next to the picture, you find the Stata command that produced the graph. This does not make use of the dialog system, but you can use his examples to enhance the graphs you make using the dialogs.

J. Scott Long published a book, *Workflow in Data Analysis Using Stata*, in 2009 that is available from Stata Press at http://www.stata-press.com/books/wdaus.html. If you are going to create datasets and manage them, this is the next book you should read. It provides sage advice from a senior scholar and expert Stata user on how to manage complex datasets. A person who masters these skills is worth his or her weight in gold to a research team. Long will take you far beyond what we have covered in this book. A great complement to *Workflow in Data Analysis Using Stata* is a book by Michael Mitchell, *Data Management Using Stata: A Practical Handbook*, which was published by Stata Press in 2010 and is available at http://www.stata-press.com/books/dmus.html.

In 2006, Stata Press published a book, *Regression Models for Categorical Dependent Variables Using Stata, Second Edition*, written by J. Scott Long and Jeremy Freese; see http://www.stata-press.com/books/regmodcdvs.html. This book builds on chapter 13 (logistic regression), providing extensions to multinomial regression, ordered regression, Poisson regression, and zero-inflated models. It is extremely accessible given the complexity of the topic. One of the most important areas of development in Stata is multilevel analysis. Although this extends beyond the scope of my book, this type of analysis is an important extension to what we did cover.

The revised edition of *Microeconometrics Using Stata* by A. Colin Cameron and Pravin K. Trivedi (2010) is also published by Stata Press and is available at http://www.stata-press.com/books/musr.html. If you are going to do more-advanced statistics than we covered in this book, Cameron and Trivedi provide a remarkably accessible guide to a wide range of procedures and show you how to use them with Stata do-files. Full understanding of econometrics requires considerable advanced background in mathematics, but most of the topics Cameron and Trivedi discuss can be understood and applied with a minimum of mathematics. Even though the statistical concepts are advanced, the elegance of Stata makes the application of them quite straightforward. You do not need to be an economist to use this book because most of what is covered is as applicable to political science, sociology, criminal justice, and human development as it is to economics.

A special strength of Stata is that users can write their own commands. An advanced user who finds himself or herself doing some combination of actions by using several commands can decide to write one command that combines these commands. Another user might write a new command to add features to an existing Stata command. For most statistical software packages, this would be a major challenge, but it is much easier to do in Stata. Christopher Baum, in addition to maintaining the Statistical Software Components website, has written a highly accessible book to assist you in writing your own commands: *An Introduction to Stata Programming*, published by Stata Press in 2009 and available at http://www.stata-press.com/books/isp.html.

A specialized topic we have not covered involves working with complex samples. Stata has strong capabilities for this type of analysis. Many national surveys use stratified or cluster sample designs and deliberately oversample certain groups that would otherwise have too few observations for analysis. If you need to use such a survey for your research, you should contact the group that designed the survey and get their recommendations on how to manage the sample design. You will need to learn how to implement their advice in Stata. A useful introduction to what you need to know is available online from the UCLA portal:

http://www.ats.ucla.edu/stat/stata/seminars/svy_stata_8/default.htm

This information is a bit dated but is still a great place to start. Once you have been through this tutorial, you should move on to the *Stata Survey Data Reference Manual*.

If you are fitting complex, multilevel, or longitudinal models, Stata has a rich collection of panel-data (longitudinal data) commands that are described in the *Stata Longitudinal-Data/Panel-Data Reference Manual*. For an example of a multilevel model, you might have individuals nested in families that are nested in communities. With such data, all individuals in a given family would have the same score on family variables, such as household income. Also, all families in a given community would have the same score on community-level variables, for example, community crime rate or unemployment rate. The resulting lack of statistical independence requires more-complex procedures than we have covered in this book. Sophia Rabe-Hesketh and Anders Skrondal's *Multilevel and Longitudinal Modeling Using Stata, Third Edition* (2012) is available from Stata Press at http://www.stata-press.com/books/mlmus3.html. It is a truly remarkable treatment of this advanced topic. Although this is an extremely advanced and innovative treatment of the topic, they use great examples from published research. They show you step by step how to apply Stata to properly analyze the data.

Consult the Stata Bookstore, http://www.stata.com/bookstore/statabooks.html, to see a list of all the available books about using Stata. These include books on many special topics.

## A.2.3   Short courses

Stata offers on-site training and public training sessions. On-site training is useful for Stata users of all levels, whether your site is new to Stata or needs an in-depth look at a specific feature of Stata. On-site training is taught at your facility by expert StataCorp personnel. If you are interested in arranging on-site training, see http://www.stata.com/training/onsite.html for additional details.

Public training courses are taught by various senior members of the StataCorp staff at sites around the country. Courses range from learning to use Stata effectively to learning to use Stata for subject-specific topics, like multilevel/mixed models and multiple imputation. To view the list of available courses, dates, and other details, see http://www.stata.com/training/public.html.

Stata provides NetCourses to help you learn to use Stata more effectively. These courses include two that are designed to introduce you to data management, analysis, and programming with Stata, and a third, more advanced programming course. You can enroll in a regularly scheduled NetCourse or in NetCourseNow, which allows you to take the course at your own pace. These courses may be a little too advanced for a true beginner, but those who have completed this book will find them useful, as will those who have expertise in a competing statistical package. These courses are useful if you already know the statistical content and if you are switching to Stata from some other package such as SYSTAT, SAS, or IBM SPSS Statistics. Now that you have completed this book, you might want to pursue these NetCourses. You will be given weekly readings and assignments, and you will have a web location where you can ask for clarification and interact with other students. For more information, consult the Stata NetCourse website at http://www.stata.com/netcourse/.

Many individual instructors provide supporting documentation for their own students, and they have placed it on the web where anybody has access to it. One such instructor is Richard Williams, an associate professor of Sociology at Notre Dame, who has data and programs that cover a one-year graduate course in social statistics; see http://www.nd.edu/~rwilliam/stats/StataHighlights.html. Williams' page is especially useful for people who have used IBM SPSS Statistics but are switching to Stata.

## A.2.4   Acquiring data

So you want to use Stata. Where can you get data? National funding organizations that support social-science research expect researchers to make the data they collect available to others. This is a professional and ethical obligation that researchers have. As a result, you can acquire many national datasets at little or no cost. When you read an article that uses a dataset, you can do a web search on the name of the dataset

with a search engine, such as Google. For example, I used a subset of variables from
NLSY97 (National Longitudinal Survey of Youth, 1997). Enter NLSY97 in your search
bar, and it will show you a link to a home page for the U.S. Department of Labor
(http://www.bls.gov/nls/nlsy97.htm). Here you can download the documentation and
data for all waves of this panel survey. They even provide an extraction-software program that will help you select variables and generate a Stata dataset. There is no
charge to do this, unless you want a hard copy of documentation (and even that charge
is minimal). Not all datasets are free, but the cost is usually minimal.

There are several clearinghouses that archive datasets. One of the best is the Inter-
University Consortium for Political and Social Research (ICPSR), located at the University of Michigan. Many universities pay an annual fee for membership in this organization. If you are affiliated with an organization that is a member, your organization has a set of numbers that identifies each computer on its network, and this
range of numbers is recorded at ICPSR. If you are eligible, go to the ICPSR webpage
(http://www.icpsr.umich.edu/). Enter a keyword and do a search. You might enter "recidivism", for example. This will give you a list of all the surveys that involve recidivism
or include questions about it. You can then download the dataset and documentation.

Although ICPSR is putting new datasets in the Stata format and updating some of
the more widely used older datasets to Stata format, this is certainly not true for all the
older datasets. SAS and IBM SPSS Statistics were used for statistical analysis for many
years before Stata was developed, and the dataset you want may be in one of those two
formats. There is a simple solution. Stat/Transfer is a program you can buy from Stata
or from Circle Systems, the publisher of Stat/Transfer (http://www.stattransfer.com/);
it converts datasets from one format to another. We discussed this in an early chapter,
and you should make use of it.

When working with national datasets, you can run into some limitations in Stata.
One possible limitation is the number of variables you can have in a dataset. With
Stata/IC, you are limited to 2,047 variables, and with Stata/SE, you are limited to
32,767 variables. Both of these numbers sound huge for any study, but some large
datasets can exceed these numbers.

## A.3  Summary

At the beginning of this book, I assumed that you had no experience using a computer
program to create and manage data or to do data analysis and graphics. If you already
had experience using another software program and had a strong statistical background,
my gentle approach may have been a bit too slow at times. I deliberately made few
assumptions about your background and decided to build from the simplest applications
to the more complex.

You may encounter specialized topics in your statistics books that I have not em-
phasized here. Having gone through this book, you may be able to figure out what to
do on your own. Suppose that you need to do what is called a path analysis. You might
see an example like figure A.2.

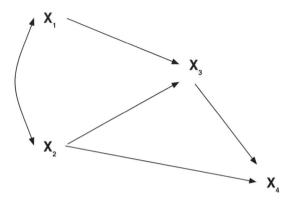

Figure A.2. A path model

Studying this figure, you see that $X_3$ depends on $X_1$ and $X_2$ and that $X_4$ depends on $X_2$ and $X_3$. Because we see a curved line with an arrowhead on both ends linking $X_1$ and $X_2$, we acknowledge that we cannot regress $X_2$ on $X_1$ or vice versa because the relationship is that they are just correlated. We are not assuming $X_1$ causes $X_2$ or that $X_2$ causes $X_1$, just that they are correlated.

Your statistics book tells you that the path coefficients are standardized $\beta$s. You might guess that we could fit this model using

```
. correlate X1 X2
. regress X3 X1 X2, beta
. regress X4 X2 X3, beta
```

The correlation would go on the curved line. The standardized $\beta$s would go on the straight lines. You would also want to report both $R^2$ values. One problem you might have is that if there are different $N$s for each variable because of missing values, you would want to do one of the following two things: 1) eliminate anybody who is missing values on any of the variables before doing the analysis, or 2) use multiple imputation, as discussed in chapter 13, to use all available data.

This example is included here to show you that you have learned a lot and you should be ready to attack specialized statistical questions. Now you can reflect on what you have learned. We covered creating graphs, charts, tables, statistical inferences for one or two variables, analyses of variance, correlations, regressions, multiple regressions, and logistic regressions; entering data; labeling variables and values; managing datasets; generating new variables; and recoding variables. This is a lot of material. At this point, you know enough to do your own study or to manage a study for a research team. Although this book has been designed more for learning Stata than as a reference manual, the extended index can help you use it in the future as a reference manual.

You now have a valuable set of skills! Congratulations!

# References

Acock, A. C. 2013. *Discovering Structural Equation Modeling Using Stata*. Rev. ed. College Station, TX: Stata Press.

Acock, A. C., and D. H. Demo. 1994. *Family Diversity and Well-Being*. Thousand Oaks, CA: Sage.

Agresti, A., and B. Finlay. 2009. *Statistical Methods for the Social Sciences*. 4th ed. Englewood Cliffs, NJ: Prentice Hall.

Allison, P. D. 2001. *Missing Data*. Thousand Oaks, CA: Sage.

Altman, D. G. 1991. *Practical Statistics for Medical Research*. London: Chapman & Hall/CRC.

Baum, C. F. 2009. *An Introduction to Stata Programming*. College Station, TX: Stata Press.

Buis, M. L. 2010. hangroot: Stata module to create a hanging rootogram comparing an empirical distribution to the best fitting theoretical distribution. Statistical Software Components, Department of Economics, Boston College. http://ideas.repec.org/c/boc/bocode/s456886.html.

Bureau of Labor Statistics. 2005. NLSY97 Questionnaires and Codebooks. http://www.bls.gov/nls/quex/y97quexcbks.htm.

Cameron, A. C., and P. K. Trivedi. 2010. *Microeconometrics Using Stata*. Revised ed. College Station, TX: Stata Press.

Carlin, J. B., J. C. Galati, and P. Royston. 2008. A new framework for managing and analyzing multiply imputed data in Stata. *Stata Journal* 8: 49–67.

Cohen, J. 1988. *Statistical Power Analysis for the Behavioral Sciences*. 2nd ed. Hillsdale, NJ: Erlbaum.

———. 1992. A power primer. *Psychological Bulletin* 112: 155–159.

Cox, N. J. 2014. egenmore: Stata modules to extend the generate function. Statistical Software Components, Department of Economics, Boston College. http://ideas.repec.org/c/boc/bocode/s386401.html.

Dattalo, P. 2008. *Determining Sample Size: Balancing Power, Precision, and Practicality*. New York: Oxford.

Day, R. D., and A. C. Acock. 2013. Marital well-being and religiousness as mediated by relational virtue and equality. *Journal of Marriage and Family* 75: 164–177.

Graham, J. W., A. E. Olchowski, and T. D. Gilreath. 2007. How many imputations are really needed? Some practical clarifications of multiple imputation theory. *Prevention Science* 8: 206–213.

Hamilton, L. C. 2013. *Statistics with Stata (Updated for Version 12)*. Boston, MA: Brooks/Cole.

Harel, O. 2009. The estimation of $R^2$ and adjusted $R^2$ in incomplete data sets using multiple imputation. *Journal of Applied Statistics* 36: 1109–1118.

Hilbe, J. 2008. oddsrisk: Stata module to convert logistic odds ratios to risk ratios. Statistical Software Components, Department of Economics, Boston College. Http://ideas.repec.org/c/boc/bocode/s456897.html.

Horton, N. J., S. R. Lipsitz, and M. Parzen. 2003. A potential for bias when rounding in multiple imputation. *American Statistician* 57: 229–232.

Jann, B. 2013. fre: Stata module to display one-way frequency table. Statistical Software Components, Department of Economics, Boston College. http://ideas.repec.org/c/boc/bocode/s456835.html.

———. 2014. center: Stata module to center (or standardize) variables. Statistical Software Components, Department of Economics, Boston College. http://ideas.repec.org/c/boc/bocode/s444102.html.

Kenward, M. G., and J. R. Carpenter. 2007. Multiple imputation: Current perspectives. *Statistical Methods in Medical Research* 16: 199–218.

Kohler, U., and F. Kreuter. 2012. *Data Analysis Using Stata*. 3rd ed. College Station, TX: Stata Press.

Landis, J. R., and G. G. Koch. 1977. The measurement of observer agreement for categorical data. *Biometrics* 33: 159–174.

Lawshe, C. H. 1975. A quantitative approach to content validity. *Personnel Psychology* 28: 563–575.

Long, J. S. 2009. *The Workflow of Data Analysis Using Stata*. College Station, TX: Stata Press.

Long, J. S., and J. Freese. 2006. *Regression Models for Categorical Dependent Variables Using Stata*. 2nd ed. College Station, TX: Stata Press.

Mitchell, M. N. 2004. *A Visual Guide to Stata Graphics*. College Station, TX: Stata Press.

———. 2010. *Data Management Using Stata: A Practical Handbook*. College Station, TX: Stata Press.

———. 2012. *A Visual Guide to Stata Graphics*. 3rd ed. College Station, TX: Stata Press.

Rabe-Hesketh, S., and A. Skrondal. 2012. *Multilevel and Longitudinal Modeling Using Stata*. 3rd ed. College Station, TX: Stata Press.

Royston, P. 2004. Multiple imputation of missing values. *Stata Journal* 4: 227–241.

———. 2009. Multiple imputation of missing values: Further update of ice, with an emphasis on categorical variables. *Stata Journal* 9: 466–477.

Royston, P., J. B. Carlin, and I. R. White. 2009. Multiple imputation of missing values: New features for mim. *Stata Journal* 9: 252–264.

Rubin, D. B. 1987. *Multiple Imputation for Nonresponse in Surveys*. New York: Wiley.

Schafer, J. L. 1997. *Analysis of Incomplete Multivariate Data*. Boca Raton, FL: Chapman & Hall/CRC.

Shultz, K. S., D. J. Whitney, and M. J. Zickar. 2014. *Measurement Theory in Action: Case Studies and Exercises*. 2nd ed. New York: Routledge.

StataCorp. 2013a. *Stata 13 Multiple-Imputation Reference Manual*. College Station, TX: Stata Press.

———. 2013b. *Stata 13 Power and Sample-Size Reference Manual*. College Station, TX: Stata Press.

Utts, J. M. 2014. *Seeing Through Statistics*. 4th ed. Belmont, CA: Brooks/Cole.

van Buuren, S., H. C. Boshuizen, and D. L. Knook. 1999. Multiple imputation of missing blood pressure covariates in survival analysis. *Statistics in Medicine* 18: 681–694.

von Hippel, P. T. 2009. How to impute interactions, squares, and other transformed variables. *Sociological Methodology* 39: 265–291.

Wang, Z. 1999. lrdrop1: Stata module to calculate likelihood-ratio test after dropping one term. Statistical Software Components, Department of Economics, Boston College. Http://ideas.repec.org/c/boc/bocode/s400901.html.

Williams, R. 2012. Using the margins command to estimate and interpret adjusted predictions and marginal effects. *Stata Journal* 12: 308–331.

# Author index

# Subject index